Topics in Applied Physics Volume 1

Topics in Applied Physics Founded by Helmut K. V. Lotsch

Volume 57 **Strong and Ultrastrong Magnetic Fields and Their Applications**
Editor: F. Herlach

Volume 58 **Hot-Electron Transport in Semiconductors**
Editor: L. Reggiani

Volume 59 **Tunable Lasers**
Editors: L. F. Mollenauer and J. C. White

Volume 60 **Ultrashort Laser Pulses** and Applications
Editor: W. Kaiser

Volume 61 **Photorefractive Materials and Their Applications I**
Fundamental Phenomena
Editors: P. Günter and J.-P. Huignard

Volume 62 **Photorefractive Materials and Their Applications II**
Survey of Applications
Editors: P. Günter and J.-P. Huignard

Volume 63 **Hydrogen in Intermetallic Compounds I**
Electronic, Thermodynamic and Crystallographic Properties, Preparation
Editor: L. Schlapbach

Volume 64 **Hydrogen in Intermetallic Compounds II**
Surface and Dynamic Properties, Applications
Editor: L. Schlapbach

Volume 65 **Laser Spectroscopy of Solids II**
Editor: W. M. Yen

Volume 66 **Light Scattering in Solids V**
Superlattices and Other Microstructures
Editors: M. Cardona and G. Güntherodt

Volumes 1–56 are listed on the back inside cover

# Dye Lasers

Edited by F. P. Schäfer

With Contributions by
K. H. Drexhage H. Gerhardt T. W. Hänsch
E. P. Ippen F. P. Schäfer
C. V. Shank B. B. Snavely J. J. Snyder

Third Enlarged and Revised Edition

With 126 Figures

Springer-Verlag Berlin Heidelberg New York
London Paris Tokyo Hong Kong

Professor Dr. Fritz P. Schäfer

Max-Planck-Institut für biophysikalische Chemie, Postfach 2841,
D-3400 Göttingen-Nikolausberg, Fed. Rep. of Germany

ISBN 3-540-51558-5 3. Auflage Springer-Verlag Berlin Heidelberg New York
ISBN 0-387-51558-5 3rd edition Springer-Verlag New York Berlin Heidelberg

ISBN 3-540-08470-3 2. Auflage Springer-Verlag Berlin Heidelberg New York
ISBN 0-387-08470-3 2nd edition Springer-Verlag New York Berlin Heidelberg



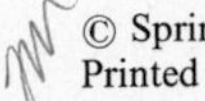



Typesetting: K + V Fotosatz, 6124 Beerfelden
Printing and binding: Brühlsche Universitätsdruckerei, 6300 Giessen
2154/3150-543210 – Printed on acid-free paper

## Preface to the Third Edition

This third edition of *Dye Lasers* appears 12 years after the second edition, which has been out of print for some years. Since Springer-Verlag has received many requests for either a simple reprint or an updated version, it was decided to find a reasonable compromise between the two. Thus, the classic chapter by B. B. Snavely on continuous-wave dye lasers has been reprinted unchanged and that by K. H. Drexhage, on structure and properties of laser dyes, appears with only minor modifications, while the chapters by F. P. Schäfer on principles of dye laser operation and by C. V. Shank and E. P. Ippen on mode-locking of dye lasers have been updated. The chapter by T. W. Hänsch on applications of dye lasers, which was written in 1973 for the first edition, when a synopsis of the relatively few dye laser applications was still possible, has been eliminated completely, since nowadays dye lasers have penetrated almost all fields of science and technology and applications have become innumerable. This made room for two new chapters, which many readers will find useful, namely the chapter by H. Gerhardt on continuous-wave dye lasers and the chapter by T. W. Hänsch and J. J. Snyder on wavemeters.

The editor is conscious of the shortcomings of such a procedure, but after the success of the first two editions it seemed sensible to try to change the character of the book as little as possible. It is hoped that readers will find this new edition useful and will add it to their personal libraries.

The volume *Tunable Lasers*, edited by L. F. Mollenauer and J. C. White and published as Volume 59 in the same series, complements this new edition, so that the two books together provide comprehensive coverage of the subject of laser tunability.

Göttingen, October 1989 *F. P. Schäfer*

# Contents

## Contributors

Drexhage, Karl H.

Physikalische Chemie, Universität Gesamthochschule, Postfach 101240, D-5900 Siegen, Fed. Rep. of Germany

Gerhardt, Harald

Laser-Laboratorium Göttingen e.V., Postfach 2619, D-3400 Göttingen, Fed. Rep. of Germany

Hänsch, Theodor W.

Max-Planck-Institut für Quantenoptik, D-8046 Garching and Universität München, Sektion Physik, Schellingstrasse 4, D-8000 München 40, Fed. Rep. of Germany

Ippen, Erich P.

Massachusetts Institute of Technology, Cambridge, MA 02139, USA

Schäfer, Fritz P.

Max-Planck-Institut für biophysikalische Chemie, Postfach 2841, D-3400 Göttingen-Nikolausberg, Fed. Rep. of Germany

Shank, Charles V.

Lawrence Berkeley Laboratory, University of California, 1 Cyclotron Rd., Berkeley, CA 94720, USA

Snavely, Benjamin B.

Eastman Kodak Company, 121 Lincoln Avenue, Rochester, NY 14653, USA

Snyder, James J.

Lawrence Livermore National Laboratory, P.O. Box 808, L-250, Livermore, CA 94551, USA

# 1. Principles of Dye Laser Operation

Fritz P. Schäfer

With 69 Figures

## Historical

Dye lasers entered the scene at a time when several hundreds of laser-active materials had already been found. Yet they were not just another addition to the already long list of lasers. They were the fulfillment of an experimenter's pipe dream that was as old as the laser itself: To have a laser that was easily tunable over a wide range of frequencies or wavelengths. Dye lasers are attractive in several other respects: Dyes can be used in the solid, liquid, or gas phases and their concentration, and hence their absorption and gain, is readily controlled. Liquid solutions of dyes are especially convenient: The active medium can be obtained in high optical quality and cooling is simply achieved by a flow system, as in gas lasers. Moreover, a liquid is self-repairing, in contrast to a solid-state active medium where damage (induced, say, by high laser intensities) is usually permanent. In principle, liquid dye lasers have output powers of the same magnitude as solid-state lasers, since the density of active species can be the same in both and the size of an organic laser is practically unlimited. Finally, the cost of the active medium, organic dyes, is negligibly small compared to that of solid-state lasers.

Early speculations about the use of organic compounds (Rautian and Sobel'mann 1961; Brock et al. 1961) produced correct expectations of the role of vibronic levels of electronically excited molecules (Broude et al. 1963), but the first experimental study that might have led to the realization of an organic laser was by Stockman et al. (1964) and Stockman (1964). Using a high-power flashlamp to excite a solution of perylene in benzene between two resonator mirrors, Stockman found an indication of a small net gain in his system. Unfortunately, he tried only the aromatic molecule perylene, which has high losses due to triplet-triplet absorption and absorption from the first excited singlet into higher excited singlet levels. Had he used some xanthene dye, like rhodamine 6G or fluorescein, we would undoubtedly have had the dye laser two years earlier. In 1966, Sorokin and Lankard (1966) at IBM's Thomas J. Watson Research Center, Yorktown Heights, were the first to obtain stimulated emission from an organic compound, namely chloro-aluminum-phthalocyanine. Strictly speaking, this dye is an organometallic compound, for its central metal atom is directly bonded to an organic ring-type molecule, somewhat resembling the compounds used in chelate lasers. In chelate lasers, however, stimulated emission originates in the central atom only, whereas in chloro-aluminum-phthalocyanine spectroscopic evidence clearly showed that the emission originated in the organic part of the molecule. Sorokin and Lankard set out to observe the resonance Raman effect in this dye, excited by a giant-pulse ruby laser (Sorokin 1969). Instead of sharp Raman lines they found a weak diffuse band at 755.5 nm, the peak of one of the fluorescence bands. They immediately suspected this might be a sign of incipient laser action and indeed, when the dye cell was incorporated into a resonator, a powerful laser beam at 755.5 nm emerged.

At that time the author, unaware of Sorokin and Lankard's work, was studying in his laboratory, then at the University of Marburg, the saturation characteristics of saturable dyes of the cyanine series. Instead of observing the saturation of the absorption, he used the saturation of spontaneous fluorescence excited by a giant-pulse ruby laser and registered by a photocell and a Tektronix 519 oscilloscope. The dye 3,3'-diethyltricarbocyanine had been chosen as a most convenient sample. This scheme worked well at very low concentrations of $10^{-6}$ and $10^{-5}$ mole/liter, but when Volze, then a student, tried to extend these measurements to higher concentrations, he obtained signals about one thousand times stronger than expected, with instrument-limited risetime that at a first glance were suggestive of a defective cable. Very soon, however, it became clear that this was laser action, with the glass-air interface of the square, all-side-polished spectrophotometer cuvette acting as resonator mirrors with a reflectance of 4%. This was quickly checked by using a snooperscope to look at the bright spot where the infrared laser beam hit the laboratory wall. Together with Schmidt, then a graduate student, we photographed the spectra at various concentrations and also with reflective coatings added to the cuvette walls. Thus we obtained the first evidence that we had a truly tunable laser whose wavelength could be shifted over more than 60 nm by varying the concentration or the resonator mirror reflectivity. This was quickly confirmed and extended to a dozen different cyanine dyes. Among these was one that showed a relatively large solvatochromic effect and enabled us to shift the laser wavelength over 26 nm merely by changing the solvent (Schäfer et al. 1966). Stimulated emission was also reported in two other cyanine dyes by Spaeth and Bortfield at Hughes Aircraft Company (Spaeth and Bortfield 1966). These authors, intrigued by Sorokin and Lankard's publication, used cryptocyanine and a similar dye excited by a giant-pulse ruby. Since their dyes had a very low quantum yield of fluorescence, they had a high-lying threshold and observed only a certain shift of laser wavelength with cell length and concentration.

Stimulated emission from phthalocyanine compounds, cryptocyanine, and methylene blue was also reported in 1967 by Stepanov and coworkers (Stepanov et al. 1967a, b). They also reported laser emission from dyes that had a quantum efficiency of fluorescence of less than one thousandth of one percent; this, however, has not yet been confirmed by others.

A logical extension of this work was to utilize shorter pump wavelengths and other dyes in the hope of obtaining shorter laser wavelengths. This was first achieved in the author's laboratory by pumping a large number of dyes, among them several xanthene dyes, by the second harmonic of neodymium and ruby lasers (Schäfer et al. 1967). Similar results were obtained independently in several other laboratories (Sorokin et al. 1967; McFarland 1967; Stepanov et al. 1967a, b; Kotzubanov et al. 1968a, b). Dye laser wavelengths now cover the whole visible spectrum with extensions into the near-ultraviolet and infrared.

Another important advance was made when a diffraction grating was substituted for one of the resonator mirrors to introduce wavelength-dependent feedback (Soffer and McFarland 1967). These authors obtained effective spectral narrowing from 6 to 0.06 nm and a continuous tuning range of 45 nm. Since then many different schemes have been developed for tuning the dye-laser wavelength; they will be discussed at length in the next chapter.

A natural step to follow was the development of flashlamps comparable in risetime and intensity to giant-pulse ruby lasers, to enable dye lasers to be pumped with a convenient, incoherent light source. This was initially achieved by techniques developed several years earlier for flash photolysis (Sorokin and Lankard 1967; W. Schmidt and Schäfer 1967). Soon after this, the author's team found that even normal linear xenon-filled flashlamps – and for some dyes even helical flashlamps – could be used, provided they were used in series with a spark gap and at sufficiently high voltage to result in a risetime of about 1 μs (Schäfer 1968). This is now standard practice for most single-shot or repetitively pumped dye lasers.

The common belief that continuous-wave (cw) operation of dye lasers was not feasible because of the losses associated with the accumulation of dye molecules in the metastable triplet state was corrected by Snavely and Schäfer (1969). Triplet-quenching by oxygen was found to decrease the steady-state population of the triplet state far enough to permit cw operation, at least in the dye rhodamine 6G in methanol solution, as demonstrated by using a lumped-parameter transmission line to feed a 500 μs trapezoidal voltage pulse to a flashlamp, giving a 140-μs dye-laser output. The premature termination of the dye laser pulse was also shown to be caused not by triplet accumulation but rather by thermal and acoustical schlieren effects. Our findings were later confirmed and extended by others who also used unsaturated hydrocarbons as triplet quenchers (Marling et al. 1970a and b; Pappalardo et al. 1970a).

The pump-power density requirements for a cw dye laser inferred from such measurements were so high that it seemed unlikely cw operation would be effected by pumping with the available high-power arc lamps. On the other hand, much higher pump-power densities can be obtained by focussing a high-power gas laser into a dye cuvette. The pumped region is thereby limited to about 50 μl, but the gain in the dye laser is usually very high.

Cw operation was a most important step towards the full utilization of the dye laser's potential; it was first achieved by O. G. Peterson et al. (1970) at Eastman-Kodak Research Laboratory. They used an argon-ion laser to pump a solution of rhodamine 6G in water with some detergent added. Water as solvent has the advantage of a high heat capacity, thus reducing temperature gradients, which are further minimized by the high velocity of the flow of dye solution through the focal region. The detergent both acts as triplet quencher and prevents the formation of non-fluorescing dimers of dye molecules, which produce a high loss in pure water solutions.

This breakthrough triggered a host of investigations and developments of cw dye lasers in the following years; these will be covered in Chap. 2 of this book.

Towards the other end of the time scale, dye lasers, because of their extremely broad spectral bandwidth, are capable of producing ultrashort pulses of smaller half-width than with any other laser. The first attempts to produce ultrashort pulses with dye lasers involved pumping a dye solution with a mode-locked pulse train from a solid-state laser; the dye cuvette sat in a resonator with a round-trip time exactly equal to, or a simple submultiple of, the spacing of the pump pulses (Glenn et al. 1968; Bradley and Durrant 1968; Soffer and Linn 1968). It subsequently proved possible to eliminate the resonator and use the superradiant traveling-wave emission from a wedged dye cuvette (Mack 1969b). The pulsewidths obtained in this way were generally some 10 to 30 ps. Self-mode-locking of flashlamp-pumped dye lasers was first achieved by W. Schmidt and Schäfer (1968), using rhodamine 6G as the laser active medium and a cyanine dye as the saturable absorber. In an improved arrangement of this type a pulsewidth of only 2 ps was reached (Bradley et al. 1970a, b). Eventually, the technique was applied to cw dye lasers to give a continuous mode-locked emission with pulses of only 1.5 ps (Dienes et al. 1972). In 1974 the first optical pulses shorter than 1 ps were generated by Shank and Ippen with a passively mode-locked dye laser. With the introduction of the colliding pulse concept by Shank and coworkers in 1981, femtosecond pulses became a reality (Fork et al. 1981). Following the initial report of 90 fs optical pulses, several laboratories developed this technique, which recently led to the production of pulses as short as 27 fs (Valdmanis et al. 1985). This is not so far from the theoretically possible limit. If mode-locking over the full bandwidth of about 250 THz were possible, the pulsewidth should be of the order of 5 fs. Another important technique for the production of ultrashort pulses was developed by Bor and coworkers using distributed feedback dye lasers (Bor 1980). Recent development of high-performance amplifier stages has brought pulse powers up to gigawatts. Chapter 4 of this book is devoted to this important field of dye laser research. [A nonlinear optical method of pulse compression (Nakatsuka and Grischkowsky 1981) actually made it possible to reach a pulsewidth of 6 fs in 1987 (Fork et al. 1987). This interesting development will not be treated here, since it is beyond the scope of this book.]

If one tries to extrapolate this historical survey and discern the outline of future developments, one can immediately foresee some quantitative improvements in several features: The wavelength coverage will be extended farther into the ultraviolet and infrared, peak pulse powers and energies will increase by several orders of magnitude, and the generation of ultrashort pulses of unprecedentedly small pulsewidth using dye mixtures might become possible. A qualitative improvement will be the incoherent pumping of cw dye lasers with specially developed high-power arc lamps that will allow a much higher output power than pumping with gas lasers. This important achievement has recently been reported by Drexhage and coworkers (Thiel et al. 1987). A most important development will concern the chemical aspect of the dye laser, namely the synthesis of special laser dyes with improved efficiency and photochemical stability (Schäfer 1983). The latter aspect will be particularly important for potential industrial applications, which at present are practically nonexistent, while applications of dye lasers in fundamental and applied research have become innumerable.

**Organization of the Book**

Chapter 1 by Schäfer is introductory. The chemical and spectroscopic properties of organic compounds are described, there is a tutorial presentation of the general principles of dye-laser operation, and aspects of the dye laser not treated in one of the special chapters are reviewed. Emphasis here is on an easy understanding of the physical and chemical principles involved, rather than on completeness, or historical or systematic presentation. The specialized chapters will build on this basis.

Chapter 2 by Snavely presents a review of cw dye lasers, with emphasis on the gain analysis. Special attention is given to the triplet problem.

Chapter 3 by Gerhardt describes the tremendous progress in cw dye lasers since 1973 in a brief overview. It does not cover frequency-stabilization techniques, since this subject has developed into a field that is beyond the scope of this book.

Mode-locking of dye lasers is the topic of Chapter 4 by Shank and Ippen. After an introduction covering methods of measuring ultrashort pulses, the authors discuss experimental methods applicable to mode-locking, pulsed and cw dye lasers, and for amplification of ultrashort pulses to very high powers.

The chemical and physical properties of dyes are discussed in Chapter 5 by Drexhage. A large amount of data on dyes is presented in support of the rules found by the author for the selection or synthesis of useful laser dyes. A list of laser dyes completes this chapter. Since this list covers only the laser dyes reported in the literature until 1973, the interested reader is referred to the books by Maeda (1984) and Brackmann (1986) for later data.

Finally, Chapter 6 by Snyder and Hänsch is a review of laser wavemeters, which have become indispensable tools in work with widely tunable dye lasers. This chapter replaces the chapter "Applications of Dye Lasers" by Hänsch in the earlier editions, since by now the number of such applications has become indeterminable.

The authors have tried to make the chapters as self-contained as possible, so that they can be read independently, albeit at the cost of some slight overlap.

## 1.1 General Properties of Organic Compounds

Organic compounds are defined as hydrocarbons and their derivatives. They can be subdivided into saturated and unsaturated compounds. The latter are

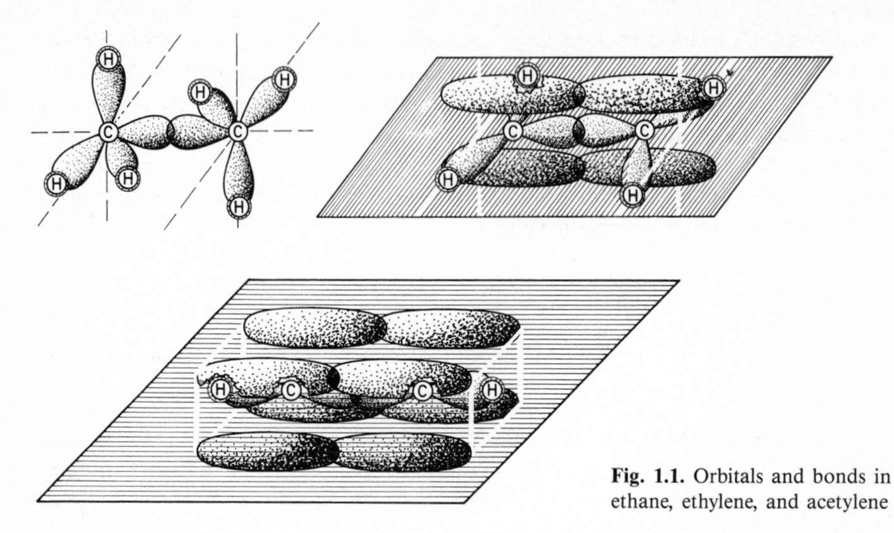

**Fig. 1.1.** Orbitals and bonds in ethane, ethylene, and acetylene

characterized by the fact that they contain at least one double or triple bond. These multiple bonds not only have a profound effect on chemical reactivity, they also influence spectroscopic properties. Organic compounds without double or triple bonds usually absorb at wavelengths below 160 nm, corresponding to a photon energy of 180 kcal/mole. This energy is higher than the dissociation energy of most chemical bonds, therefore photochemical decomposition is likely to occur, so such compounds are not very suitable as the active medium in lasers. In unsaturated compounds all bonds are formed by $\sigma$ electrons; these are characterized by the rotational symmetry of their wave function with respect to the bond direction, i.e. the line connecting the two nuclei that are linked by the bond. Double (and triple) bonds also contain a $\sigma$ bond, but in addition use $\pi$ electrons for bonding. The $\pi$ electrons are characterized by a wave function having a node at the nucleus and rotational symmetry along a line through the nucleus and normal to the plane subtended by the orbitals of the three $\sigma$ electrons of the carbon or heteroatom (Fig. 1.1). A $\pi$ bond is formed by the lateral overlap of the $\pi$-electron orbitals, which is maximal when the symmetry axes of the orbitals are parallel. Hence, in this position, bond energy is highest and the energy of the molecule minimal, thus giving a planar molecular skeleton of high rigidity. If two double bonds are separated by a single bond, as in the molecule butadiene,

```
H              H
 \            /
  C=C—C=C        ,
 /            \
H              H
```

the two double bonds are called *conjugated*. Compounds with conjugated double bonds also absorb light at wavelengths above 200 nm. All dyes in the proper sense of the word, meaning compounds having a high absorption in the visible part of the spectrum, possess several conjugated double bonds. The basic mechanism responsible for light absorption by compounds containing conjugated double bonds is the same, in whatever part of the spectrum these compounds have their longest wavelength absorption band, whether near-infrared, visible, or near-ultraviolet. We thus use the term *dye* in the wider sense as *encompassing all substances containing conjugated double bonds.* Whenever the term *dye* is used in this book, it will have this meaning.

For the remainder of this chapter we restrict our discussion of the general properties of organic compounds to dyes, since for the foreseeable future these are the only organic compounds likely to be useful laser-active media in the near-ultraviolet, visible and near infrared.

The thermal and photochemical stability of dyes is of utmost importance for laser applications. These properties, however, vary so widely with the almost infinite variety of chemical structure, that practically no general valid rules can be formulated. Thermal stability is closely related to the long-wavelength limit of absorption. A dye absorbing in the near-infrared has a low-lying excited singlet state and, even slightly lower than that, a metastable triplet state. The triplet state has two unpaired electrons and thus, chemically speaking, biradical character. There is good reason to assume that most of the dye molecules that reach this highly reactive state by thermal excitation will react with solvent molecules, dissolved oxygen, impurities, or other dye molecules to yield decomposition products. The decomposition would be of pseudofirst order with a reaction constant $k_1 = A \exp(-E_A/RT)$, where $A$ is the Arrhenius constant and has most often a value of $10^{12}\,\mathrm{s}^{-1}$ for reactions of this type (ranging from $10^{10}$ to $10^{14}\,\mathrm{s}^{-1}$), $E_A$ is the activation energy, $R$ is the gas constant and $T$ the absolute temperature. The half-life of such a dye in solution then is $t_{1/2} = \ln 2/k_1$. Assuming as a minimum practical lifetime one day, the above relations yield an activation energy of 24 kcal/mole, corresponding to a wavelength of 1.2 µm. If $A = 10^{10}\,\mathrm{s}^{-1}$, this shifts the wavelength to 1.7 µm, and with $A = 10^{14}\,\mathrm{s}^{-1}$ it would correspond to 1.1 µm. If we assume that a year is the minimum useful half-life of the dye (and $A = 10^{12}\,\mathrm{s}^{-1}$), we get a wavelength of 1.0 µm.

Obviously, it becomes more and more difficult to find stable dyes having the maximum of their long-wavelength band of absorption in the infrared beyond 1.0 µm, and there is little hope of ever preparing a dye absorbing beyond 1.7 µm that will be stable in solution at room temperature. Thus, dye-laser operation at room temperature in the infrared will be restricted to wavelengths not extending far beyond 1.0 µm. This limit seems to have been reached with some new laser dyes synthesized by Drexhage and coworkers (Polland et al. 1983) which have their peak absorption wavelength at up to 1.45 µm, thus enabling dye laser action to be extended to 1.85 µm.

The short-wavelength limit of dye-laser operation, already mentioned implicitly, is given by the absorption of dyes containing only two conjugated double bonds and having their long-wavelength absorption band at wavelengths of about 220 nm. Since the fluorescence, and hence the laser emission, is always red-shifted, dye lasers can hardly be expected to operate at wavelengths below about 250 nm. Even if we were to try to use compounds possessing only one double bond, like ethylene, absorbing at 170 nm, we could at best hope to reach 200 nm in laser emission. At this wavelength, however, photochemical decomposition already competes effectively with radiative deactivation of the molecule, since the energy of the absorbed quantum is higher than the energy of any bond in the molecule. In addition, at shorter wavelengths the probability of excited state absorption, cf. p. 29, can become higher than that of stimulated emission, thus preventing laser action. This seems to be the main reason why at present, after years of attempts to extend dye laser action further into the ultraviolet, the shortest dye laser wavelength still remains fixed at 308.5 nm (Zhang and Schäfer 1981).

Another important subdivision of dyes is into ionic and uncharged compounds. This feature mainly determines melting point, vapor pressure, and solubility in various solvents. An uncharged dye already mentioned is butadiene, $CH_2{=}CH{-}CH{=}CH_2$; other examples are most aromatics: anthracene, pyrene, perylene, etc. They usually have relatively low-lying melting points, relatively high vapor pressures, and good solubility in nonpolar solvents, like benzene, octane, cyclohexane, chloroform, etc. Cationic dyes include the large class of cyanine dyes, e.g. the simple cyanine dye

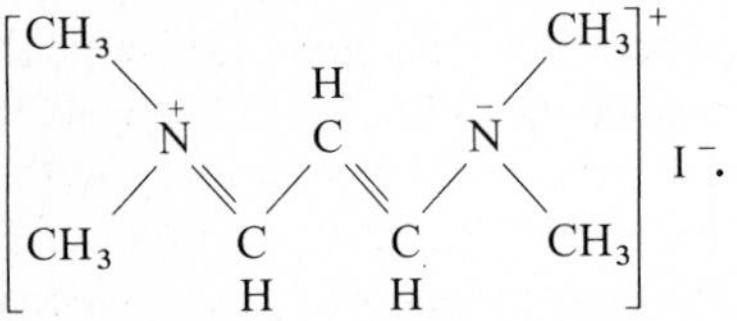

$$\left[(CH_3)_2\overset{+}{N}{=}CH{-}CH{=}CH{-}N(CH_3)_2\right]^+ I^-.$$

These compounds are salts, consisting of cations and anions, so they have high melting points, very low vapor pressure, good solubility in more polar solvents like alcohols, and only a slight solubility in less polar solvents. Similar statements can be made for anionic dyes, e.g. for the dye

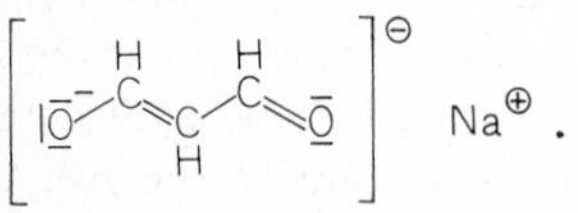

$$\left[{}^{-}O{-}CH{=}CH{-}CH{=}O\right]^{\ominus} Na^{\oplus}.$$

Many dyes can exist as cationic, neutral and anionic molecules depending on the pH of the solution, e.g. fluorescein:

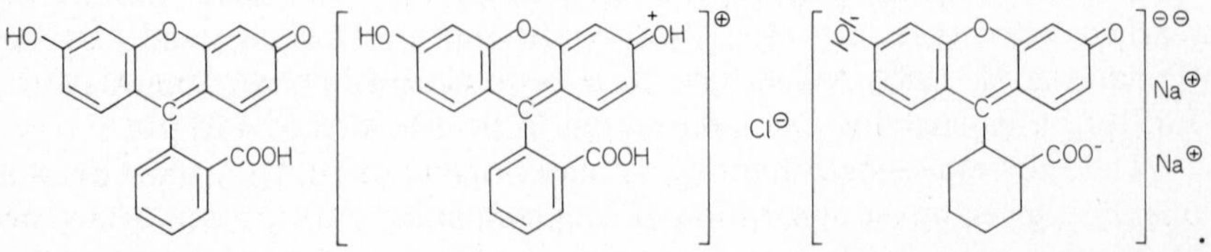

HO … O … O; C; COOH — Neutral form (alcoholic solution)

[HO … O … $\overset{+}{O}H$; C; COOH]$^{\oplus}$ $Cl^{\ominus}$ — Cationic form (hydrochloric acid solution)

[$^{-}O$ … O … O; C; $COO^-$]$^{\ominus\ominus}$ $Na^{\oplus}$ $Na^{\oplus}$ — Di-anionic form sodium hydroxyde solution

It should be stressed here that dyes are potentially useful as laser-active media in the solid, liquid and vapor phases. Since most dyes form good single crystals, it might be attractive to use them directly in this form. There are two main obstacles to their use in crystal form: The extremely high values attained by the extinction coefficient in dyes, which prevents the pump-light from exciting more than the surface layer a few microns thick; and the concentration quenching of fluorescence that usually sets in whenever the dye molecules approach each other closer than about 10 nm. Doping a suitable host crystal with a small fraction (one thousandth or less) of dye circumvents these difficulties. On the other hand, solid solutions of many different types can be used. For instance, one can dissolve a dye in the liquid monomer of a plastics material and then polymerize it; or one can dissolve it in an inorganic glass (e.g. boric acid glass) or an organic glass (e.g. sucrose glass) or some semirigid material like gelatine or polyvinylalcohol. The utilization of some of these techniques for dye lasers will be discussed later, together with the use of dyes in the liquid and vapor phases.

## 1.2 Light Absorption by Organic Dyes

The light absorption of dyes can be understood on a semiquantitative basis if we take a highly simplified quantum-mechanical model, such as the free-electron gas model (Kuhn 1959). This model is based on the fact that dye molecules are essentially planar, with all atoms of the conjugated chain lying in a common plane and linked by $\sigma$ bonds. By comparison, the $\pi$ electrons have a node in the plane of the molecule and form a charge cloud above and below this plane along the conjugated chain. The centers of the upper and lower lobes of the $\pi$-electron cloud are about one half bond length distant from the molecular plane. Hence, the electrostatic potential for any single $\pi$ electron moving in the field of the rest of the molecule may be considered constant, provided all bond lengths and atoms are the same (Fig. 1.2). Assume that the conjugated chain which extends approximately one bond length to the left and right beyond the terminal atoms has length $L$. Then the energy $E_n$ of the $n$th eigenstate of this electron is given by $E_n = h^2 n^2/8mL^2$, where $h$ is Planck's constant, $m$ is the mass of the electron, and $n$ is the quantum number giving the number of antinodes of the eigenfunction along the chain. According to the Pauli principle, each state can be occupied by two electrons. Thus, if we have $N$ electrons, the lower $(1/2)N$ states are filled with two electrons each, while all higher states are empty (provided $N$ is an even number; this is usually the case in stable molecules since only highly reactive radicals possess an unpaired electron). The absorption of one photon of energy $\Delta E = hc_0/\lambda$ (where $\lambda$ is the wavelength of the absorbed radiation and $c_0$ is the velocity of light) raises one electron from an occupied to an empty state. The longest wavelength absorption band then corresponds to a transition from the highest occupied to the lowest empty state with

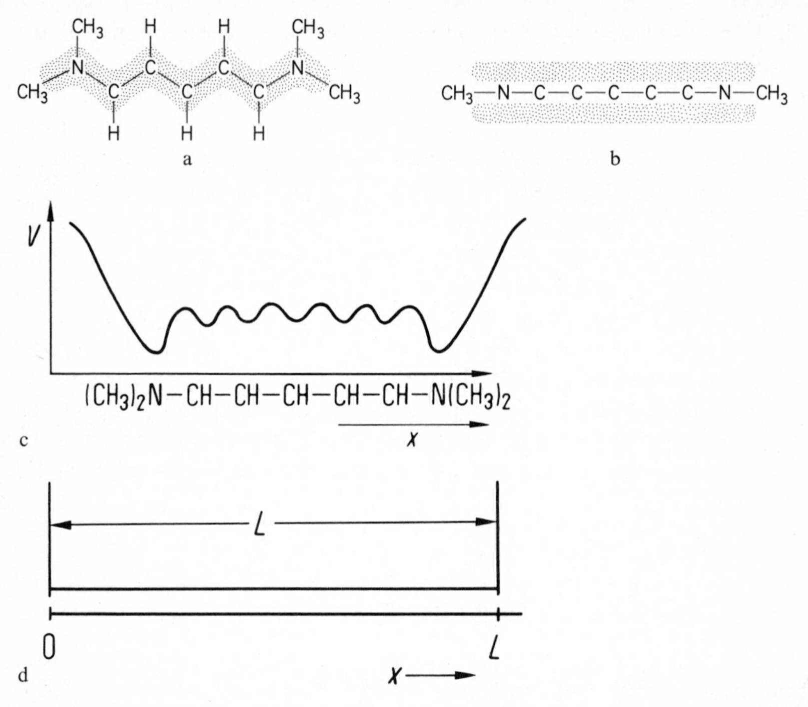


**Fig. 1.2.** **(a)** $\pi$-Electron cloud of a simple cyanine dye seen from above the molecular plane; **(b)** the same as seen from the side; **(c)** potential energy $V$ of a $\pi$ electron moving along the zig-zag chain of carbon atoms in the field of the rump molecule; **(d)** simplified potential energy trough; $L$ = length of the $\pi$-electron cloud in a as measured along the zig-zag chain. (From Försterling and Kuhn 1971)

$$\Delta E_{\min} = \frac{h^2}{8mL^2}(N+1) \quad \text{or} \quad \lambda_{\max} = \frac{8mc_0}{h}\frac{L^2}{N+1} \; .$$

This indicates that to first approximation the position of the absorption band is determined only by the chain length and by the number of $\pi$ electrons $N$. Good examples of this relation are the symmetrical cyanine dyes of the general formula

$$R\overset{\oplus}{N}(R_1)C\!-\!(CH{=}CH)_{j-2}\!-\!CH{=}C\,N(R_1)\,R \longleftrightarrow R\,N(R_1)\,C{=}CH\!-\!(CH{=}CH)_{j-2}\!-\!C\,\overset{\oplus}{N}(R_1)R$$

where $j$ is the number of conjugated double bonds, $R_1$ a simple alkyl group like $C_2H_5$, and R indicates that the terminal nitrogen atoms are part of a larger group, as e.g. in the following dye (homologous series of thiacyanines):

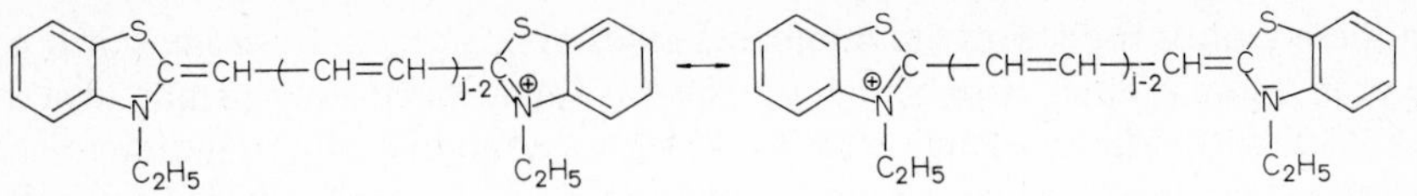

The double-headed arrow means that the two formulae are limiting structures of a resonance hybrid. The $\pi$ electrons in the phenyl ring can be neglected in first approximation, or treated as a polarizable charge cloud, leading to an apparent enlargement of the chain $L$. In the case of the last-mentioned dye, good agreement is found between calculated and experimental absorption wavelength, when the chain length $L$ is assumed to extend 1.3 bond lengths (instead of 1.0 bond length as above) beyond the terminal atoms. The bond length in cyanines is 1.40 Å. The good agreement between the results of this simple calculation and the experimental data for the above thiacyanines is shown by the following comparison:

Wavelength (in nm) of absorption maximum for thiacyanines

| | Number of conjugated double bonds $j$ = | | | |
|---|---|---|---|---|
| | 2 | 3 | 4 | 5 |
| Calculated | 395 | 521 | 649 | 776 |
| Experimental | 422 | 556 | 652 | 760 |

Similarly good agreement can be found for all the other homologous series of symmetrical cyanines, once the value of the end length extending over the terminal N atoms is found by comparison with the experimentally observed absorption wavelength for one member of the series. Absorption wavelengths have been reported for a large number of cyanines (Miyazoe and Maeda 1970).

The following nomenclature is customarily used for cyanine dyes. If $j = 2$, then the dye is a cyanine in the narrower sense or monomethine dye, since it contains one methine group, $-CH-$, in its chain; if $j = 3$, the dye is a carbocyanine or trimethine dye; if $j = 4, \ldots, 7$, the dye is a di-, tri-, tetra-, or pentacarbocyanine or penta-, hepta-, nona- or undecamethine dye. The heterocyclic terminal groups are indicated in an abridged notation; thus "thia" stands for the benzthiazole group:

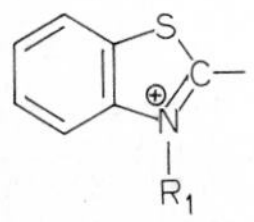

and "oxa" for the benzoxazole group

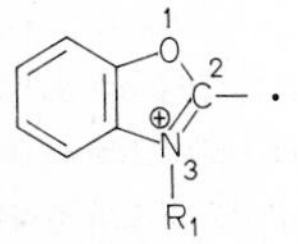

In these groups the atoms are numbered clockwise, starting from the sulfur or oxygen atom. Thus, the name of the last-mentioned thiacyanine dye is 3,3′-diethyl-thiadicarbocyanine for $j = 4$. If the terminal group is quinoline, the syllable "quinolyl" indicative of this end group is frequently omitted from the name. In this case, however, the position of linkage of the polymethine chain to the quinoline ring must be given, e.g.

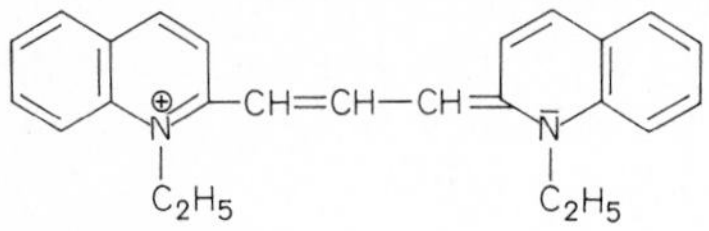

is 1,1′-diethyl-2,2′-carbocyanine (trivial name: pinacyanol), while

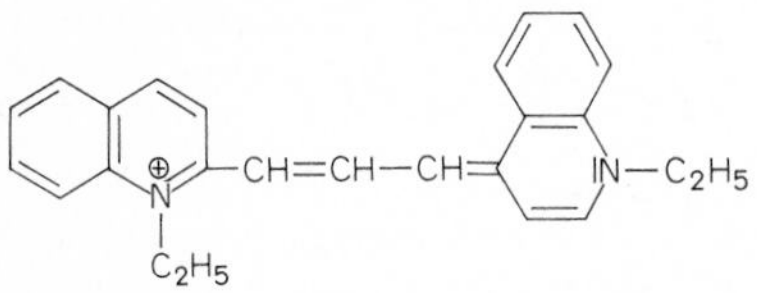

is 1,1′-diethyl-2,4′-carbocyanine (trivial name: dicyanine), and

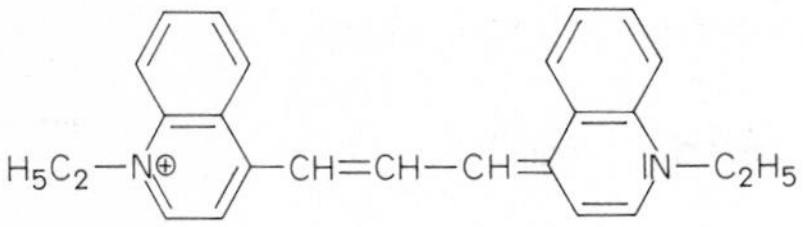

is 1,1′-diethyl-4,4′-carbocyanine (trivial name: cryptocyanine). Where there are two different end groups, these are named in alphabetical order, e.g. 3,3′-diethyl-oxa-thiacarbocyanine is

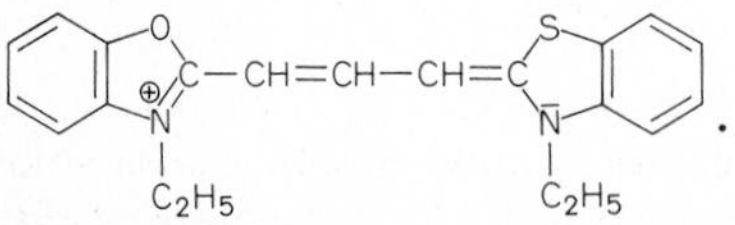

It is easy to eliminate the above assumption of identical atoms in the conjugated chains. If, for instance, one CH group is replaced by an N atom, the higher electronegativity of the nitrogen atom adds a small potential well to the constant potential; a simple perturbation treatment shows that every energy level is shifted by $\varepsilon = -B\Psi^2$, where $B$ is a constant characteristic of the electronegativity of the heteroatom (in the case of $=\mathrm{N}-$, $B = 3.9 \times 10^{-20}$ erg cm) and $\Psi$ is the value of the normalized wave function at the heteroatom. This means that the shift of an energy level is zero if the wave function has a node at the heteroatom, and maximal if it has an antinode there. The change in absorption wavelength with heterosubstitution calculated in this way is in good agreement with the experimentally observed values. The second of the above assumptions, that of equal bond lengths along the chain, can be eliminated in a similar way. An illustrative example is offered by the polyenes of the general

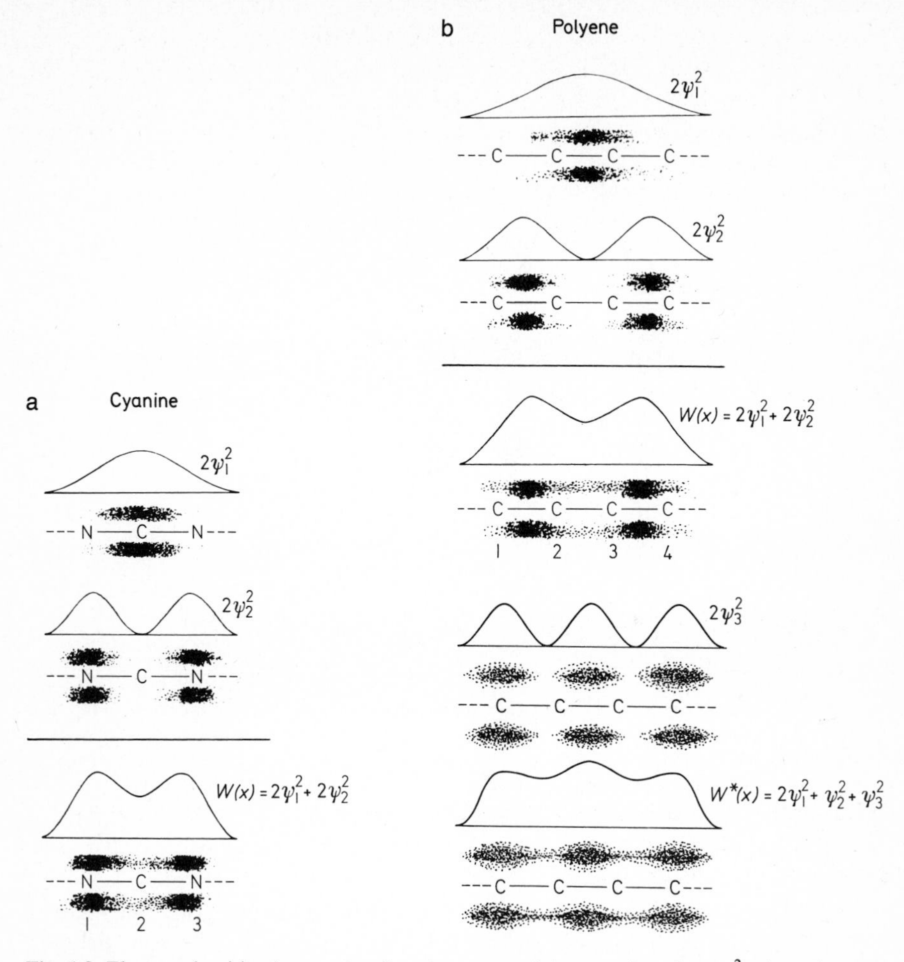


**Fig. 1.3.** Electron densities (proportional to the square of the wave function, $\psi^2$) along the conjugated chain of (**a**) a cyanine, (**b**) a polyene (butadiene). The symbols $\psi_1$ and $\psi_2$ denote the wave functions of the first and second filled $\pi$-electron molecular orbital, and $\psi_3$ that of the lowest orbital normally empty, which is occupied by one electron when excited. $W(x)$ is the total $\pi$-electron density for a molecule in the ground state, and $W^*(x)$ is the same for butadiene in the first excited state

formula $R-(CH=CH)_j-R$. These compounds have an even number of atoms. The total charge distribution results in localized double and single bonds (Fig. 1.3) and hence in alternating short and long bond lengths (1.35 Å and 1.47 Å, respectively). A $\pi$ electron moving along the chain will therefore experience greater attraction to the neighboring atoms in the middle of a double bond than in the middle of a single bond. This fact may be represented by a periodic perturbing potential with minima at the center of the double bonds and maxima at the center of the single bonds (Fig. 1.4). Assuming a sinusoidal potential with an amplitude of 2.4 eV gives good agreement between the

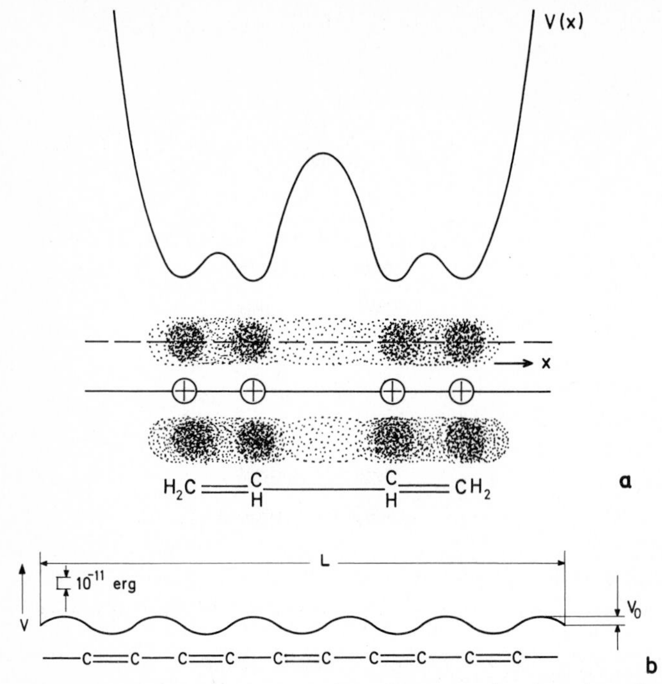


**Fig. 1.4.** (**a**) Potential energy $V(x)$ of a $\pi$ electron moving along the carbon atom chain in the field of the rump molecule of butadiene; (**b**) simplified potential energy trough of a long polyene molecule with perturbing sinusoidal potential of amplitude $V_0 = 2.4$ eV. The energy difference between highest filled and lowest empty orbital is given by $\Delta E = (h^2/8mL^2)(2j+1)+0.83(1-1/2j) \times V_0$, where $j$ is the number of conjugated double bonds. (From Kuhn 1959)

calculated and experimental values of the absorption wavelengths. The relation between chain length and absorption wavelength is found to be very different in the cyanines and the polyenes. In symmetrical cyanines the absorption wavelength is shifted by a constant amount of roughly 100 nm on going from one member of a series to the next higher which has one more double bond. In polyenes this shift decreases with increasing number of double bonds. A similar treatment can be applied to unsymmetrical cyanines in which the difference in electronegativity of the different end groups gives a similar polyene-type perturbation. The perturbation is less pronounced when the difference in electronegativity of the end groups is less.

Many dyes containing a branched chain of conjugated double bonds can, in a first approximation, be classified either as symmetrical cyanine-like or polyene-like substances, or as intermediate cases. For example, Michler's hydrol blue

$(H_3C)_2\bar{N}$–$C_6H_4$–$CH$=$C_6H_4$=$\overset{\oplus}{N}(CH_3)_2$ ⟷ $(H_3C)_2\overset{\oplus}{N}$=$C_6H_4$=$CH$–$C_6H_4$–$\bar{N}(CH_3)_2$

absorbs at practically the same wavelength as the symmetrical cyanine

$$(H_3C)_2\bar{N}(CH{=}CH)_4CH{=}\overset{\oplus}{N}(CH_3)_2$$

$$\leftrightarrow (H_3C)_2\overset{\oplus}{N}{=}CH(CH{=}CH)_4\bar{N}(CH_3)_2.$$

Hence it seems justified to treat it like a cyanine by neglecting the lower branches in the formula of Michler's hydrol blue. If, however, additional branching is introduced by connecting positions 4 and 8 through a $\left(\begin{array}{c}=N-\\ |\\ CH_3\end{array}\right)$ bridge, this gives acridine orange

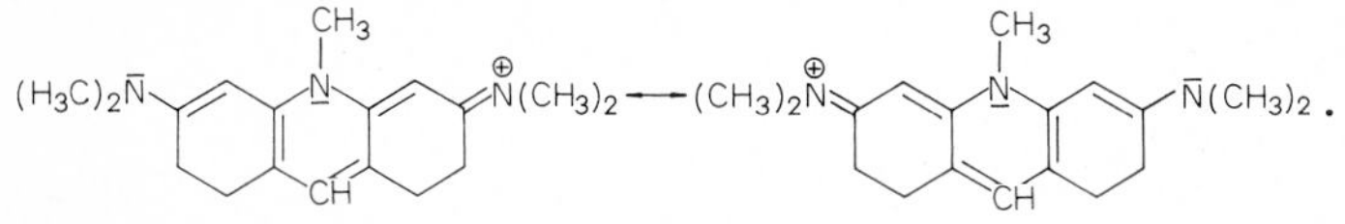

$(H_3C)_2\bar{N}$–[acridine ring, N–$CH_3$, CH]=$\overset{\oplus}{N}(CH_3)_2$ ⟷ $(CH_3)_2\overset{\oplus}{N}$=[acridine ring, N–$CH_3$, CH]–$\bar{N}(CH_3)_2$ .

The absorption wavelength is shifted from 603 nm to 491 nm. It indicates that in this case branching cannot be neglected even in a first approximation.

With an O atom as bridging group instead of the $N-CH_3$ group, the xanthylium dye pyronine G with an absorption wavelength of 550 nm is obtained:

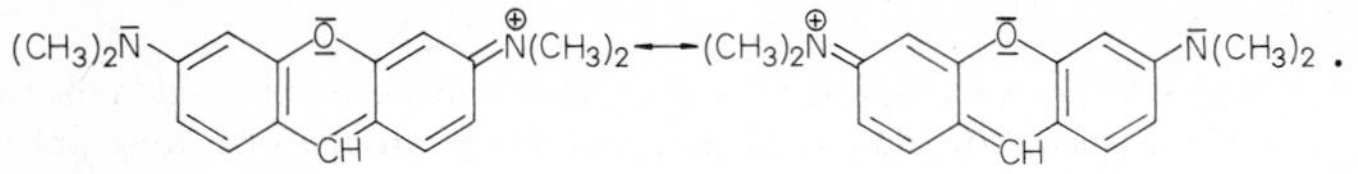

$(CH_3)_2\bar{N}$–[xanthene ring, $\bar{O}$, CH]=$\overset{\oplus}{N}(CH_3)_2$ ⟷ $(CH_3)_2\overset{\oplus}{N}$=[xanthene ring, $\bar{O}$, CH]–$\bar{N}(CH_3)_2$ .

For a detailed treatment of molecules containing such a branched free-electron gas, the reader is referred to the literature (Kuhn 1959).

Another important class of dyes in which the branching of the conjugated chain can be neglected to first approximation, are the phthalocyanines and similar large-ring molecules. In first approximation the benzene rings are separated resonance systems, leaving the 16-membered ring indicated in Fig. 1.5 by heavy lines. Now there are $18\,\pi$ electrons on a ring of circumference $L = 16l$. In the lowest state ($n = 0$) the eigenfunction has no nodes, for $n \neq 0$ there are two degenerate eigenfunctions for every $n$, corresponding to the sine and cosine in the constant-potential approximation. A large perturbation is, however, introduced by the nitrogen atoms, which remove the degeneracy. This effect is most pronounced for $n = 4$, since one eigenfunction then has its nodes at the N atoms, the other its antinodes. The lowest unfilled levels are again degenerate because of equal perturbation energy. The absorption wavelengths for the two transitions from $n = 4$ to $n = 5$ are then readily calculated (with the above value for $B$): $\lambda_1 = 690$ nm and $\lambda_2 = 340$ nm (Fig. 1.5). The experimental values are 674 nm and 345 nm.

The free-electron model, even in its simplest form, gives very satisfactory agreement between calculated and experimental values of the absorption wavelength for large dye molecules. Nevertheless, its one-electron functions are not sufficient for a quantitative description of light absorption by small molecules of high symmetry, like benzene, naphthalene and similar molecules,

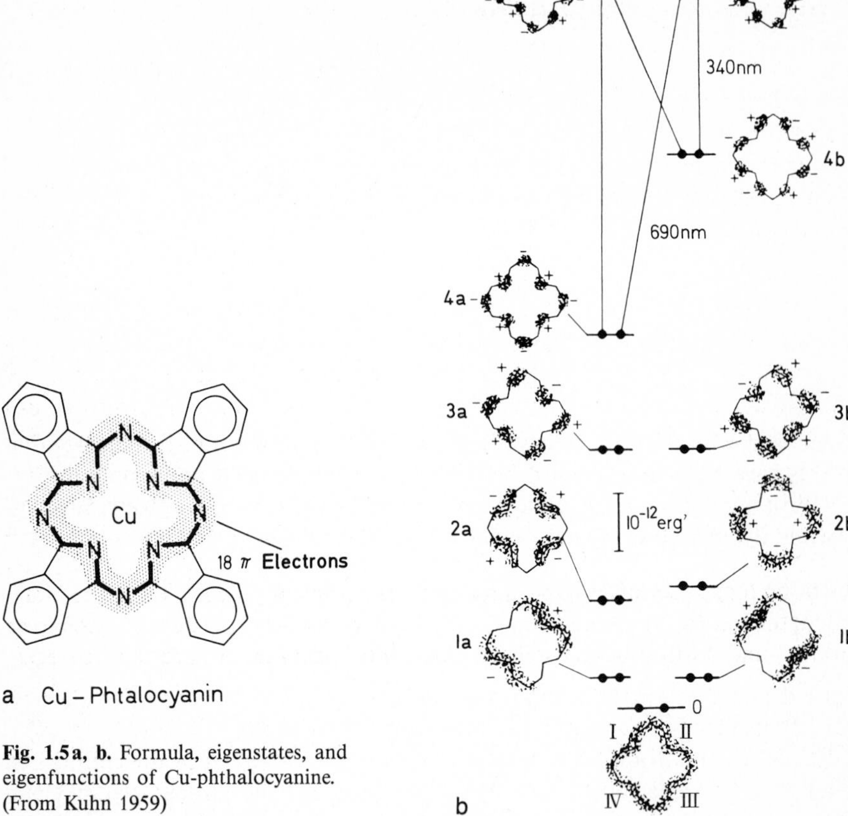


**Fig. 1.5a, b.** Formula, eigenstates, and eigenfunctions of Cu-phthalocyanine. (From Kuhn 1959)

since here the repulsion between the $\pi$ electrons plays an important role. For the inclusion of electron correlations into the free electron model, the reader is referred to Försterling et al. (1966).

The oscillator strength of the absorption bands can also be calculated easily by the free-electron model. Transition moments $X$ and $Y$ along and normal to the long molecular axis are connected with the oscillator strength $f$ of the absorption band by

$$f = 2\,\frac{8 m_0 \pi^2}{3 h^2}\,\Delta E (X^2 + Y^2)\ ,$$

and can be calculated using the electron gas wave functions $\psi_a$ and $\psi_b$ of the eigenstates between which the transition occurs:

$$X = \int\limits_{\text{molecule}} \psi_a x \psi_b dx\ , \qquad Y = \int\limits_{\text{molecule}} \psi_a y \psi_b dy\ .$$

The relative strengths of the $x$ and $y$ components also yield the orientation of the transition moment in the molecule. One finds good agreement on comparing the $f$ values obtained from the absorption spectrum of the dye using the relation

$$f = 4.32 \times 10^{-9} \int_{\substack{\text{absorption}\\ \text{band}}} \varepsilon(\tilde{\nu})\, d\tilde{\nu} \,,$$

where $\varepsilon$ is the numerical value of the molar decadic extinction coefficient measured in liter/(cm mole), and $\tilde{\nu}$ is the numerical value of the wave number measured in $cm^{-1}$.

A peculiarity of the spectra of organic dyes as opposed to atomic and ionic spectra is the width of the absorption bands, which usually covers several tens of nanometers. This is immediately comprehensible when one recalls that a typical dye molecule may possess fifty or more atoms, giving rise to about 150 normal vibrations of the molecular skeleton. These vibrations, together with their overtones, densely cover the spectrum between a few wave numbers and $3000\ cm^{-1}$. Many of these vibrations are closely coupled to the electronic transitions by the change in electron densities over the bonds constituting the conjugated chain. After the electronic excitation has occurred, there is a change in bond length due to the change in electron density. If, e.g. in Fig. 1.3, the bond length connecting atoms 1 and 2 is $r$ (typically 1.35 Å) and this bond is lengthened in the excited molecule, the new equilibrium distance being $r^*$, atoms 1 and 2 will start to oscillate, classically speaking, around this new position with an amplitude $r^* - r$ (typically 0.02 Å) after the electronic transition has occurred. A molecular skeletal vibration is excited in this way. The new equilibrium total $\pi$ electron density $W^*(x)$ for the excited state is given in Fig. 1.3c demonstrating the large increase in $\pi$ electron density over the bond connecting atoms 2 and 3. Quantum-mechanically this means that transitions have occurred from the electronic and vibrational ground state $S_0$ of the molecule

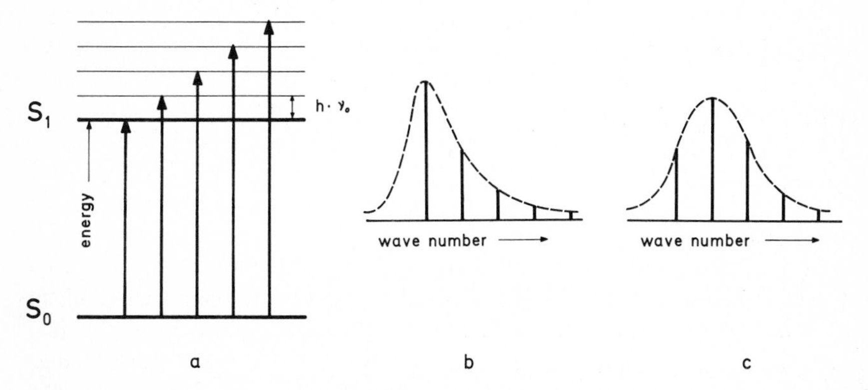


**Fig. 1.6.** **(a)** Electronic and vibronic energy levels of a dye molecule; $S_0$: ground state, $S_1$: excited state, and absorptive transitions from $S_0$ to $S_1$; **(b)** and **(c)** two possible forms of concomitant spectra

to an electronically and vibrationally excited state $S_1$, as depicted in Fig. 1.6. This results in spectra like Fig. 1.6b or c, depending on how many of the vibrational sublevels, spaced at $h\nu_0\,(v+1/2)$, with $v = 0, 1, 2, 3 \ldots$, are reached and what the transition moments to these sublevels are.

In the general case of a large dye molecule, many normal vibrations of differing frequencies are coupled to the electronic transition. Furthermore, collisional and electrostatic perturbations, caused by the surrounding solvent molecules, broaden the individual lines of such vibrational series as that given in Fig. 1.6. As a further complication, every vibronic sublevel of every electronic state, including the ground state, has superimposed on it a ladder of rotationally excited sublevels. These are extremely broadened because of the frequent collisions with solvent molecules which hinder the rotational movement so that there is a quasicontinuum of states superimposed on every electronic level. The population of these levels in contact with thermalized solvent molecules is determined by a Boltzmann distribution. After an electronic transition, which, as described above, leads to a nonequilibrium state (Franck-Condon state) the approach to thermal equilibrium is very fast in liquid solutions at room temperature. The reason is that a large molecule experiences at least $10^{12}$ collisions/s with solvent molecules, so that equilibrium is reached in a time of the order of one picosecond. Thus the absorption is practically continuous all over the absorption band. The same is true for the fluorescence emission corresponding to the transition from the electronically excited state of the molecule to the ground state. This results in a mirror image of the absorption band displaced towards lower wave numbers by reflection at the wave number of the purely electronic transition. This condition exists, since the emissive transitions start from the vibrational ground state of the first excited electronic state $S_1$ and end in vibrationally excited sublevels of the electronic ground state. The resulting typical form of the absorption and fluorescence spectra of an organic dye is given in Fig. 1.7.

Further complications of dye spectra arise from temperature and concentration dependence and acid–base equilibria with the solvent. If the temperature of a dye solution is increased, higher vibrational levels of the ground state are populated according to a Boltzmann distribution and more and more transitions occur from these to higher sublevels of the first excited singlet state. Consequently, the absorption spectrum becomes broader and the superposition of so many levels blurs most of the vibrational fine structure of the band, while cooling of the solution usually reduces the spectral width and enhances any vibrational features that may be present. Thus, spectra of solid solutions of dyes in EPA, a mixture of 5 parts ethyl ether, 5 parts isopentane, and 5 parts ethanol that forms a clear organic glass when cooled down to 77 K, are often used for comparison with calculated spectra because of their well-resolved vibrational structure. Further cooling below the glass point, when the free movement of solvent molecules or parts thereof is inhibited, usually brings about no further sharpening of the spectral features (Fig. 1.8). The many possible different configurations of the solvated molecule in the cage of solvent molecules cannot be attained when the temperature is lowered because the ac-

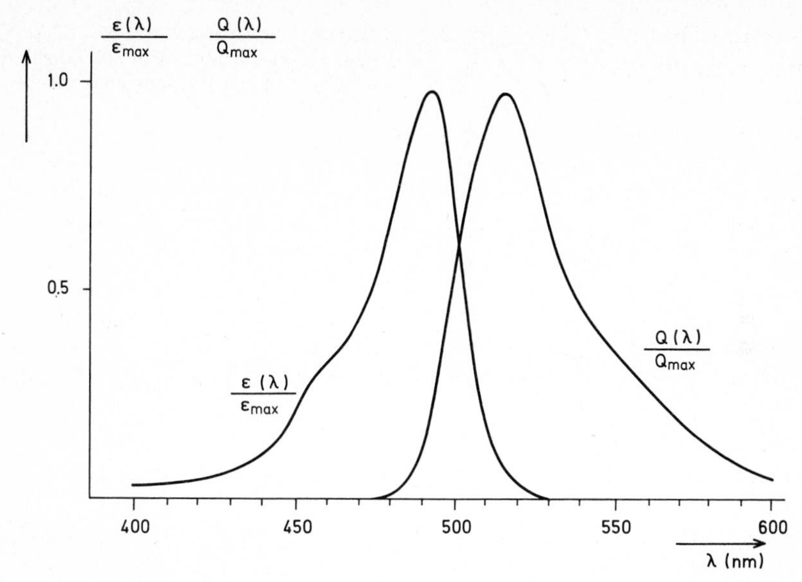


**Fig. 1.7.** Absorption spectrum, $\varepsilon(\lambda)/\varepsilon_{max}$, and fluorescence spectrum, $Q(\lambda)/Q_{max}$, of a typical dye molecule (fluorescein-Na in water)

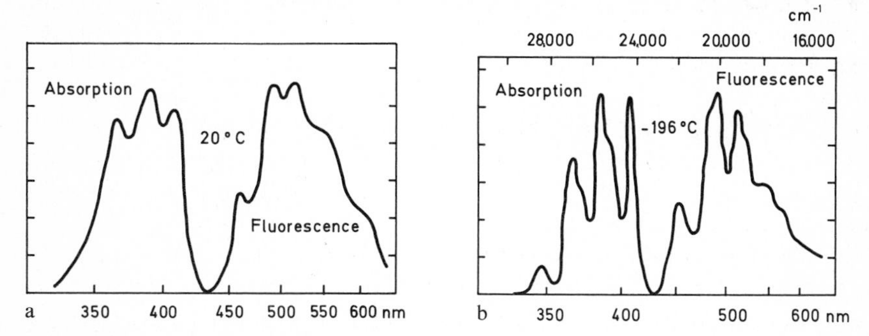


**Fig. 1.8.** Absorption and fluorescence spectra of diphenyloctatetraene in xylene. (**a**) at 20 °C, (**b**) at −196 °C. (From Hausser et al. 1935)

tivation energy required to reach the new equilibrium positions is too high. A very special case is the Shpolski effect (Shpolski 1962): This refers to the appearance of very sharp, line-like spectra (often termed quasi-line spectra in the Russian literature) of about one $cm^{-1}$ width instead of the usual diffuse band spectra of dye molecules (most often aromatics) in a matrix of *n*-paraffins at low temperature (usually below 20 K). Evidently this is because there are only a few different possibilities of solvation of the molecule in that matrix, and each of the different sites causes a series of spectral lines in absorption as well as in emission. Analyzing these series gives the energy difference of the different sites, usually a few hundred $cm^{-1}$ (Fig. 1.9). The Shpolski effect has

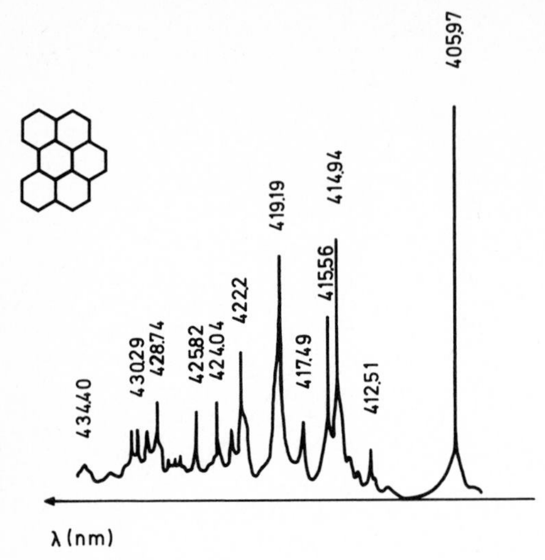


Fig. 1.9. Fluorescence spectrum of 1,12-benzoperylene in n-hexane at 77 K (Shpolski effect). (From Personov and Solochmov 1967)

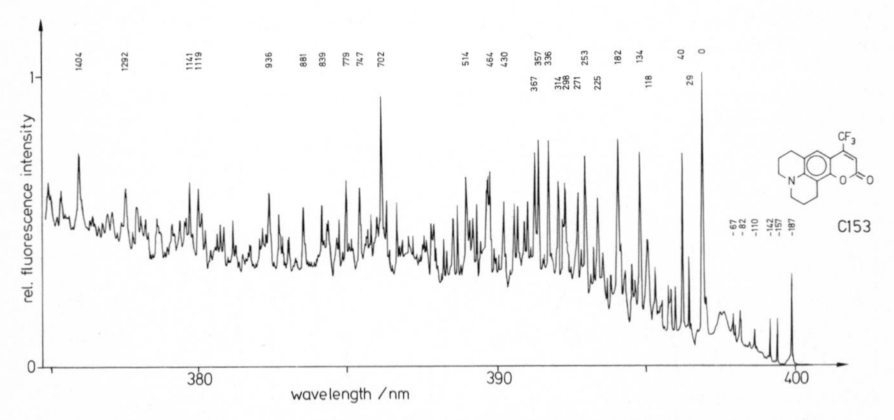


Fig. 1.10. Excitation spectrum of jet-cooled laser dye C 153 with "rigidized" amino group. (From Ernsting et al. 1982)

also been called the optical analog of the Mössbauer effect (Rebane and Khizhnyakov 1963), since here also the recoil of the interacting photons seems to be taken up by the matrix as a whole.

The highest resolution absorption and fluorescence spectra of laser dyes can be obtained using the relatively new tool of cold beam spectroscopy. For this purpose a stream of noble gas is passed over a heated sample of the dye and then expanded through a small orifice into a vacuum. The adiabatic expansion cools the molecules quickly down to a few degrees Kelvin after a flight over a few millimeters. The density of the isolated cool molecules is then still very high (up to $10^{14}\,cm^{-3}$), so that absorption and fluorescence spectra can easily be obtained. An example of such a spectrum is shown in Fig. 1.10.

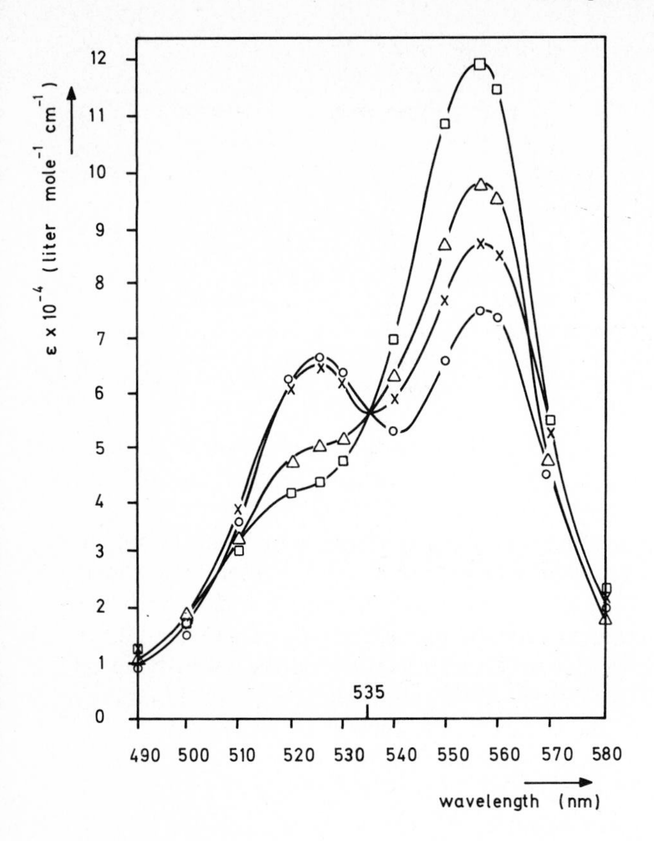


**Fig. 1.11.** Absorption spectra of aqueous solutions of rhodamine B at 22 °C, (○) 1.5 $\times 10^{-3}$ M, (×) 7.6$\times 10^{-4}$ M, (△) 1.5$\times 10^{-4}$ M, (□) 3.0$\times$ $10^{-6}$ M. (From Selwyn and Steinfeld 1972)

The concentration dependence of dye spectra is most pronounced in solutions where the solvent consists of small, highly polar molecules, notably water. Dispersion forces between the large dye molecules tend to bring the dye molecules together in a position with the planes of the molecules parallel, where the interaction energy usually is highest. This is counteracted by the repulsive Coulomb forces if the dye molecules are charged. In solvents of high dielectric constants this repulsion is lowered and the monomer–dimer equilibrium is far to the side of the dimer. The equilibrium constant of the dimerization process monomer + monomer ⇌ dimer is $K = [\text{dimer}]/[\text{monomer}]^2$ and can easily be obtained from spectra of differently concentrated dye solutions (Fig. 1.11). Usually the polymerization process stops at this stage, as evidenced by at least one isosbestic point (535 nm in Fig. 1.11). However, in some cases reversible polymerization occurs to a degree of polymerization of up to 1 million dye molecules (Scheibe 1941; Bücher and Kuhn 1970). In this case, a very sharp, high-intensity absorption and emission line becomes prominent in the spectra. The spectral differences between monomer and dimer, or even polymer, dye molecules can easily be understood at least qualitatively (Förster 1951). Figure 1.12 shows schematically the two main electronic energy levels of two distant dye molecules, and the splitting of the degenerate levels

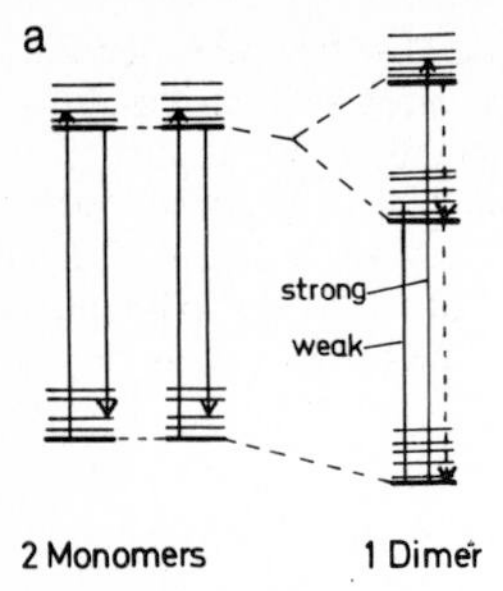


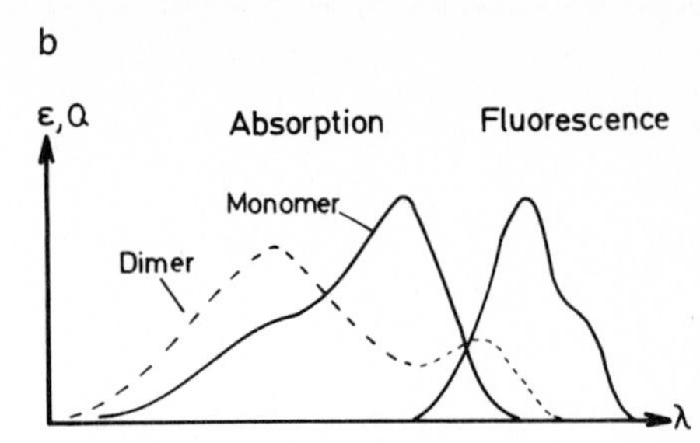


**Fig. 1.12.** **(a)** Energy levels of two monomers and the dimer molecule formed by them; **(b)** resulting spectra

if they come close enough to each other to experience interaction energy and the concomitant formation of a dimer. Now two transitions are possible, and these in general possess different transition moments depending on the wave functions of the dimer. Most often the long-wavelength transition has a practically vanishing transition moment, so that only one absorption band of the dimer is observed, lying to the short-wavelength side of the monomer band. This has an important consequence for the fluorescence of such dimers. Since the upper level from which the fluorescence starts is always the lowest-lying excited electronic level, and since a small transition moment is coupled to a long lifetime of the excited state, these dimers would show a very slow decay of their fluorescence. This, however, makes them susceptible to competing quenching processes, which in liquid solution are generally diffusion-controlled and hence very fast processes. Consequently, in most of these cases the fluorescence of the dimers is completely quenched and cannot be observed.

This is the reason why dimers constitute an absorptive loss of pump power in dye lasers and must be avoided by all means. There are several ways in which this may be done. One is to use a less polar solvent, like alcohol or chloroform. There are very few dyes which show dimerization in alcohol at the highest concentrations and at low temperatures. Another possibility is to add a detergent to the aqueous dye solution, which then forms micelles that contain one dye molecule each (Förster and Selinger 1964). This method is of prime importance for cw dye lasers and is discussed in Chap. 5, on laser dyes.

Acid–base equilibria have already been mentioned above for the case of fluorescein. Very often these equilibria are less obvious. For example, the dye 3,6-bis-dimethylaminoacridine can be protonated at the central nitrogen atom and then exhibits an absorption band shifted to longer wavelengths by about 55 nm (J. Ferguson and Mau 1972b), as shown in Fig. 1.13. Several other cases of acid–base equilibria of laser dyes will be discussed in detail in Chap. 5.

In addition to above-mentioned transitions between $\sigma$ orbitals ($\sigma \rightarrow \sigma^*$ transitions) and the extensively discussed transitions between $\pi$ orbitals ($\pi \rightarrow \pi^*$ transitions), we should also mention the transitions from the orbital of a lone-electron pair (so-called n orbitals), e.g. in a keto group, C=O, to a $\pi$ orbital. Since the n orbitals have either spherical or rotational symmetry, with the sym-

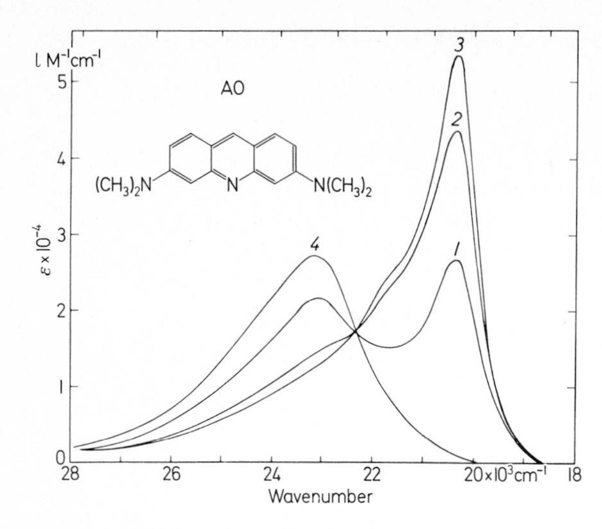


**Fig. 1.13.** Absorption spectra of 3,6-bis-dimethylamino-acridine (AO) in ethanol, (*1*) $3\times10^{-5}$ M, (*2*) 2% water added, (*3*) bubbled with $CO_2$ gas, (*4*) anhydrous potassium carbonate added. (From J. Ferguson and Mau 1972b)

metry axis lying in the molecular plane, the overlap with $\pi$ orbitals, and hence the transition moments, is very small. Correspondingly, the molar decadic extinction coefficient $\varepsilon$ (as defined by Beer's law, $I = I_0 \cdot 10^{-\varepsilon cd}$, where $I$ and $I_0$ are the transmitted and incident light intensity, $c$ is the concentration in mole/liter of the absorbing species, and $d$ is the thickness of the absorbing layer in cm) of absorption bands caused by $n \to \pi^*$ transitions usually ranges from 1 to $10^3$ l/(mole cm), while $\pi \to \pi^*$ transitions exhibit values of $\varepsilon$ lying between $10^3$ and $10^6$ l/(mole cm).

The free-electron model can also provide a simple explanation for another important property of the energy levels of organic dyes, namely the position of the triplet levels relative to the singlet levels. In the ground state of the dye molecule, which in the free-electron model is the state with the $(1/2)N$ lowest levels filled, the spins of two electrons occupying the same level are necessarily antiparallel, resulting in zero spin. However, there is also the possibility of a parallel arrangement of the spins if one of the electrons is raised to a higher level. The resulting spin $S = 1$ can place itself either parallel, antiparallel or orthogonal with respect to an external magnetic field. The parallel arrangement of the spins of the two most energetic electrons thus gives a triplet state of the same energy as the singlet state with zero spin within the framework of one-electron functions. The Dirac formulation of the Pauli exclusion principle states the total wave function, including the spin function, must be antisymmetric with respect to the exchange of any two electrons. In the two-electron case considered this means that the following four antisymmetrical product wave functions can be used. Here spin $+1/2$ is described by the spin function $\alpha$, spin $-1/2$ by the spin function $\beta$:

$$\psi_s = \{\psi_m(1)\,\psi_n(2) + \psi_n(1)\,\psi_m(2)\}\{\alpha(1)\beta(2) - \alpha(2)\beta(1)\} ,$$

$$\psi_{T,+1} = \{\psi_m(1)\,\psi_n(2) - \psi_n(1)\,\psi_m(2)\}\{\alpha(1)\alpha(2)\} ,$$

$$\psi_{T,0} = \{\psi_m(1)\psi_n(2) - \psi_n(1)\psi_m(2)\}\{\alpha(1)\beta(2) + \alpha(2)\beta(1)\} ,$$

$$\psi_{T,-1} = \{\psi_m(1)\psi_n(2) - \psi_n(1)\psi_m(2)\}\{\beta(1)\beta(2)\} ,$$

where $\psi_s$ is the singlet, $\psi_{T,+1}$, $\psi_{T,-1}$, $\psi_{T,0}$ are the three triplet wave functions, and the argument 1 or 2 refers to electron no. 1 or no. 2. Because of the symmetry of the spin factor, the spatial factor of these functions is symmetric for the singlet wave function and antisymmetric for the triplet wave functions. These spatial factors of one-dimensional two-electron functions can be interpreted in terms of two-dimensional one-electron functions

$$\psi_{m,n}(s_1, s_2) + \psi_{n,m}(s_1, s_2)$$

for the singlet case and

$$\psi_{m,n}(s_1, s_2) - \psi_{n,m}(s_1, s_2)$$

for the triplet case.

In a crude approximation, one can think of electron 1 as travelling in the upper lobe of the electron cloud along the molecular chain, its coordinate being $s_1$, and electron 2 in the lower lobe, its coordinate being $s_2$, as depicted for three relative positions in the right half of Fig. 1.14a. In the left half of this figure these three positions A, B, C are plotted in a plane determined by $s_1$ and $s_2$ as cartesian coordinates.

For every configuration of the two electrons we can give the repulsion energy of the two electrons

$$V = \frac{e_0^2}{Dr} = \frac{e_0^2}{D[(s_1 - s_2)^2 + d^2]^{1/2}} ,$$

where $r$ is the distance between the two electrons, $d = 0.12$ nm is the distance between the centers of the upper and lower lobes, and $D$ denotes the dielectric constant in which the two electrons are imbedded.

The potential energy profile has a crest along the symmetry axis $s_1 = s_2$ (Fig. 1.14b). Since the spatial factor of the singlet function is symmetric with respect to this axis, it must have antinodes there. By contrast, the antisymmetric spatial factor of the triplet wave functions has a nodal line along $s_1 = s_2$. Consequently, the mean potential energy of the electrons in the singlet state with quantum numbers $n$, $m$ is higher than those in the triplet state with the same quantum numbers. This crude model gives the most important result that *for every excited singlet state there exists a triplet state of somewhat lower energy.* In addition the numerical calculation usually yields a value of the energy difference between corresponding singlet and triplet states that is correct within a factor of two or three.

Direct observation of absorptive transitions from the singlet ground state into triplet states is very difficult since the transitions are spin-forbidden and

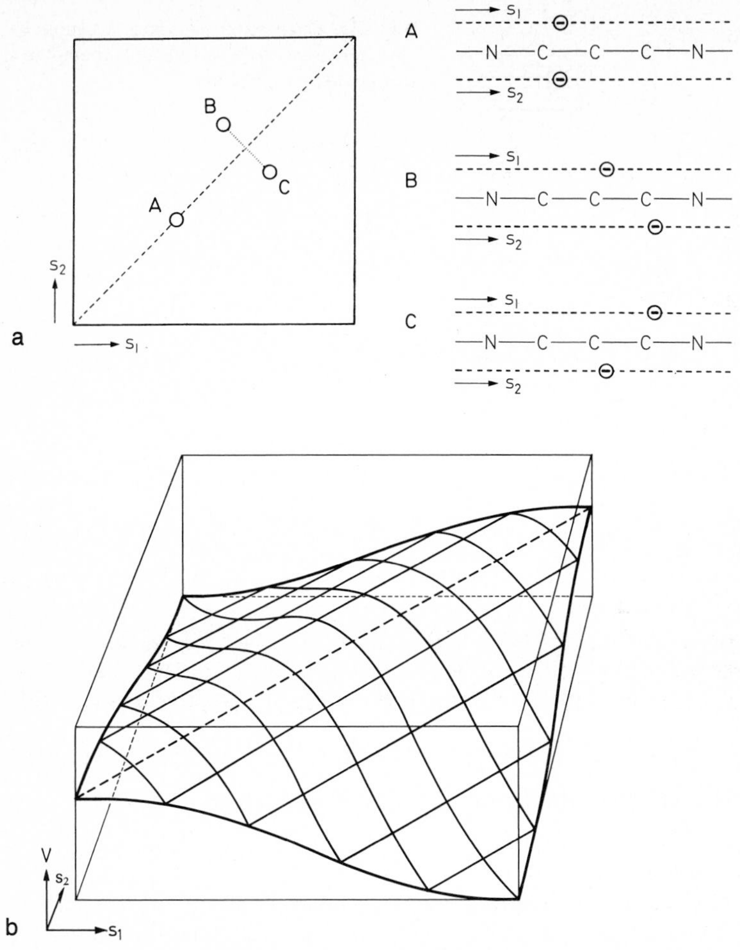


**Fig. 1.14.** **(a)** Three configurations of a two-electron system. The three points A, B, C in the cartesian coordinate system on the left belong to the three two-electron configurations pictured on the right. **(b)** Potential energy $V(s_1, s_2)$ of the two-electron system. (From Kuhn 1955)

the corresponding extinction coefficient at least a factor of $10^6$ lower than that in spin-allowed $\pi \rightarrow \pi^*$ transitions.

Proceeding finally from the two-electron wave function to the *N*-electron wave function, the following picture of the eigenstates of the dye molecule is obtained (Fig. 1.15). There is a ladder of singlet states $S_i (i = 1, 2, 3, \ldots)$ containing also the ground state *G*. Somewhat displaced towards lower energies there is the ladder of triplet states $T_i (i = 1, 2, 3, \ldots)$. The longest wavelength absorption is from G to $S_1$, the next absorption band from G to $S_2$, etc. By contrast the absorption from G to $T_i$ is spin-forbidden. The absorptions $G - S_1$, $G - S_2$, etc., will usually have differing transition moments. The fate of a molecule after it has undergone an absorptive transition and finds itself in an excited state $S_1$ or $S_2$ is discussed in the next section.

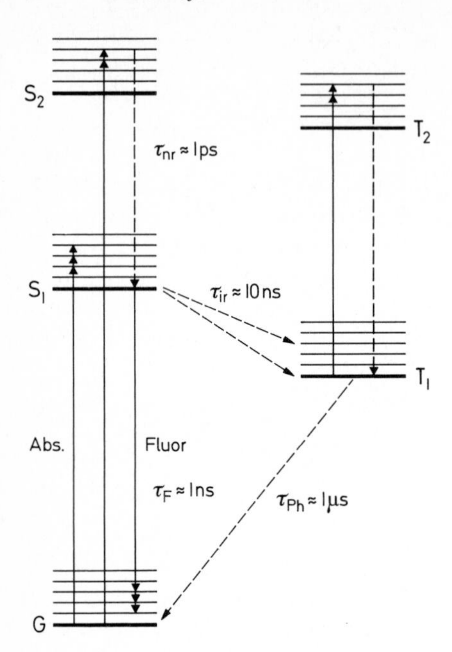


**Fig. 1.15.** Eigenstates of a typical dye molecule with radiative (*solid lines*) and nonradiative (*broken lines*) transitions

Recently a first example of dye laser emission of a radical was recorded (Ishizaka and Kotani 1985). The transient radical $H_2N$–⟨benzene ring⟩–S was photochemically produced by irradiation of a liquid solution of $H_2N$–⟨benzene ring⟩–S–S–⟨benzene ring⟩–$NH_2$ by a XeCl laser and excited by the absorption of one or more additional photons of XeCl laser radiation. In this case of radicals with an odd number of $\pi$ electrons, we have doublet and quartet instead of singlet and triplet states. Since the ground state is a doublet, and the quartet states, being doubly excited, lie higher than the corresponding doublet states, there is no intersystem crossing and consequently no quartet–quartet absorption with concomitant losses in dye laser operation.

## 1.3 Deactivation Pathways for Excited Molecules

There are very many processes by which an excited molecule can return directly or indirectly to the ground state. Some of these are schematically depicted in Fig. 1.15. It is the relative importance of these which mainly determines how useful a dye will prove in dye lasers. The one process that is directly used in dye lasers is the radiative transition from the first excited singlet state $S_1$ to the ground state G. If this emission, termed "fluorescence", occurs spontaneously, its radiative lifetime $\tau_{rf}$ is connected with the Einstein coefficient for spontaneous emission $A$ and the oscillator strength $f$ of the pertinent absorption band by

$$A \equiv 1/\tau_{rf} = (8\pi^2\mu^2 e_0^2/m_0 c_0)\,\tilde{\nu}^2 f\ ,$$

where $\mu$ is the refractive index of the solution, $e_0$ is the charge and $m_0$ is the mass of the electron, $c_0$ is the velocity of light, and $\tilde{\nu}$ denotes the wave number of the center of the absorption band. This relation is valid if the half-width of the emission band is small and its position is not shifted significantly from that of the absorption band. Since the $f$ values of the transition are near unity in most dyes, the radiative lifetime $\tau_{rf}$ is typically of the order of a few nanoseconds. Generally, however, the fluorescence spectrum is broad and shows considerable Stokes shift. In this case the radiative lifetime can be computed with the aid of the following relation (Strickler and Berg 1962):

$$\frac{1}{\tau_{rf}} = 2.88\times 10^{-9}\mu^2 \frac{\int F(\tilde{\nu})\,d\tilde{\nu}}{\int \tilde{\nu}^{-3}F(\tilde{\nu})\,d\tilde{\nu}} \int\limits_{\substack{\text{longest}\\ \text{wavelength}\\ \text{absorption band}}} \frac{\varepsilon(\tilde{\nu})\,d\tilde{\nu}}{\tilde{\nu}}$$

where $F(\tilde{\nu}) = dQ/d\tilde{\nu}$ is the fluorescence spectrum (in quanta $Q$ per wave number) and $\varepsilon(\tilde{\nu})$ is the molar decadic extinction coefficient. There also exists the possibility that the energy hypersurface of the excited state will approach closely enough to the ground state for the molecule to tunnel through the barrier between them. It is then found in a very highly excited vibronic level of the ground state. This process is generally termed "internal conversion" and its rate constant is $k_{SG}$. There is very little theoretical knowledge about this and other related radiationless processes (Robinson and Frosch 1963; Nitzan and Jortner 1973a,b; E.J. Heller and Brown 1983; Kommandeur 1983). Experimental results (which range over several orders of magnitude of $k_{SG}$) will be discussed in detail in Chap. 5.

The difficult problem of the measurement of picosecond vibrational relaxation times of dye molecules in liquid solution in the ground and excited singlet states has been solved with different methods (Malley and Mourou 1974; Mourou and Malley 1974; Magde and Windsor 1974a; Ricard and Ducuing 1974; Ricard et al. 1972; Lin and Dienes 1973a; Mourou 1975; Laubereau et al. 1975; Ricard and Ducuing 1975; Shilov et al. 1976; Penzkofer and Falkenstein 1976; Penzkofer et al. 1976; Shank et al. 1977). Especially noteworthy is the measurement of vibrational relaxation times in a large dye molecule (coumarin 6) in the vapor phase under collisionless conditions, which showed that even in this case vibrational relaxation times are still only a few picoseconds (Maier et al. 1977) or even less (Erskine et al. 1984; A.J. Taylor et al. 1984; Rosker et al. 1986; Ackerman et al. 1982; Rentzepis and Bondybey 1984). A direct measurement of the intermolecular energy transfer between dye and solvent molecules is reported by Seilmeier et al. (1984).

The internal conversion between $S_2$ (and higher excited states) and $S_1$ is usually extremely fast, taking place in less than $10^{-11}$ s. This is the reason why fluorescence quantum spectra of dyes generally do not depend on the excitation wavelength. The only known exception to these general rules was for a long time the aromatic molecule azulene and ist derivatives. The fluorescence of this molecule is directly from $S_2$ to the ground state, while internal conversion from $S_1$ to the ground state is so fast that no fluorescence originating in $S_1$ can be detected except with very high peak power laser pulses (Birks 1972). Recently several other examples of a weak fluorescence from $S_2$ to the ground state have been discovered (Steer and Ramamurthy 1988).

The radiationless transition from an excited singlet state to a triplet state can be induced by internal perturbations (spin-orbit coupling, substituents containing nuclei with high atomic number) as well as by external perturbations (paramagnetic collision partners, like $O_2$ molecules in the solution or solvent molecules containing nuclei of high atomic number). These radiationless transitions are usually termed "intersystem crossing" and have the rate constant $k_{ST}$.

If some fluorescence quenching agent Q is present in a solution or gas mixture in a known concentration [Q], its contribution to the radiationless deactivation can be separated and expressed by a rate constant $k'_{SG} = k_Q[Q]$. Since for most quenchers quenching occurs at every encounter with an excited dye molecule, $k_Q$ equals the diffusion-controlled bimolecular rate constant, given by $k_Q = 8RT/2000\eta$, where $R$ is the gas constant, $T$ the absolute temperature and $\eta$ the viscosity of the solvent (Osborne and Porter 1965).

The quantum yield of fluorescence, $\phi_f$, is then defined as the ratio of radiative and nonradiative transition rates

$$\phi_f = \frac{1/\tau_{rf}}{1/\tau_{rf} + k_{ST} + k_{SG} + k_Q[Q]} .$$

The denominator is the reciprocal of the observed fluorescence lifetime $\tau_f$. The quantum yield of fluorescence can thus be calculated from the measured fluorescence lifetime $\tau_f$ and the radiative lifetime $\tau_{rf}$, obtained through the absorption spectrum: $\phi_f = \tau_f/\tau_{rf}$. This method usually gives good results, whereas the direct experimental measurement of absolute quantum yields is very difficult (see Chap. 5).

The radiative transition from $T_1 \rightarrow G$ is termed "phosphorescence". In principle, one should be able to estimate the radiative lifetime $\tau_{rp}$ of the phosphorescence on the basis of the $G \rightarrow T_1$ absorption. Since this transition is normally completely obscured by absorption due to impurities, the phosphorescence radiative lifetime can be obtained from the quantum yield of phosphorescence, $\phi_p$, and the observed phosphorescence lifetime $\tau_p$, as for fluorescence: $\tau_p = \phi_p \tau_{rp}$. As expected for spin-forbidden transitions, it is extremely long, ranging from milliseconds to many seconds for molecules with small spin-orbit coupling. Consequently, even relatively slow quenching processes can lead to radiationless deactivation in liquid solution. Hence the observed $\tau_p$ is generally very low in liquid solution, becoming appreciable only at low temperatures, e.g. 77 K in solid solutions. (A similar comment can be made regarding the long-lived emission resulting from the above-mentioned $\pi^* \leftarrow n$ transitions, which are usually very weak, if any fluorescence is observed at all.) Note that $\phi_p$ is the true quantum yield of phosphorescence. It is defined as the ratio of the phosphorescence rate to the total rate of radiative and radiationless deactivation

$$\phi_p = \frac{1/\tau_{rp}}{1/\tau_{rp} + k_{TG} + k_Q[Q]} .$$

For comparison, one often quotes the apparent quantum yield of phosphorescence $\phi_p^+$, which relates the rate of phosphorescence emission to the rate of light absorption. The relation between the two is $\phi_p^+ = \phi_p \phi_T$. Here $\phi_T$ is the triplet formation efficiency, i.e. the ratio of the intersystem crossing rate $k_{ST}$ to the total rate of deactivation of the fluorescent level:

$$\phi_T = \frac{k_{ST}}{1/\tau_{rf} + k_{ST} + k_{SG} + k_Q \cdot [Q]} .$$

Usually no measurable phosphorescence can be observed in liquid solutions. Here $\tau_p$, which is identical with the triplet lifetime $\tau_T$, is often determined by flash spectroscopy from the vanishing of the triplet–triplet absorption bands. Assuming the value of the radiative lifetime at elevated temperatures is the same as at low temperatures, the quantum yield $\phi_p$ in liquid solutions can be obtained. To give a typical example, for the eosin-di-anion in methanol ($7 \times 10^{-5}$ mole/l) the phosphorescence lifetime at room temperature is $\tau_p = 1.5$ ms, the apparent quantum yield of phosphorescence $\phi_p^+ = 3.9 \times 10^{-3}$, the triplet quantum yield $\phi_T = 0.45$, hence the radiative lifetime $\tau_{rp} = 0.2$ s (Parker and Hatchard 1961).

The lowest singlet and triplet states can also be deactivated by a long-range radiationless energy transfer to some other dye molecule. For this to happen, the absorption band of the latter, "acceptor" molecule must overlap the fluorescence or phosphorescence band of the former, "sensitizer" molecule. The so-called critical distance, at which the quantum yield is reduced by a factor of two, can reach values of 10 nm for the case of singlet–singlet energy transfer. For details of these and some other processes resulting in "delayed fluorescence", which is less important in the present context, the reader is referred to Parker (1968).

Other deactivation pathways for the excited molecule include reversible and irreversible photochemical reactions, e.g. protolytic reactions, cis-trans-isomerizations, dimerizations, ring-opening reactions and many others. Some of these which are of major importance in dye lasers are discussed in Chap. 5.

The absorptive transitions of excited molecules into higher excited states can constitute a loss mechanism for the pump as well as the dye-laser radiation. It thus is of utmost importance for dye-laser operation.

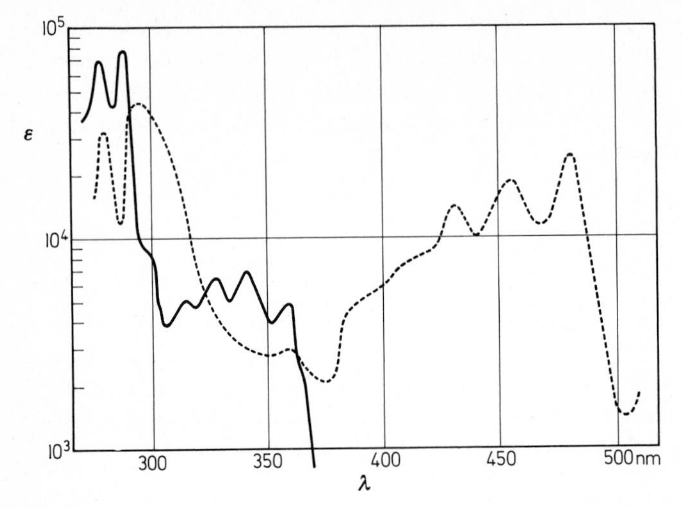


**Fig. 1.16.** Triplet–triplet absorption spectrum (*broken line*) of 1,2-benzanthracene in hexane (*solid line*: absorption spectrum from ground to excited states). (From Labhart 1964)

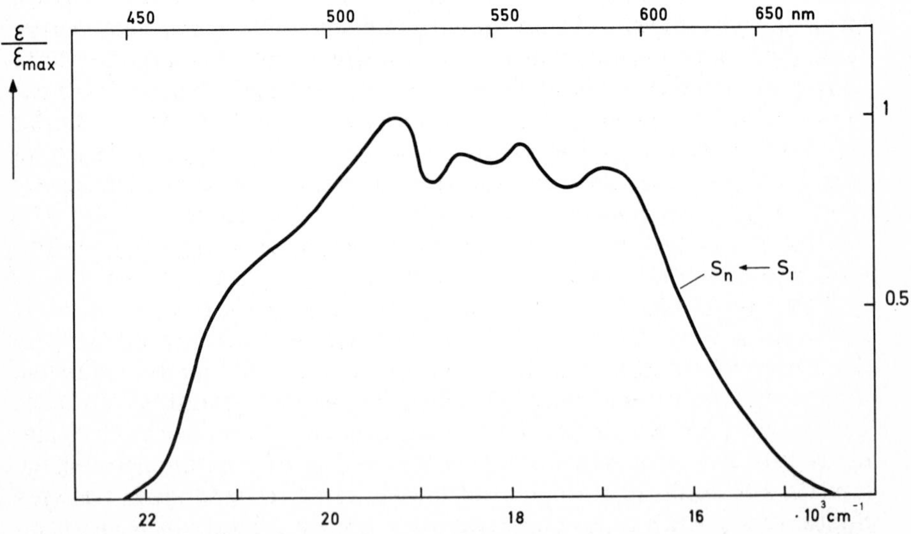


**Fig. 1.17.** $S_1 \rightarrow S_n$-absorption spectrum of 1,2-benzanthracene in ethanol. (From Müller and Sommer 1969)

Triplet–triplet absorption is measured by flash photolysis as well as by photostationary methods (Porter and Windsor 1958; Labhart 1964; Zanker and Miethke 1957b). An example of dye spectra thus obtained is given in Fig. 1.16. It is much more difficult to measure the absorption from the short-lived excited singlet state and has only become possible through the use of laser spectroscopy methods (Novak and Windsor 1967; Müller 1968; Bonneau et al.

1968). So far spectra for only about 20 molecules have been recorded (Birks 1970). An example is given in Fig. 1.17.

## 1.4 Laser-Pumped Dye Lasers

### 1.4.1 Oscillation Condition

From the foregoing discussion of the spectroscopic properties of dyes one might come to the conclusion that there are two possible ways, at least in principle, of using an organic solution as the active medium in a laser: One might utilize either the fluorescence or the phosphorescence emission. At first sight the long lifetime of the triplet state makes phosphorescence look more attractive. On the other hand, due to the strongly forbidden transition, a very high concentration of the active species is required to obtain an amplification factor large enough to overcome the inevitable cavity losses. In fact, for many dyes this concentration would be higher than the solubility of the dyes in any solvent. A further unfavorable property of these systems is that there will almost certainly be losses due to triplet–triplet absorption. It must be remembered that triplet–triplet absorption bands are generally very broad and diffuse and the probability they will overlap the phosphorescence band is high. Because of these difficulties no laser using the phosphorescence of a dye has yet been reported [a preliminary report (Morantz et al. 1962 and Morantz 1963) was evidently in error]. The possibility cannot be excluded, however, that further study of phosphorescence and triplet–triplet absorption in molecules of different types of chemical constitution might eventually lead to a triplet dye laser operating, for example, at the temperature of liquid nitrogen. On the other hand, the probability for this seems low at present and these systems will not be considered here. For a more detailed discussion of phosphorescent systems the reader is referred to Lempicki and Samelson (1966).

If the fluorescence band of a dye solution is utilized in a dye laser, the allowed transition from the lowest vibronic level of the first excited singlet state to some higher vibronic level of the ground state will give a high amplification factor even at low dye concentrations. The main complication in these systems is the existence of the lower-lying triplet states. The intersystem crossing rate to the lowest triplet state is high enough in most molecules to reduce the quantum yield of fluorescence to values substantially below unity. This has a twofold consequence: Firstly, it reduces the population of the excited singlet state, and hence the amplification factor; and secondly, it enhances the triplet–triplet absorption losses by increasing the population of the lowest triplet state. Assume a light-flux density which slowly rises to a level $P$ [quanta $s^{-1}$ $cm^{-2}$], a total molecular absorbing cross-section $\sigma$ [$cm^2$][1], a quantum yield $\phi_T$ of

[1] The absorption cross-section $\sigma$ can be determined from the molar decadic extinction coefficient $\varepsilon$ by $\sigma = 0.385 \times 10^{-20} \varepsilon$. Here $\sigma$ is given in $cm^2$, if $\varepsilon$ is measured in liter/(mole cm).

triplet formation, a triplet lifetime $\tau_T$, populations of the triplet and ground state of $n_T$ and $n_0$ [$cm^{-3}$], respectively, and, neglecting the small population of the excited singlet state, a total concentration of dye molecules of $n = n_0 + n_T$. A steady state is reached when the rate of triplet formation equals the rate of deactivation:

$$P\sigma n_0 \phi_T = n_T/\tau_T \ . \tag{1.1}$$

Thus, the fraction of molecules in the triplet state is given by

$$n_T/n = P\sigma\phi_T\tau_T/(1+P\sigma\phi_T\tau_T) \ . \tag{1.2}$$

Assuming some typical values for a dye, $\sigma = 10^{-16}\,cm^2$, $\phi_T = 0.1$ (corresponding to a 90% quantum yield of fluorescence), and $\tau_T = 10^{-4}\,s$, the power to maintain half of the molecules in the triplet state is $P_{1/2} = 10^{21}$ quanta $s^{-1}\,cm^{-2}$, or an irradiation of only 1/2 kW $cm^{-2}$ in the visible part of the spectrum. This is much less than the threshold pump power calculated below. Hence a slowly rising pump light pulse would transfer most of the molecules to the triplet state and deplete the ground state correspondingly. On the other hand, the population of the triplet level can be held arbitrarily small, if the pumping light flux density rises fast enough, i.e. if it reaches threshold in a time $t_r$ which is small compared to the reciprocal of the intersystem crossing rate $t_r \ll 1/k_{ST}$. Here $t_r$ is the risetime of the pump light power, during which it rises from zero to the threshold level. For a typical value of $k_{ST} = 10^7\,s^{-1}$, the risetime should be less than 100 ns. This is easily achieved, for example with a giant-pulse laser as pump light source, since giant pulses usually have risetimes of 5–20 ns. In such laser-pumped dye-laser systems one may neglect all triplet effects in a first approximation.

We thus can restrict our discussion to the singlet states. Molecules that take part in dye-laser operation have to fulfill the following cycle (Fig. 1.18): Absorption of pump radiation at $\tilde{\nu}_p$ and with cross-section $\sigma_p$ lifts the molecule from the ground state with population $n_0$ into a higher vibronic level of the first (or second) excited singlet state $S_1$ (or $S_2$) with a population $n'_1$ (or $n'_2$). Since the radiationless deactivation to the lowest level of $S_1$ is so fast (see Sect. 1.3), the steady-state population $n'_1$ is negligibly small, provided the temperature is not so high that this vibronic level is already thermally populated by the Boltzmann distribution of the molecules in $S_1$. At room temperature, $kT = 200\,cm^{-1}$, so that this is not the case. Stimulated emission then occurs from the lowest vibronic level of $S_1$ to higher vibronic levels of G. Again the population $n'_0$ of this vibronic level is negligible since the molecules quickly relax to the lowest vibronic levels of G.

It is easy then to write down the oscillation condition for a dye laser (Schäfer 1968; Schäfer et al. 1968). In its simplest form a dye laser consists of a cuvette of length $L$ [cm] with dye solution of concentration $n$ [$cm^{-3}$] and of two parallel end windows carrying a reflective layer each of reflectivity $R$ for the laser resonator. With $n_1$ molecules/$cm^3$ excited to the first singlet state, the

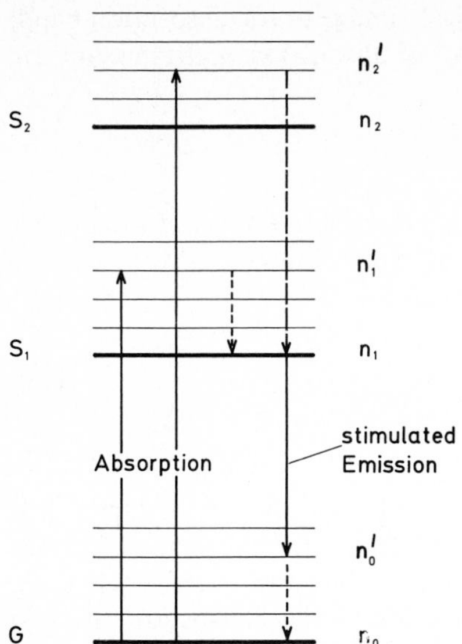


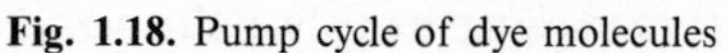

**Fig. 1.18.** Pump cycle of dye molecules

dye laser will start oscillating at a wave number $\tilde{\nu}$, if the overall gain is equal to or greater than one:

$$\exp\left[-\sigma_\mathrm{a}(\tilde{\nu})n_0 L\right] R \exp\left[+\sigma_\mathrm{f}(\tilde{\nu})n_1 L\right] \geqq 1 \ . \tag{1.3}$$

Here $\sigma_\mathrm{a}(\tilde{\nu})$ and $\sigma_\mathrm{f}(\tilde{\nu})$ are the cross-sections for absorption and stimulated fluorescence at $\tilde{\nu}$, respectively, and $n_0$ is the population of the ground state. The first exponential term gives the attenuation due to reabsorption of the fluorescence by the long-wavelength tail of the absorption band. The attenuation becomes the more important, the greater the overlap between the absorption and fluorescence bands. The cross-section for stimulated fluorescence is related to the Einstein coefficient $B$

$$\sigma_\mathrm{f}(\tilde{\nu}) = g(\tilde{\nu}) B h \tilde{\nu}/c_0 \qquad \int\limits_{\substack{\text{fluorescence}\\ \text{band}}} g(\tilde{\nu})d\tilde{\nu} = 1 \ . \tag{1.4}$$

Substituting the Einstein coefficient $A$ for spontaneous emission according to

$$B = \frac{1}{8\pi\tilde{\nu}^2} A \frac{1}{h\tilde{\nu}} \tag{1.5}$$

and realizing that $g(\tilde{\nu}) A \phi_\mathrm{f} = Q(\tilde{\nu})$, the number of fluorescence quanta per wave number interval, one obtains

$$\sigma_\mathrm{f}(\tilde{\nu}) = \frac{1}{8\pi c_0 \tilde{\nu}^2} \frac{Q(\tilde{\nu})}{\phi_\mathrm{f}} \ . \tag{1.6}$$

Since the fluorescence band usually is a mirror image of the absorption band, the maximum values of the cross-sections in absorption and emission are found to be equal:

$$\sigma_{f,\max} = \sigma_{a,\max} \ . \tag{1.7}$$

Taking the logarithm of (1.3) and rearranging it leads to a form of the oscillation condition which makes it easier to discuss the influence of the various parameters:

$$\frac{S/n+\sigma_a(\tilde{\nu})}{\sigma_f(\tilde{\nu})+\sigma_a(\tilde{\nu})} \leqq \gamma(\tilde{\nu}) \ , \tag{1.8}$$

where $S = (1/L)\ln(1/R)$ and $\gamma(\tilde{\nu}) = n_1/n$.

The constant $S$ on the left-hand side of (1.8) only contains parameters of the resonator, i.e. the active length $L$, and reflectivity $R$. Other types of losses, like scattering, diffraction, etc., may be accounted for by an effective reflectivity, $R_{\text{eff}}$. The value $\gamma(\tilde{\nu})$ is the minimum fraction of the molecules that must be raised to the first singlet state to reach the threshold of oscillation. One may then calculate the function $\gamma(\tilde{\nu})$ from the absorption and fluorescence spectra for any concentration $n$ of the dye and value $S$ of the cavity. In this way one finds the frequency for the minimum of this function. This frequency can also be obtained by differentiating (1.8) and setting $d\gamma(\tilde{\nu})/d\tilde{\nu} = 0$. This yields

$$\frac{\sigma_a'(\tilde{\nu})}{\sigma_f'(\tilde{\nu})+\sigma_a'(\tilde{\nu})}(\sigma_f(\tilde{\nu})+\sigma_a(\tilde{\nu})) = \frac{S}{n} \tag{1.9}$$

(prime means differentiation with respect to $\tilde{\nu}$) from which the start-oscillation frequency can be obtained.

Figure 1.19 represents a plot of the laser wavelength $\lambda$ (i.e. the wavelength at which the minimum of $\gamma(\tilde{\nu})$ occurs) versus the concentration with $S$ a fixed parameter. Similarly, Fig. 1.20 shows the laser wavelength versus the active length $L$ of the cuvette with dye solution, with the concentration of the dye as a parameter. Both figures apply to the dye 3,3′-diethylthiatricarbocyanine (DTTC). These diagrams demonstrate the wide tunability range of dye lasers

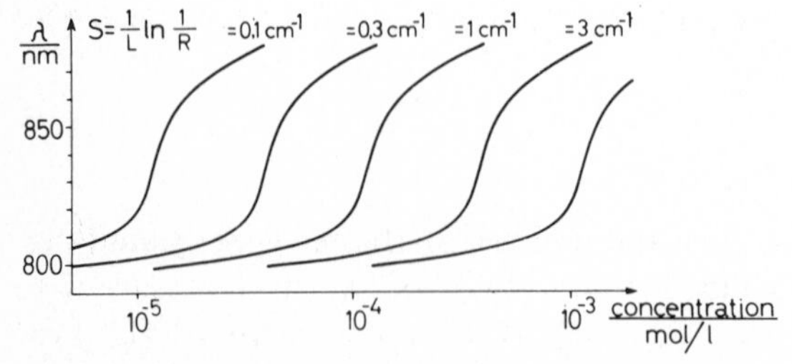


**Fig. 1.19.** Plot of calculated laser wavelength vs concentration of the laser dye 3,3′-diethylthiatricarbocyanine bromide, with $S$ as a parameter. (From Schäfer 1968)

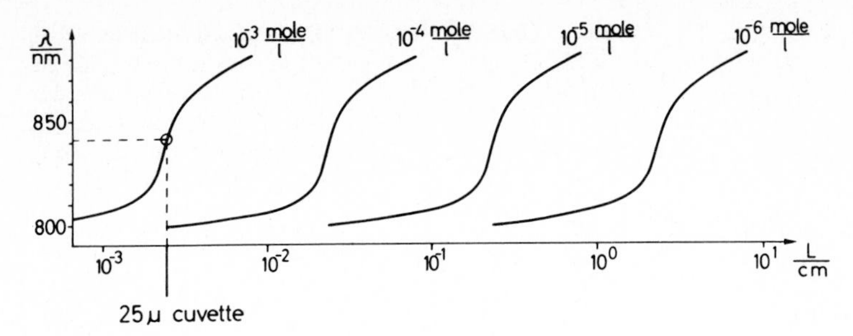


**Fig. 1.20.** Plot of calculated laser wavelength vs active length of the laser cuvette, with the concentration of the solution of the dye 3,3′-diethylthiatricarbocyanine bromide as a parameter. (From Schäfer 1968)

which can be induced by changing the concentration of the dye solution, or length, or $Q$ of the resonating cavity. They also show the high gain which permits the use of extremely small active lengths.

The absorbed power density $W$ necessary to maintain a fraction $\gamma$ of the molecular concentration $n$ in the excited state is

$$W = \gamma n h c_0 \tilde{\nu}_p / \tau_f \ , \tag{1.10}$$

and the power flux, assuming the incident radiation is completely absorbed in the dye sample,

$$P = W/n\sigma = \gamma h c_0 \tilde{\nu}_p / \tau_f \sigma \ , \tag{1.11}$$

where $\tilde{\nu}_p$ is the wave number of the absorbed pump radiation. If the radiation is not completely absorbed, the relation between the incident power $W_{in}$ and the absorbed power is $W = W_{in}[1 - \exp(-\sigma_p n_0 L)]$. Since in most cases $n \approx n_0$, this reduces for optically thin samples to $W = W_{in}\sigma_p n L$. The threshold incident power flux, $P_{in}$, then is

$$P_{in} = \gamma h c_0 \tilde{\nu}_p / \tau_f \sigma_p \ .$$

In the above derivation of the oscillation condition and concentration dependence of the laser wavelength, broad-band reflectors have been assumed. The extension to the case of wavelength-selective reflectors and/or dispersive elements in the cavity is straightforward and will not be treated here.

### 1.4.2 Practical Pumping Arrangements

A simple structure for a dye-solution laser was used in many of the early investigations. It is still useful for exploratory studies of new dyes. It consists of a square spectrophotometer cuvette filled with the dye solution (Fig. 1.21) which is excited by the beam from a suitable laser. As shown in the figure, the resonator is formed by the two glass-air interfaces of the polished sides of the

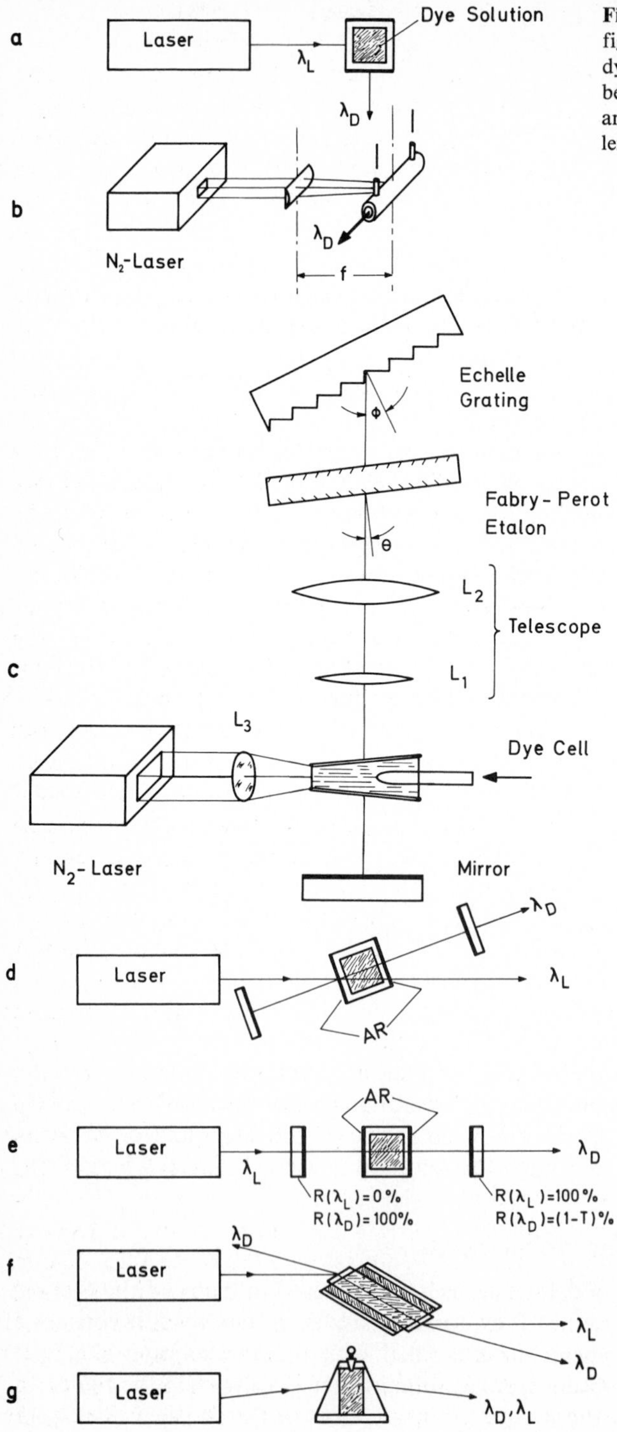


**Fig. 1.21a–g.** Resonator configurations for laser-pumped dye lasers. ($\lambda_L$ pumping-laser beam, $\lambda_D$ dye-laser beam, AR antireflection coating, $f$ focal length of cylindrical lens

cuvette (heavy lines in Fig. 1.21 a) and the exciting laser and dye-laser beams are at right angles. Reflective coatings on the windows consisting of suitable metallic or multiple dielectric layers enhance the $Q$ of the resonator. It is also possible to use antireflective coatings or Brewster windows at the cuvette and separate resonator mirrors. In the transverse pumping arrangement (Fig. 1.21 a) the population inversion in the dye solution is nonuniform along the exciting laser beam, since the exciting beam is attenuated in the solution. Consequently, at a higher concentration, the threshold might only be reached in a thin layer directly behind the entrance window of the exciting beam. This gives rise to large diffraction losses and beam divergence angles. If the concentration is not too high, say, less than $10^{-4}$ molar, the measured threshold pump power in such an arrangement can be used to test the value calculated from the oscillation condition. Using (1.8) and (1.11), known spectral data, and the observed spontaneous decay time of fluorescence $\tau_f = 2$ ns, one computes a threshold pump power flux of $P = 1.2 \times 10^5$ W/cm$^2$ for the exciting giant pulse ruby laser and a $10^{-4}$ molar solution of 3,3′-diethylthiatricarbocyanine bromide in methanol in a 1 cm square spectrophotometric cuvette. The experimental value is $3.0 \times 10^5$ W/cm$^2$, while the calculated and observed values for a $10^{-3}$ molar solution are $1.2 \times 10^4$ W/cm$^2$ and $1.2 \times 10^5$ W/cm$^2$, respectively (Volze 1969). The higher value of the observed threshold for the $10^{-4}$ molar solution is easily accounted for by neglection of triplet losses. The risetime of the exciting pulse was 8 ns in these experiments, while the spontaneous lifetime of the dye is 2 ns. An estimate of the quantum yield of fluorescence gives $\phi_f = 0.4$. Hence, an appreciable fraction of the molecules has already been converted to the triplet state at threshold. The larger discrepancy between calculated and observed values for the $10^{-3}$ molar solution can be traced to the high diffraction losses. In this experiment the exciting laser beam is attenuated to $1/e$ of incident intensity in a depth of only 85 μm. A similar transverse arrangement is often used for nitrogen laser-pumped dye lasers (Fig. 1.21 b). Here the emission of a nitrogen laser is focused by a cylindrical lens into a line that coincides with the axis of a quartz capillary of inside diameter $d \approx 1$ mm. The transmitted pump radiation is reflected by an aluminum coating at the back surface of the capillary tube. If the beam divergence of the nitrogen laser is $\alpha$ and the length of the cylindrical lens $f$, then the width of the focal line is $H = f\alpha$ and is best so chosen that $H$ is about one fourth of the inside diameter of the tube. The dye concentration $c$ is then adjusted so that the absorption $A = 2\varepsilon c d \approx 1$. The endfaces of the tube can either be normal to the axis and act directly as resonator mirrors, or set at the Brewster angle for use with external mirrors.

Another arrangement developed early for the pumping of dye lasers with a nitrogen laser is shown in Fig. 1.21 c (Hänsch 1972). The dye cell is 10 mm long and is made of 12-mm diameter Pyrex tubing. Antireflection-coated quartz windows are sealed to the ends of the cell under a wedge angle of about 10 degrees to avoid internal cavity effects. The dye solution is transversely circulated by means of a small centrifugal pump at a flow speed of about 1 m/s in the active region. The nitrogen laser emission is focused by a spherical lens of 135 mm focal length into a line of about 0.15 mm width at the inner cell wall. To provide a near-circular active cross-section, the dye concentration is adjusted so that the penetration of the pump light is also of the order of 0.15 mm. A plane outcoupling mirror is mounted at one end of the optical cavity at a distance of 50 mm. Because of the large ratio of this distance to the diameter of the active region, the latter acts as an active pinhole and the emerging beam is essentially diffraction-limited and has a beam divergence (at 600 nm emission wavelength) of 2.4 mrad. The other end of the cavity contains an inverted telescope, consisting of two lenses $L_1$ and $L_2$ of focal lengths $f_1 = 8.5$ mm and $f_2 = 185$ mm (antireflection-coated multielement systems corrected for spherical aberration and coma within $\lambda/8$) and a separation between cell and lens $L_1$ of 75 mm. The lenses are used slightly off-axis to avoid back reflection and etalon effects. The initial waist size of 0.08 mm is thus enlarged to 4 mm and the beam divergence is reduced to 0.05 mrad. This is especially important if a diffraction grating in Littrow mounting is to be used instead of a second mirror. In this case, if the beam were unexpanded, it would eventually burn the grating and the spectral resolution would be very low.

A better configuration for laser-pumped dye lasers is the longitudinal arrangement (Sorokin et al. 1966b) of Fig. 1.21e. In this arrangement the exciting laser beam passes through one of the resonator mirrors of the dye laser. For best effect this is coated by a dielectric multilayer mirror with a very low reflection coefficient at the exciting-laser wavelength and a high one at the dye-laser wavelength. Concentration and depth of the solution should be adjusted so that the excitation power is very nearly uniform over the whole volume of the cuvette. This results in a much lower beam divergence. Typical values of 3–5 mrad were reported. It is often more convenient to orient the exciting-laser beam a few degrees to the normal of the mirror for spatial separation of the exciting-laser and the dye-laser beams. If the dimension of the cuvette is much smaller than the cavity length, the exciting-laser beam may pass by the side of the laser mirror (Fig. 1.21d). This eliminates the requirement for a special mirror system that must withstand the full power of the exciting laser.

Another advantage of the longitudinal pumping arrangement is the possibility of using an extremely small depth of dye solution. Volze (1969) described a dye laser consisting of two mirrors in direct contact with the dye solution and kept apart by spacers ranging from 100 μm down to 5 μm. The mirrors had 80% transmission at the exciting ruby-laser wavelength, and a constant 98% reflection over the range of the dye-laser emission. Using a $10^{-3}$ molar solution of 3,3′-diethylthiatricarbocyanine bromide (or other anion), the observed threshold for a 5 μm spacer is 350 kW/cm$^2$, while the computed value is 120 kW/cm$^2$. This is about the same error factor as in the above transverse pumped case with the $10^{-4}$ molar solution. Here, too, the error should be due mainly to neglect of the triplet losses. The operation of a dye laser of only 5 μm width demonstrates the extremely high gain in the dye solution. This experiment indicates an amplification factor of 1.02 per 5 μm which implies a gain of $G = 170$ dB/cm.

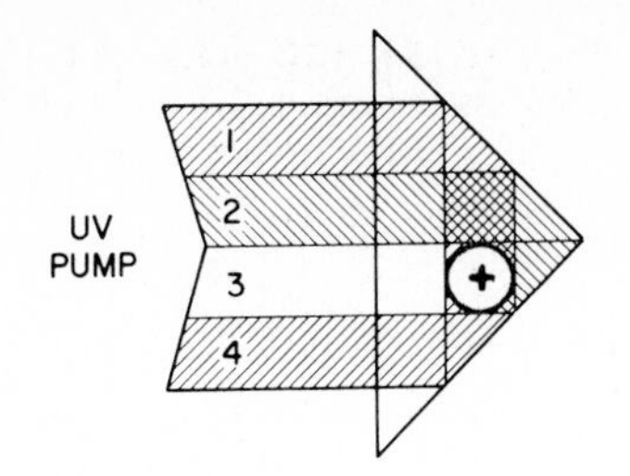


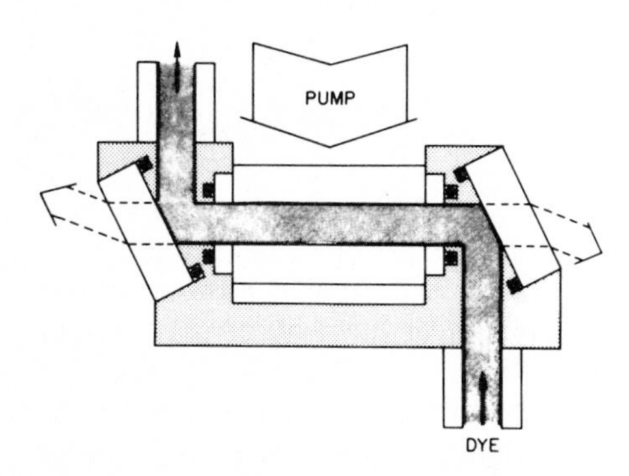


**Fig. 1.22.** Cross-sections through a Bethune-type cell. (From Bethune 1981)

An ingenious improvement of transversal pumping was reported by Bethune (1981). It takes advantage of a dye cell in the form of a roof-top prism with a longitudinal bore, as shown in Fig. 1.22. If a pump beam enters the hypotenuse face normally, the four partial beams 1 – 4, indicated by different hatching in the figure, illuminate the bore from four sides, thus creating a more uniform illumination of the dye solution in the bore than can be obtained with the normal transverse pumping using only a cylindrical lens. A quantitative treatment shows that this solution is still not completely satisfactory, since the pump intensity of each partial beam decreases towards the center by the absorption in the dye solution, resulting in a minimum of the intensity distribution near the axis, thus creating a ring-shaped inversion for any value of absorption coefficient that gives reasonable pump-light absorption.

An improvement circumventing the described disadvantage was described by Schäfer (1986). It makes use of a 45° glass cone with an axial bore, as shown in cross-section in Fig. 1.23. A pump beam entering normal to the cone base is totally reflected at the conical surface and impinges normally on the bore containing the dye solution. Since now the pump radiation is focused on the axis, one can easily calculate that the intensity has a maximum on the axis and perfect rotational symmetry is obtained.

Superradiant dye lasers can be pumped either by a normal giant-pulse laser or by ultrashort pulses. In the first case (Fig. 1.21 f), the exciting-laser pulse passes through a cuvette of small optical length compared to the length of the exciting pulse.

Typically, the solution is contained in a 5-cm Brewster cuvette with quartz tubing and windows (Volze 1969). The solvent DMSO has the same refractive

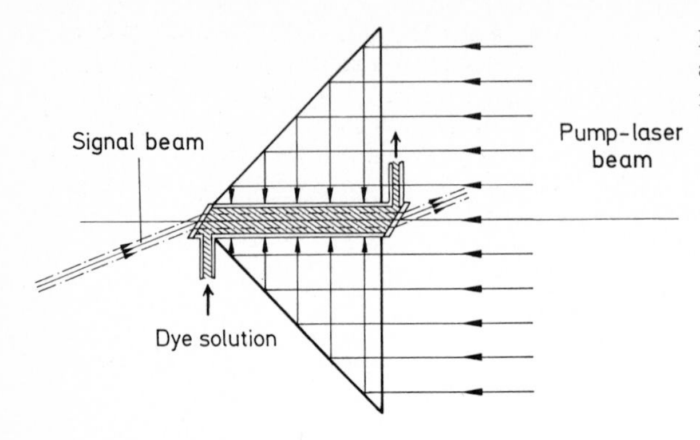


**Fig. 1.23.** Cross-section through an axicon dye cell. (Adapted from Schäfer 1986)

index ($n = 1.48$) as the cuvette walls. The cuvette tubing is surrounded by a cylinder containing a suitable dye solution in a solvent of higher refractive index so that fluorescence light incident on the cuvette walls is absorbed. This eliminates the unwanted closed-feedback path. Near the peak of the exciting-laser pulse the spatial distribution of pump light in the cuvette is nearly stationary. With this assumption, the amplification can be computed for spontaneously emitted fluorescence quanta that originate at one window and travel along the optical axis towards the other window. As expected, in this model the superradiant dye-laser intensities are almost equal in both forward and backward directions, and the beam divergence is determined by the aspect ratio of the cuvette. A plot of the superradiant versus the exciting intensity for this configuration with a $3 \times 10^{-5}$ molar solution of DTTC is shown in Fig. 1.24.

For traveling-wave excitation of a superradiant dye laser, the solution is contained in a wedged (10°) cell typically 2 cm in depth (Mack 1969). A methanol solution of DTTC or some similar dye is pumped by a ruby-laser pulse of a few ps half-width and 5 GW peak power. This results in an almost complete inversion of the dye solution at the position of the exciting pulse with correspondingly high gain. For a pulsewidth of 5 ps, corresponding to an inverted region of 2 mm, spontaneously emitted fluorescence quanta passing through this region would experience a maximum gain of 50 dB in a $10^{-4}$ molar solution of DTTC. It is assumed that no saturation effects occur and that the relaxation time from the initial Franck-Condon state to the upper laser level is negligible compared to the duration of the exciting pulse. Since the vibrational relaxation time of dyes in solution is generally of the order of a few ps, and since saturation certainly must occur as the exciting pulse travels through the cuvette, the actual gain is less than the above value. The traveling-wave nature of the superradiant dye-laser emission may be verified by measuring the forward to backward ratio of the dye-laser emission. It turned out to be 100 to 1 in this experiment, and the beam divergence was 15 mrad.

The polarization of the dye laser beam in the longitudinal and transverse arrangements is determined mainly by the polarization of the exciting laser beam, the relative orientation of the transition moments in the dye molecule for the pumping and laser transitions, and the rotational diffusion–relaxation

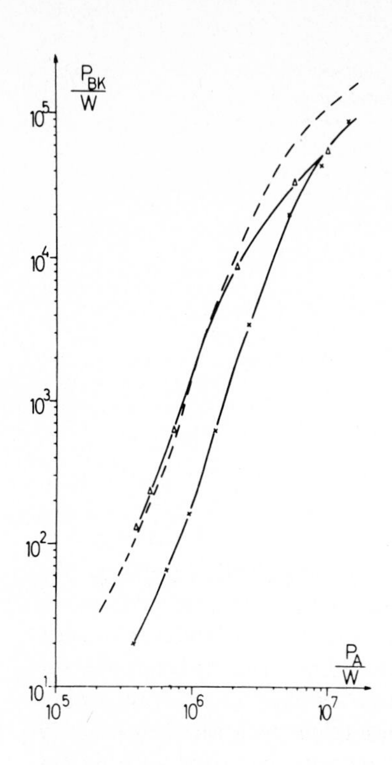


**Fig. 1.24.** Superradiant power $P_{BK}$ from a Brewster angle cuvette vs pumping laser power $P_A$ for a $3 \times 10^{-5}$ M solution of 3,3′-diethylthiacarbocyanine bromide. (Triangles: emission in backward direction; crosses: emission in forward direction. The broken line gives the theoretical expectation). (From Volze 1969)

time. The latter is determined by solvent viscosity, temperature and molecular size. The direction of the transition moments of the fluorescence and the long-wavelength absorption is identical, since the same electronic transition is involved in both processes. For rotational diffusion–relaxation times which are long compared to the exciting pulse, the following table gives the theoretically expected relative orientation of the pumping and dye-laser polarization. In two cases one obtains a depolarized dye-laser beam. There the emissive transition moments of the excited molecules are directed along the resonator axis. Hence, the molecules must first rotate before they can radiate along the resonator axis. Since this rotation is effected by diffusion, the resulting angular orientation distribution is essentially isotropic and the laser output is depolarized. The same is true when the rotational diffusion-relaxation time is short compared to the pumping pulse. A special case is that of large, ring-type molecules like the phthalocyanines. These have two degenerate transitions with orthogonal transition moments in the plane of the ring. The polarization of dye lasers with these compounds is given in brackets in the following table. Very few such possibilities have been experimentally investigated and verified (Sorokin et al. 1967; Sevchenko et al. 1968a, b; McFarland 1967). In addition, the polarization of dye laser beams can, of course, be manipulated in obvious ways by introducing into the resonator polarizing elements like Brewster windows.

The above techniques employ cells for the containment of liquid dye solutions, but even simpler arrangements are possible with solid solutions of dyes or dye crystals. Various dyes have been incorporated in polymethylmethacry-

| Relative orientation of pumping and dye-laser polarizations | Relative orientation of transition moments in absorption and emission | |
|---|---|---|
| | ‖ to each other | ⊥ to each other |
| Longitudinal pumping | ‖ (depol.) | ⊥ (depol.) |
| Transverse pumping, pump laser polar. ⊥ to P | ‖ (depol.) | ⊥ (⊥) |
| Transverse pumping, pump laser polar. ‖ to P | depol. (⊥) | depol. (depol.) |

P = plane containing exciting and dye-laser beams. Data in brackets refer to large, ring-type molecules, e.g. phthalocyanines.

late and other plastics (Soffer and McFarland 1967). If two parallel faces of such a sample are optically polished, the Fresnel reflection of the air-plastics interface can be utilized as the air-glass reflection in the spectrophotometer cuvette of Fig. 1.21 a. A number of dyes have been dissolved in gelatin. The molten gelatin was poured onto a glass slide in a layer a few microns thick. After hardening, this thin film could be stripped off the glass slide. It showed strong superradiant emission when the emission of a nitrogen laser was focused on it with a cylindrical lens (Hänsch et al. 1971 a).

Free jets of flowing dye solutions have completely replaced the easily damaged dye cells for cw dye lasers. Since they are not often used with pulsed dye lasers, they will be described in Chap. 3.

Organic single crystals, e.g. fluorene crystals containing $2\times10^{-3}$ anthracene, which grow and cleave well, can also be pumped by a nitrogen laser, using the two cleaved faces of the laser crystal itself as the optical resonator (Karl 1972). In the case of the fluorene crystal containing anthracene as laser dye, the crystals had a thickness of 0.3 – 0.7 mm and absorbed the pump light almost completely. Laser emission was found at 408 nm.

The vapor pressure of many nonionic laser dyes is quite high at relatively low temperatures. For example, the dye POPOP (1,4-di-[2-phenyloxazolyl]benzene) has a vapor pressure of 20 Torr at a temperature of 410 °C, corresponding to a concentration of $4\times10^{-4}$ mole/l, similar to the concentrations used in dye solution lasers. Using exactly these conditions and a 2-cm cell with 20% reflectivity coatings at 400 nm, the first purely vapor-phase dye laser emission was observed when the power of the pumping nitrogen laser was raised above 1 MW (Steyer and Schäfer 1974 a). An intermediate case between vapor phase and solution is obtained when high pressures (usually some tens of atmospheres) of solvent vapor, e.g. isopentane or ether, are used, which enhances the volatility and thermal stability of the dye molecules (Borisevich et al. 1973). Using polar solvents above the critical point, one could even expect vapor-phase dye laser emission from ionic dyes, which, however, has not yet been attempted. A fairly recent review of vapor-phase dye lasers (Stoilov 1984) lists 57 suitable laser dyes, some of which have shown efficiencies of up to 42%.

Pump lasers used so far with one or the other of the above arrangements include neodymium (1.06 μm) and its harmonics (532 nm, 354 nm, and 266 nm), ruby (694 nm) and its harmonic (347 nm), nitrogen (337 nm), excimer lasers (XeCl, 308 nm; KrF, 248 nm; XeF, 351 nm, etc.), copper-vapor lasers (510 and 578 nm), xenon-ion lasers (multiline from 430 to 596 nm), and other dye lasers.

If the repetition rate of the pump pulses is increased above about one shot per minute, the heat generated via radiationless transitions in the dye molecules and energy transfer to the solvent can cause schlieren in the cuvette, which reduce the dye-laser output. It is thus advisable to use a flow system so that the cuvette contains fresh solution for every shot. Obviously, this is also the way to deal with problems of photochemical instability.

### 1.4.3 Time Behavior and Spectra

Most often the pump pulse is of approximately Gaussian shape and its full width at half maximum power is less than the reciprocal of the intersystem-crossing rate constant $k_{ST}$. Thus, we can expect the following time behavior, neglecting finer detail for the moment. Shortly after the pump pulse reaches threshold level, dye-laser emission starts. The dye-laser output power closely follows the pump power till it drops below threshold, when dye-laser emission stops. The dye-laser pulse shape should thus closely resemble that of the part of the pump pulse above the threshold level. This behavior was indeed observed by Schäfer et al. (1966) pumping transversely a methanol solution of the dye DTTC in a square spectrophotometer cuvette of $1 \times 1$ cm dimensions, polished on all sides. Figure 1.25 shows oscillograms of this pump and dye-laser pulse at different peak pump power levels. From an evaluation of these oscillograms the input–output plot shown in Fig. 1.26 is obtained. It clearly shows how the dye fluorescence intercepted by the photocell is increased by several orders of magnitude from the spontaneous level, while the pump power is increased slightly above threshold. When the threshold is reached at the foot of the pump pulse, so that pump and dye-laser pulse forms are practically identical, the expected linear relationship between input and output obtains. The efficiency in this case was found to be 10%.

A more detailed treatment of the time behavior of the laser-pumped dye laser was given by Sorokin et al. (1967) who solved the rate equations for the excited state population in the dye and the photons in the cavity

$$dN_1/dt = W(t) - (N_1/N_t)(Q/t_c) - N_1/\tau_f$$

$$dQ/dt = (Q/t_c)(N_1/N_t - 1) \ .$$

The quantities appearing in these equations are as follows: $N_1$ is the excited-state population, $N_t$ is the threshold inversion, $Q$ is the number of quanta in the cavity, $t_c$ is the resonator lifetime, $\tau_f$ is the fluorescence lifetime, and $W(t)$ denotes the pumping pulse which was assumed to have a Gaussian distribution with half-width at half-power points equal to $T_1$, i.e.

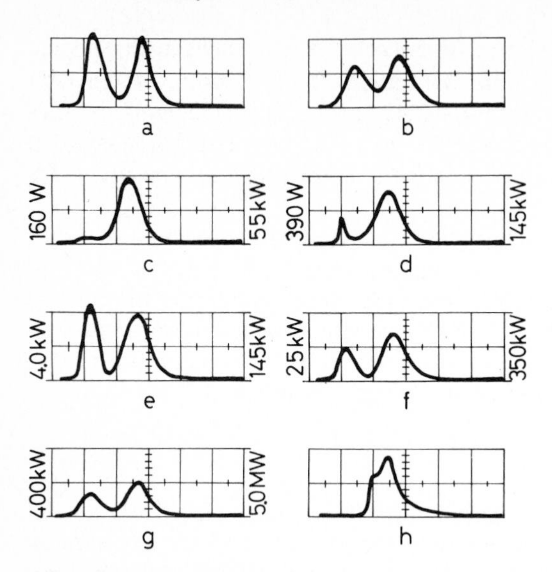


**Fig. 1.25 a–h.** Oscillograms of fluorescence, dye-laser and pumping-laser pulses. (Sweep speed: 10 ns/division for (**a**–**g**), and 2 ns/div. for (**h**). (**a**) Pumping pulse and delayed reference pulse, (**b**) spontaneous fluorescence of a $10^{-5}$ M solution of the dye 3,3′-diethylthiatricarbocyanine bromide in methanol and delayed reference pulse, (**c**–**g**) dye-laser pulse from a $10^{-3}$ M solution of the dye and delayed reference pulse. The power corresponding to two major scale divisions is annotated to the right of each oscillogram for the pumping ruby laser pulse, to the left for the dye-laser pulse. (**h**) Dye-laser pulse just above threshold. (From Schäfer et al. 1966)

$$W(t) = W_{\max} \exp[-(t\sqrt{\ln 2}/T_1)^2]$$

and normalized to $\int_{-\infty}^{+\infty} W(t)\,dt = N_{\text{pump}}$, the total number of pumping photons. Digital computer solutions were obtained for a large number of parameter combinations, many of which agreed well with the experimentally observed time behavior even in such details as risetime and initial overshoot of the dye-laser pulse. A sample computer solution and an oscillogram of the dye-laser output for one set of parameters are reproduced from this work in Fig. 1.27.

For many applications it is advantageous to have a simple method for obtaining dye laser pulses that are much shorter than the pump laser pulses. This was first achieved by Lin and Shank (1975) using the method of pulse shortening by resonator transients, which had been developed earlier for solid-state lasers (Roess 1966). Pumping a 1-cm dye cell with a grating in Littrow mounting as the rear reflector and one cell window as the front reflector just above threshold with a 10-ns nitrogen laser pulse, they obtained 600-ps pulses with rhodamine 6G and 900-ps pulses with 7-dimethylamino-4-methylcoumarin. Further description of this technique can be found in (Lin 1975a and 1975b). By reducing the cavity lifetime using cells of only 10–500 μm width and pumping longitudinally with harmonics of a mode-locked Nd laser, pulses as short as 2 ps could be achieved (Fan and Gustafson 1976; Cox et al. 1982). Many different designs of ultrashort cavity dye lasers have been published since then

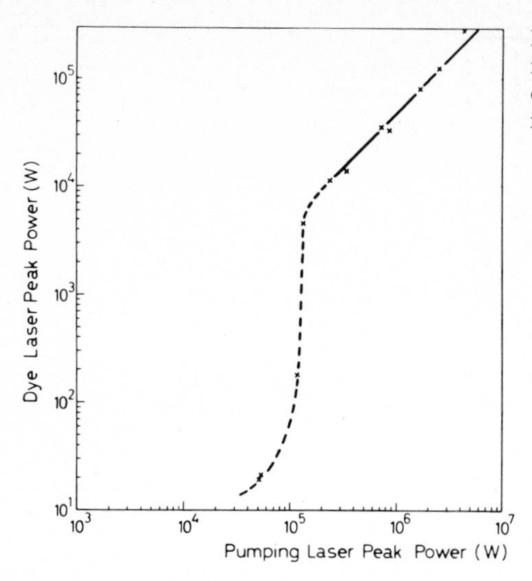


**Fig. 1.26.** Dye-laser peak power vs pumping laser peak power for a $10^{-3}$ M solution of 3,3'-diethylthiatricarbocyanine bromide in methanol. (From Schäfer et al. 1966)

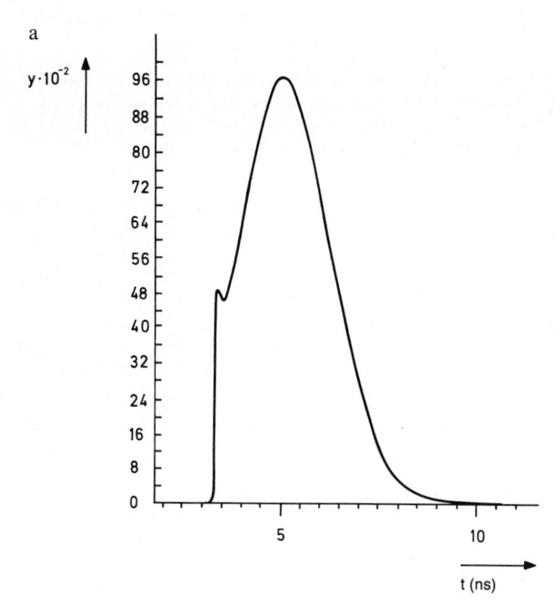


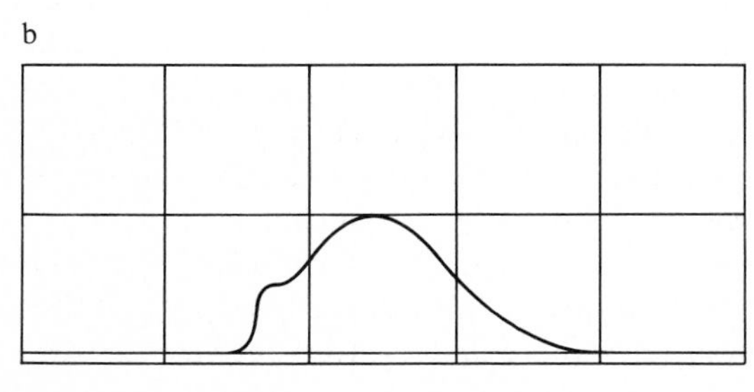


**Fig. 1.27 a, b.** Comparison of a computer solution of the rate equations and an actual oscillogram. (**a**) Computer solution for $N_{pump}/N_t = 350$, $\tau/t_c = 3.0$, $T_1/t_c = 6.0$, $y = Q/(t W_{max})$; (**b**) oscillogram of a chloro-aluminum-phthalocyanine laser. (Mirror spacing: 5 cm, mirror reflectivities: 80% and 98%, sweep speed: 20 ns/division) (From Sorokin et al. 1967)

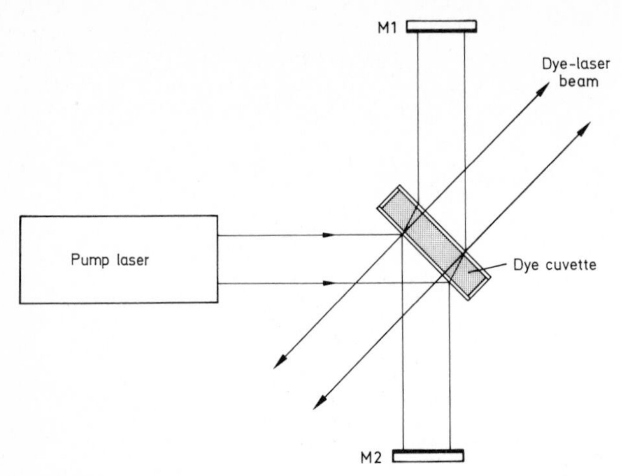


**Fig. 1.28.** Scheme for producing short dye laser pulses using two crossed resonators with a common dye cell. (Adapted from Eranian et al. 1973)

(Aussenegg and Leitner 1980; Yao 1982; Scott et al. 1984; Chiu et al. 1984; Ishida et al. 1985; Yamada et al. 1986; S. C. Hsu and Kwok 1986).

Another method of producing short dye laser pulses in a simple way makes use of two resonators with one dye cell as the common active medium, as shown in Fig. 1.28 (Eranian et al. 1973). While the short resonator is made up of the Fresnel reflections from the 2 mm dye cell, the long resonator consists of two mirrors of maximum reflectivity. Even though the threshold of the short resonator is high because of its low feedback, it reaches threshold first due to its short round-trip time. Its emission is, of course, normal to the cell windows. After the long resonator has reached threshold, the inversion in the dye solution drops to the critical inversion determined by the high Q of the long resonator. Since this inversion is much lower than the threshold inversion needed for the short resonator with its low Q, the dye laser emission from the short resonator is stopped. In this way a pump pulse of 20 ns produced a dye laser pulse of only 2 ns.

Similar arrangements of quenched dye lasers were described by Andreoni et al. (1975) and Braverman (1975). While both methods, resonator transients and quenching, produced dye laser pulses that were shorter than the pump pulse by a factor of about 10, the combination of both methods increased this pulse shortening factor to over 100 (Schäfer et al. 1983). This technique, called by the authors the method of quenched resonator transients, allows strong pumping of the dye cell with a steeply rising pump pulse, so that a first resonator transient of very short duration is produced, which would normally be followed by a damped transient oscillation. This is prevented, however, by careful adjustment of the length of the long resonator, which quenches the dye laser emission right after the end of the first resonator transient. Dye laser pulses of down to 100 ps were easily produced with excimer laser pump pulses of 20 ns duration. The dye laser pulsewidth could even be decreased to less

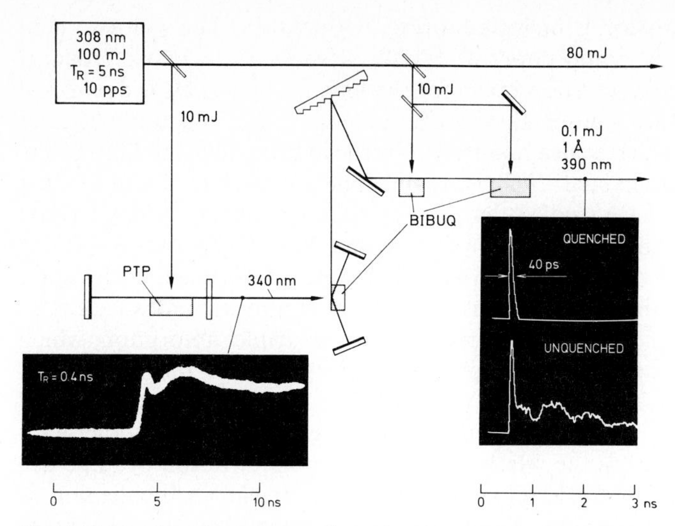


**Fig. 1.29.** Dye laser system to produce ultrashort pulses with long excimer laser pulse pumping. (From Bor and Rácz 1985)

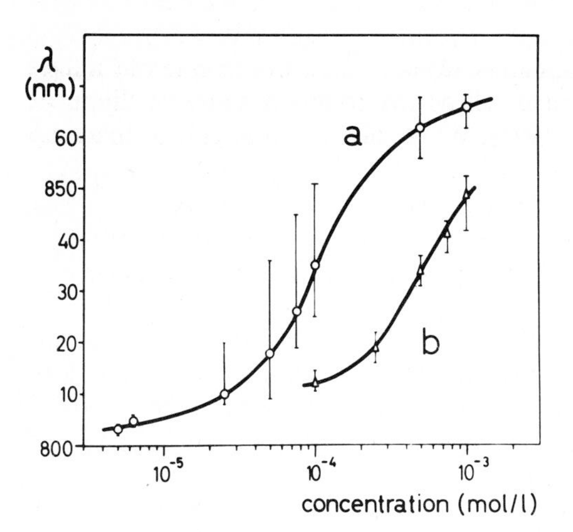


**Fig. 1.30.** Wavelength shift of the dye laser with concentration of a solution of 3,3′-diethylthiatricarbocyanine bromide in methanol, curve a for a silvered and curve b for an unsilvered cuvette. The circles and triangles give the wavelengths of the intensity maxima, the bars show the bandwidth of the laser emission. (From Schäfer et al. 1966)

than 30 ps by sharpening the front of the pump pulse by first pumping a standard broadband dye laser with the excimer laser in order to produce a steeply rising dye laser pulse which then pumped the quenched resonator transient arrangement, as shown in Fig. 1.29. A theoretical treatment of this method and a further experimental refinement was reported recently by Simon et al. (1986).

As explained above, it is not difficult to calculate the oscillation wavelength of a laser-pumped dye laser near threshold. Figure 1.30 gives the observed dye-laser wavelength as a function of DTTC dye concentration for a low and a high value of cavity Q (Schäfer et al. 1966). These curves show satisfactory agreement with the computed curves of Fig. 1.19. The time-integrated spectral width of the dye-laser emission is indicated in Fig. 1.30 by bars. The width is measured with a constant pump power of 5 MW. It is seen to be rather broad in the middle of the concentration range for the high Q curve. This is due to the simultaneous and/or consecutive excitation of many longitudinal and transverse modes, which is also manifested by the increase in beam divergence. Several authors (Farmer et al. 1968; Bass and Steinfeld 1968; Gibbs and Kellog 1968; Vrehen 1971) have studied the wavelength sweep of the dye-laser emission. The results obtained are not easy to interpret. Frequently, there is a sweep from short to long wavelength of up to 6 nm, in other cases the sweep direction is reversed towards short wavelengths after reaching a maximum wavelength. Probably part of this sweep can be traced to triplet–triplet absorption, which should increase with time as the population density grows in the triplet level. Another part might be attributable to thermally induced phase gratings and similar disturbances of the optical quality of the resonator. Still another contributing factor might be the relatively slow rotational diffusion of large dye molecules in solvents like DMSO or glycerol. Inhomogeneous broadening of the ground state, on the other hand, although suggested as an explanation (Bass and Steinfeld 1968), may be excluded on this time scale as a reason for the observed sweep. This explanation assumes that vibrational excitation of the ground state of the dye molecule has a lifetime of at least a few ns in the solution but, in fact, it is deactivated in a few ps through collisions with the solvent molecules.

The dependence of the laser wavelength on the resonator Q, i.e. mainly on cell length and mirror reflectivities, can also be deduced from the start oscillation condition as described in Sect. 1.4.1. The variation with Q was first noted by Schäfer et al. (1966) when they put silver reflectors on the faces of the dye cell. This is shown in Fig. 1.30. The dependence on the cell length was experimentally observed by Farmer et al. (1969).

The temperature dependence of the dye-laser wavelength can be similarly explained. Schappert et al. (1968) found that the laser emission of DTTC in ethanol shifted towards shorter wavelengths with decreasing temperature. This is caused by the narrowing of the fluorescence and the absorption band with decreasing temperature, which results in higher gain and fewer reabsorption losses near the fluorescence peak. The solvent has a most important influence on the wavelength and efficiency of the dye-laser emission.

The first observation of large shifts of the dye-laser emission with changing solvents is reproduced in the following table for a $10^{-4}$ molar solution of the dye 1,1′-diethyl-$\gamma$-nitro-4,4′-dicarbocyanine tetrafluoroborate (Schäfer et al. 1966):

| Solvent | Laser wavelength (in nm) | Solvent | Laser wavelength (in nm) |
|---|---|---|---|
| Methanol | 796 | Dimethylformamide | 815 |
| Ethanol | 805 | Pyridine | 821 |
| Acetone | 814 | Benzonitrile | 822 |

In this case the nitro group, as a highly polar substituent in the middle of the polymethine chain, interacts with the dipoles of the solvent to shift the energy levels of the dye. This type of solvent shift is especially large for dyes whose dipole moments differ appreciably in the ground and excited states. The transition from ground to excited state by light absorption is fast compared with the dipolar relaxation of the solvent molecules. Hence the dye molecule finds itself in a nonequilibrium Franck-Condon state following light absorption, and it relaxes to an excited equilibrium state within about $10^{-13}$ to $10^{-11}$ s. Similarly, the return to the equilibrium ground state is also via a Franck-Condon state, followed by dipolar relaxation:

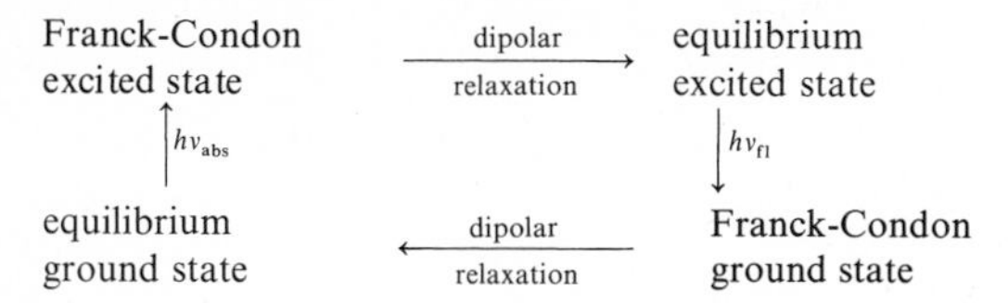


Consider the typical example of *p*-dimethylaminonitrostilbene. It has a dipole moment of 7.6 debye in the ground state and 32 debye in the excited state (Lippert 1957). Therefore, it is clear that in a polar solvent like methanol the equilibrium excited state is lowered even more strongly than the equilibrium ground state by dipole–dipole interaction. In the present case, for example, the absorption maximum occurs at 431 nm and the fluorescence maximum at 757 nm in i-butanol, while the corresponding values for cyclohexane are 415 nm and 498 nm, respectively. Similar shifts are found in laser emission.

Gronau et al. (1972) report laser emission for the similar case of the dye 2-amino-7-nitrofluorene dissolved in 1,2-dichlorobenzene.

Another property of the solvent that can have an important influence on the dye-laser emission is the acidity of the solvent relative to that of the dye. As already mentioned in Sect. 1.2, many dyes show fluorescence as cations, neutral molecules, and anions. Correspondingly, the dye laser emission of such molecules usually changes with the pH of the solution, since generally the different ionization states of the molecule fluoresce at different wavelengths. (A more detailed discussion is to be found in Chap. 5.)

An important subdivision of these dyes is that of molecules, whose acidity in the excited state is considerably different from that in the ground state due to changes of the $\pi$-electron distribution with excitation. This might cause the molecule to lase or take up a proton from the solvent. After return to the ground state, the reverse action occurs, leaving the original molecule in the ground state. Schematically the reaction is

$$\begin{array}{ccc} (AH)^* & \xrightarrow[+H^{\oplus}]{-H^{\oplus}} & (A^{\ominus})^* \\ \uparrow h\nu_{abs} & & \downarrow h\nu_{fl} \\ AH & \xleftarrow[-H^{\oplus}]{+H^{\oplus}} & A^{\ominus} \end{array} \quad \text{or} \quad \begin{array}{ccc} A^* & \xrightarrow[-H^{\ominus}]{+H^{\oplus}} & (AH^{\oplus})^* \\ \uparrow h\nu_{abs} & & \downarrow h\nu_{fl} \\ A & \xleftarrow[+H^{\ominus}]{-H^{\oplus}} & AH^{\oplus}\,. \end{array}$$

A well-known example is acetylaminopyrenetrisulfonate, which loses a proton from the amino group to form the tetra-anion in the excited state (Weller 1958).

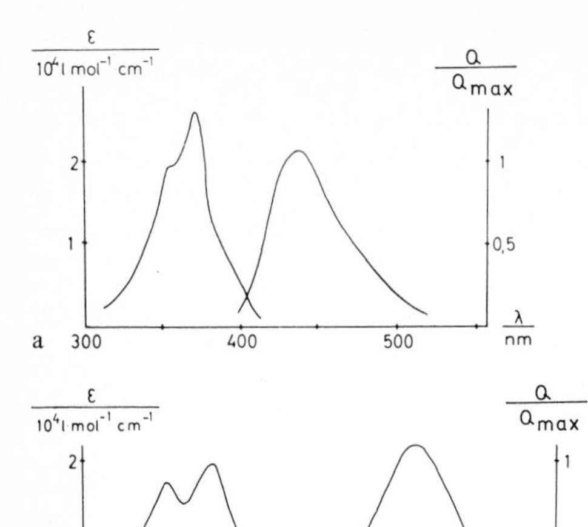


**Fig. 1.31.** Absorption and fluorescence spectra of acetylaminopyrene-trisulfonate, **(a)** neutral solution, **(b)** alkaline solution (1 N NaOH)

Absorption and fluorescence spectra of the neutral molecule, recorded in slightly acid solution, and of the tetra-anion, recorded in alkaline solution, are reproduced in Fig. 1.31. Many examples of this type can be found among the numerous known fluorescence indicators which show a change in fluorescence wavelength and/or intensity with a change of pH of the solution. While the first example of a dye laser utilizing such a protolytic reaction in the excited state (Schäfer 1968) was the above-mentioned acetylaminopyrenetrisulfonate pumped by a flashlamp, the first example used in a nitrogen-laser pumped dye laser and showing an especially large shift of the emission wavelength with pH was given by Shank et al. (1970a, b) and Dienes et al. (1970). They used two coumarins which can form a protonated excited form that fluoresces at longer wavelengths. The chemistry of these dyes is discussed in detail in Chap. 5. By judicious adjustment of the acidity of an ethanol/water mixture, neutral and protonated forms of these dyes could be made to lase. The tuning extended over 176 nm (from 391 to 567 nm). The term "exciplex", which these authors use for the excited protonated form, should be reserved for cases where an excited complex is formed with another solute molecule, not merely with a proton.

Donor-acceptor charge-transfer complex formation between a dye and a solvent molecule can occur in the ground state as well as in the excited state. The latter case is especially interesting, since the ground state of the complex can have a repulsion energy of a few kcal, while in the excited state the enthalpy of formation is comparable to that of a weak to moderately strong chemical bond (Knibbe et al. 1968). Thus the complex is stable in the excited state only and cannot be detected in the absorption spectrum. The rate of complex formation is limited by the diffusion of the two constituents. It is thus proportional to the product of the concentrations and strongly depends on the viscosity of the solution.

$$
\begin{array}{ccccc}
A^* + D & \rightleftarrows & (A^{\ominus} D^{\oplus})^* & & \\
\uparrow h\nu_{abs} & & \downarrow h\nu_{fl} & & \\
A + D & \leftrightarrows & (A^{\ominus} D^{\oplus}) & \longrightarrow A^{\ominus} + D^{\oplus} & \longrightarrow \text{reaction products.}
\end{array}
$$

If the excited complex $(A^{\ominus} D^{\oplus})^*$ is nonfluorescing, the addition of D to A results merely in quenching of the fluorescence of A. This is generally believed to be a possible mechanism of the so-called dynamic quenching of fluorescence. On the other hand, if the excited complex is fluorescent, a new fluorescence band appears, while the original fluorescence disappears with increasing concentration of D, which might be one of the constituents of a solvent mixture. An example of this behavior is the complex formed from dimethylaniline and anthracene in the excited state (Knibbe et al. 1968). While this principle has not yet been utilized in laser-pumped lasers, a methanol solution of a pyrylium dye in a flashlamp-pumped laser showed a significant lowering of the threshold and displacement of the laser wavelength by 10 nm to the red on addition of a small quantity of dimethylaniline, evidently as a result of the formation of a charge-transfer complex in the excited state.

Assuming complete absorption of the pump radiation for a specific dye (and a constant temperature), the energy conversion efficiency is strongly dependent on the solvent. This is evidenced by the results of Sorokin et al. (1967) given in the following table of conversion efficiencies of an end-pumped DTTC laser with the concentration adjusted to absorb 70% of the ruby-laser pump radiation.

| Solvent | Conversion efficiency, % |
|---|---|
| Methyl alcohol | 9.0 |
| Acetone | 7.0 |
| Ethyl alcohol | 6.5 |
| 1-propanol | 11.5 |
| Ethylene glycol | 14.0 |
| DMF | 15.5 |
| Glycerin | 15.5 |
| Butyl alcohol | 9.0 |
| DMSO | 25.0 |

Similar tabulations of conversion efficiencies for excimer-laser pumping of various dyes in a number of solvents are reported in the literature (V. S. Antonov and Hohla 1963; Cassard et al. 1981). In general, for dye lasers pumped not too high above threshold, the conversion efficiencies are roughly proportional to the corresponding quantum yields of spontaneous fluorescence. For higher pump intensities, however, the influence of excited state absorption on the pump light and on the dye-laser emission becomes more pronounced. These effects have been studied in detail (Speiser and Shakkour 1985).

We have already seen that the deactivation pathway provided by intersystem crossing can be neglected for sufficiently short pump pulses, resulting in correspondingly higher conversion efficiencies. Using picosecond pump pulses, it has recently become possible to neglect even the very fast internal conversion processes. This was done in a travelling-wave pumping arrangement described at the end of Chap. 4. In such an arrangement, molecules that have just been transferred to the first excited singlet state by the absorption of a pump-light photon are immediately brought back down to the ground state by stimulated emission at a rate much faster than the competing deactivating radiationless processes.

In this way it was possible to achieve conversion quantum efficiencies of 10%–20%, even using dyes with fluorescence quantum yields of only $5 \times 10^{-4}$ (Polland et al. 1983). A demonstration experiment for easy visual observation of this effect is described by Szatmári and Schäfer (1984).

## 1.5 Flashlamp-Pumped Dye Lasers

### 1.5.1 Triplet Influence

In the case of flashlamp-pumped lasers, triplet effects become important because of the long risetime or duration of the pump light pulse. Figure 1.32 shows the time dependence of the population density $n_1$ in the excited singlet state (Schmidt and Schäfer 1967). The solid curve applies to a slowly rising, and the broken curve to a fast rising pump light pulse. Both are computed for two different values of the quantum yield of fluorescence $\phi_f$. It is assumed that there is no direct radiationless internal conversion between the first excited singlet and the ground state, so that a fraction $\varrho = 1 - \phi_f$ of excited molecules in the singlet state make an intersystem crossing to the triplet state. It is further assumed that the triplet state has a very long lifetime compared to the singlet state. It is seen that there is a maximum in the excited singlet state population

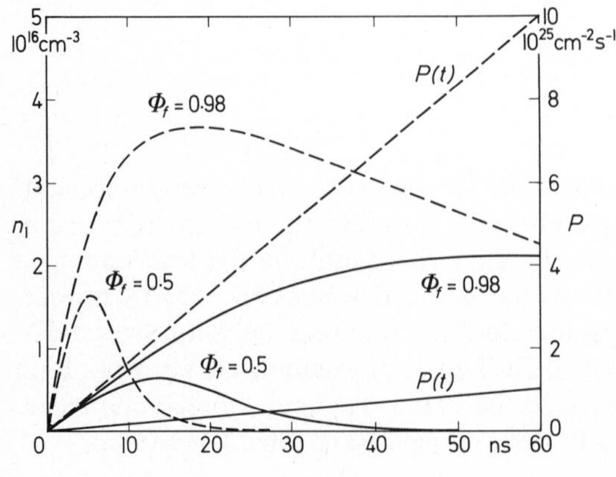


**Fig. 1.32.** Analog computer solutions of the population of the first excited singlet state $n_1$, for different pump powers $P(t)$ and quantum efficiencies of fluorescence, $\phi_f$. (From Schmidt and Schäfer 1967)

density $n_1$, which is reached while the pump light intensity is still rising. This maximum is higher for a faster rise of the pump light and also for a higher quantum yield of fluorescence. Despite the continuing rise of the pump light intensity, $n_1$ falls to a low value after passing the maximum since the ground state becomes depleted and virtually all of the molecules accumulate in the triplet state. It should be pointed out that this behavior was computed for the spontaneous fluorescence of an optically thin layer of dye solution whose intensity is proportional to $n_1$. Obviously, this type of behavior would remain practically unaltered when stimulated emission is taken into account. Thus a dye laser may be pumped above threshold by a fast-rising light source. On the other hand, the necessary population density in the excited singlet state may not be achieved with a slowly rising light source, even if the asymptotic pump level is the same for both situations. In any case, the laser emission would be extinguished after some time. Even more important than this depletion of the ground state, however, are additional losses due to molecules accumulated in the triplet state. These give rise to triplet–triplet absorption spectra, as described in Sect. 1.3, which very often extend into the region of fluorescence emission. Therefore they can lead to losses that often are higher than the gain of the laser and eventually prevent laser emission.

Experimental evidence of the premature stopping of laser action due to triplet losses can be found in almost any flashlamp-pumped dye laser. A very convincing demonstration of the importance of triplet–triplet absorption in flashlamp-pumped dye lasers is given by Schäfer et al. (1978). They compare the output energies and pulse forms of a mixture of the laser dyes p-terphenyl and dimethyl-POPOP with those of a bifluorophoric dye, which was synthesized by linking the chromophores of these two dyes by a saturated hydrocarbon chain, which in this example was a single $CH_2$ group. In addition to the pump-light photons absorbed by the longer-wavelength-absorbing dimethyl-POPOP moiety in the bifluorophoric dye, those absorbed by the p-terphenyl moiety give rise to a very efficient intramolecular radiationless energy transfer to the dimethyl-POPOP moiety in less than 1 ps (B. Kopainsky et al. 1978) so that the output energy is increased in comparison to a dye solution of dimethyl-POPOP. In a mixture of equal concentrations ($10^{-4}$ molar each) of the two dyes, however, the great intermolecular distances reduce the energy transfer from the p-terphenyl to the dimethyl-POPOP molecules to a negligible fraction. This has the important consequence that most of the p-terphenyl molecules can accumulate in the triplet state during the long pump pulse duration and prevent lasing of the dimethyl-POPOP by their triplet–triplet absorption, which is known to be strong at the laser wavelength of dimethyl-POPOP. Only at very high pumping is a weak output in a short pulse observed. These facts are clearly seen in the input–output curves of Fig. 1.33 and the oscillograms of the pulse forms given in Fig. 1.34.

It must be stressed that in the bifluorophoric dye just described, the triplet–triplet absorption of the shorter-wavelength-absorbing (donor) chromophore is eliminated by fast intramolecular singlet-singlet energy transfer, whereas the triplet–triplet absorption of the longer-wavelength-absorbing (acceptor)

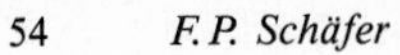

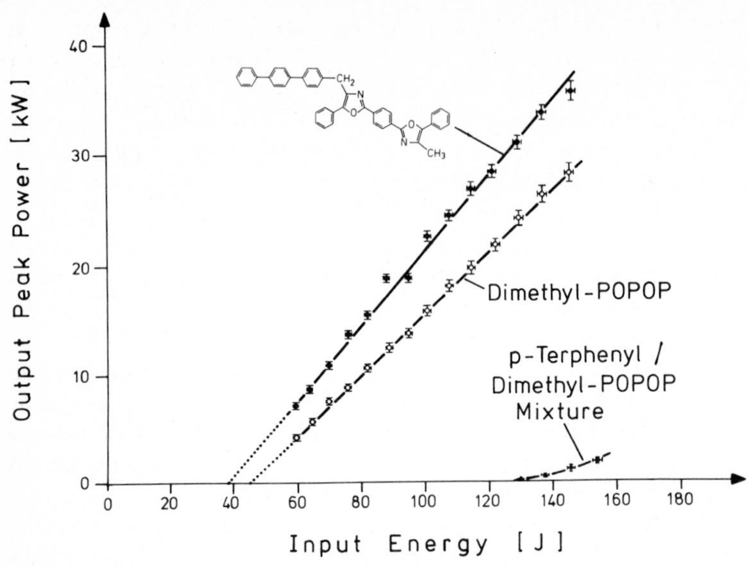


**Fig. 1.33.** Input–output curves of a flashlamp-pumped dye laser for a dye with intramolecular energy transfer, for dimethyl-POPOP, and for a dye mixture. (From Schäfer et al. 1978)

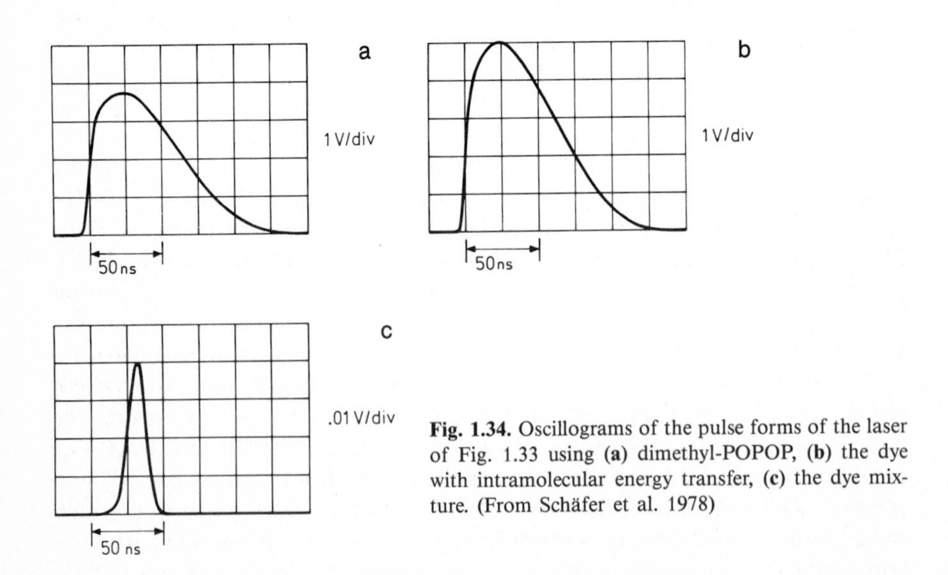


**Fig. 1.34.** Oscillograms of the pulse forms of the laser of Fig. 1.33 using (**a**) dimethyl-POPOP, (**b**) the dye with intramolecular energy transfer, (**c**) the dye mixture. (From Schäfer et al. 1978)

chromophore is still effective and is the reason for the relatively low overall efficiency of this laser dye and its relatively short (compared to the pump) pulse duration. In a later part of this chapter possibilities of external as well as internal quenching of the deleterious triplet state will be described.

Several authors have treated the kinetics of dye-laser emission by a set of coupled rate equations including terms to account for triplet–triplet absorp-

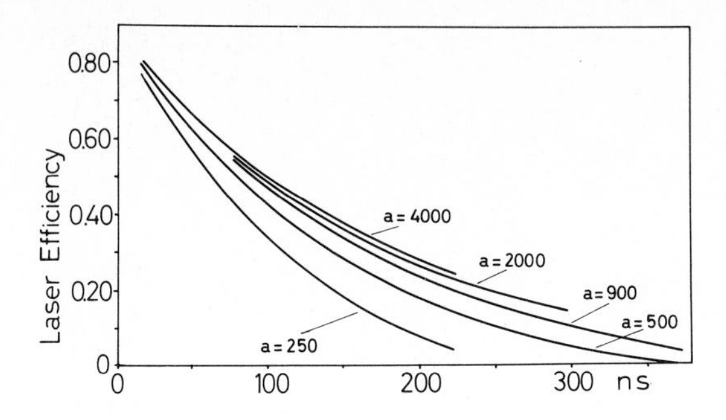


**Fig. 1.35.** Dye-laser efficiency vs half-width of pumping pulse at constant pulse energy (parameter $a$). (From Sorokin et al. 1968)

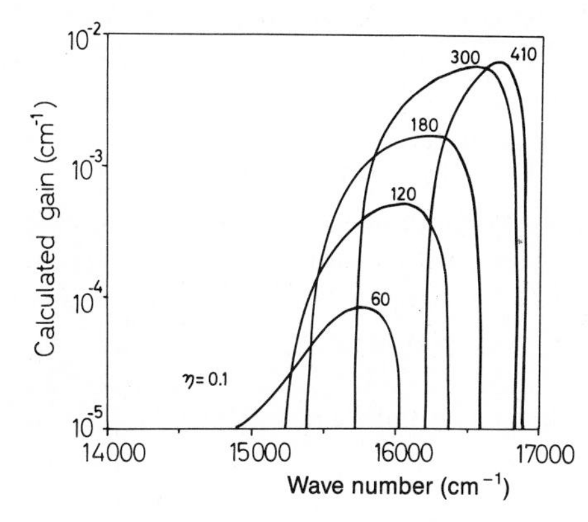


**Fig. 1.36.** Gain vs wavelength calculated for a rhodamine B laser with time (in ns) after initiation of the pump pulse as a parameter. (From M. J. Weber and Bass)

tion losses. Since the triplet losses are time-dependent, they affect the efficiency as well as the emission wavelength of the dye laser. Figure 1.35 gives the dye laser efficiency as a function of the excitation pulsewidth (Sorokin et al. 1968). The parameter $a$ is the ratio of the total number of photons in the excitation pulse to the number of photons at threshold. The figure applies to the case $\phi_f = 0.88$ and $\sigma_a = 10\sigma_T$ and $\tau_T = \infty$, where $\sigma_T$ is the cross-section for the triplet–triplet absorption band. Figure 1.36 is a plot of gain vs wavelength calculated for a rhodamine B laser with time (in ns) after initiation of the flashlamp pulse as a parameter (M. J. Weber and Bass 1969). The results of such calculations apply only to the specific light pulse form considered. It it therefore worthwhile to have an expression connecting population densities in the ground, lowest excited singlet, and triplet states with the laser wavelength and cavity parameters.

The following simple modification of the oscillation condition of Sect. 1.4.1 takes into accout triplet–triplet absorption losses, characterized by a cross-section $\sigma_T$. Let $\alpha = n_0/n$ be the normalized ground-state population density, and $\beta = n_T/n$ the triplet-state population density. Then the oscillation condition, its derivative with respect to wave number, and the balance of population densities give the following three equations for $\alpha$, $\beta$, $\gamma$ (the prime denotes differentiation with respect to wave number):

$$-\sigma_a\alpha - \sigma_T\beta + \sigma_f\gamma = S/n \ , \tag{1.12a}$$

$$-\sigma_a'\alpha - \sigma_T'\beta + \sigma_f'\gamma = 0 \ , \tag{1.12b}$$

$$\alpha + \beta + \gamma = 1 \ . \tag{1.12c}$$

From these equations and the observed wavelength, the population densities in the ground, excited singlet, and triplet states can be obtained. As an example of this procedure the case of rhodamine 6G in a cw laser will be discussed in detail in Chap. 2. The ratio $\beta/\gamma$, obtained from the above relations, and the time $t_0$ to reach threshold can be used for an estimate of $k_{ST}$, if one assumes $t_0 \ll \tau_T$ and a linearly rising pump light source. Then we have $d\gamma/dt$ = constant and $d\beta/dt = k_{ST}\beta$, which yields after integration $\beta/\gamma = \frac{1}{2}k_{ST}t_0$. On the other hand, the observed $t_0$ can give an upper limit for $\sigma_T$ at the laser wavelength for known $k_{ST}$, since laser emission can only be achieved for $\gamma\sigma_f > \beta\sigma_T$, yielding $\sigma_T < 2\sigma_f/t_0 k_{ST}$ (Sorokin et al. 1968).

For pulses that are long compared with $\tau_T$ or for continuous operation, a steady state must be reached. Then the triplet production rate $k_{ST}\gamma$ equals the deactivation rate $\beta/\tau_T$, so that the ratio $\beta/\gamma = k_{ST}\tau_T$. Setting this value equal to that obtained above from the observed laser wavelength, an estimate of $\tau_T$ can be obtained.

To obviate the need for a rapidly rising pump light intensity in dye lasers and to achieve cw emission, one must reduce the triplet population density to a sufficiently low level by reducing $\tau_T$. One way of achieving this is by adding to the dye solution suitable molecules that enhance the intersystem-crossing rate $k_{TG} = 1/\tau_T$ from the triplet to the ground state by the effects described in Sect. 1.3. The improvement in dye-laser performance possible with this method was shown by Keller (1970), who solved the complete rate equations of Sorokin et al. (1968) after adding a term $n_T/\tau_T$ for the deactivation of the triplet state to the equation for the triplet-state population density.

A molecule that has long been known to quench triplets is $O_2$. At the same time, however, it also increases $k_{ST}$ and thus also quenches the fluorescence of a dye. The relative importance of triplet and fluorescence quenching then determines whether or not oxygen will improve dye-laser performance. This is shown in Fig. 1.37, which gives the output from a long-pulse rhodamine 6G laser as a function of the partial pressure of oxygen in the atmosphere in contact with the dye solution in the reservoir (Schäfer and Ringwelski 1973). While the output first rises steeply with increasing oxygen content, it levels off after about 20% oxygen content, showing that the positive effect of the reduction in triplet lifetime is offset by the detrimental effect of fluorescence quenching. This can be understood quantitatively, if one assumes that the laser output is proportional to the ratio $\gamma/\beta$. In the presence of an oxygen concentration $[O_2]$ the triplet production rate is $(k_{ST}+k_{QS}[O_2])n_1$ and is set equal to the deactivation rate $(k_{TG}+k_{QT}[O_2])n_T$, where $k_{QS}$ and $k_{QT}$ are the quenching constants for the singlet and triplet states, respectively. This yields $\gamma/\beta = (k_{TG}+k_{QT}[O_2])/(k_{ST}+k_{QS}[O_2])$. Using $k_{ST} = 2\times10^7\ \mathrm{s}^{-1}$, $k_{QS} =$

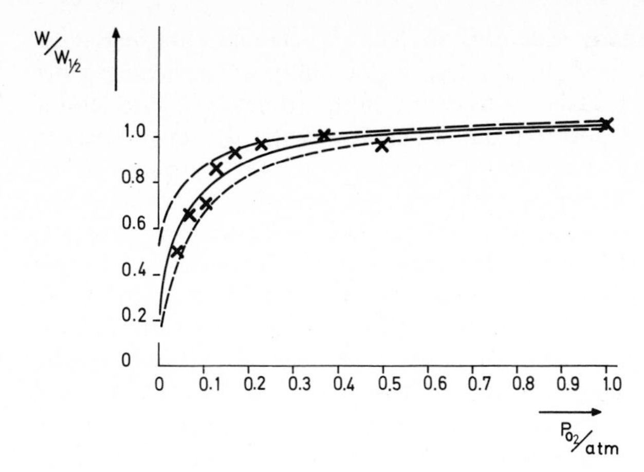


**Fig. 1.37.** Normalized laser output $W/W_{1/2}$ vs oxygen partial pressure $P_2$ in the gas mixture over a solution of rhodamine 6G. (*Crosses*: measurements; *solid line*: theoretical result for $k_{TG} = 5\times10^5\,s^{-1}$; *upper dashed curve*: same, but for $k_{TG} = 10^6\,s^{-1}$; *lower dashed curve*: same for $k_{TG} = 10^4\,s^{-1}$). (From Schäfer and Ringwelski 1973)

$3\times10^{10}\,s^{-1}\,l\,mole^{-1}$, and $k_{QT} = 3.3\times10^9\,s^{-1}\,l\,mole^{-1}$, we find good agreement with the experimental results for $k_{TG} = 5\times10^5\,s^{-1}$. The triplet lifetime of rhodamine 6G in carefully deaerated methanol solutions at room temperature is thus $\tau_T = 2\,\mu s$, while the effective triplet lifetime with 20% oxygen content in the atmosphere above the dye is $\tau_{Teff} = 1/(k_{TG} + k_{QT}[O_2])$ or 140 ns. Marling et al. (1970a) measured the effect of molecular oxygen at different partial pressures on the dye laser emission of a number of dyes. They found several dyes where the fluorescence quenching of oxygen was much stronger than its triplet quenching effect, e.g. the dye brilliant sulphaflavine, whose laser emission was extinguished when the argon atmosphere in the reservoir was replaced by one atmosphere of oxygen.

It is thus seen to be more appropriate not to use a paramagnetic gas like oxygen, which enhances both $k_{ST}$ and $k_{TG}$, but rather to apply energy transfer from the dye triplet so some additive molecule with a lower-lying triplet that can act as acceptor molecule. Triplet–triplet energy transfer was shown by Terenin and Ermolaev (1956) to occur effectively with unsaturated hydrocarbons. This scheme was used by Pappalardo et al. (1970b), who used cyclooctatetraene as acceptor molecule and obtained a dye-laser output pulse from rhodamine 6G as long as 500 µs, demonstrating that the triplet lifetime had been reduced by the quenching action of cyclooctatetraene below the steady-state value necessary for cw-laser operation. This compound is still the most effective triplet quencher known for rhodamine 6G, although a number of others have been tested and several quite effective ones found (Marling et al. 1970b), e.g. N-aminohomopiperidine, 1,3-cyclooctadiene, and the nitrite ion, $NO_2^-$.

It must be stressed that this energy transfer occurs in such a way that energy as well as spin is exchanged between dye and acceptor molecule, leaving the acceptor in the triplet state, according to $^3D + {}^1A \rightarrow {}^1D + {}^3A$. Marling et al. (1970b) has pointed out that there is also a possibility of quenching a dye triplet by collision with a quencher molecule in the triplet state through the process of triplet–triplet annihilation, which leaves the dye in the excited singlet state and the quencher molecule in the ground state: $^3D + {}^3A \rightarrow {}^1D^* + {}^1A$ with subsequent fluorescence emission or radiationless deactivation of the excited singlet state. In this case, the quencher molecules must first be excited to the first excited singlet state and pass by intersystem crossing to the triplet state, before they can become effective. The first excited singlet state of quencher molecules must always lie higher than that of the dye, to that no pump radiation useful for pumping the dye is absorbed and no fluorescence quenching energy transfer can occur from the first excited singlet state of the dye so that of the quencher molecule. This means that, for efficient quencher molecule triplet production, the pump light source must have a high proportion of ultraviolet light output. Usually this means at the same time a higher photodegradation rate for the dye, which also absorbs part of the ultraviolet pump light. Because of these disadvantages and its generally lower effectiveness, this indirect method of triplet quenching seems less attractive. The distinction between direct and indirect triplet quenching can, however, clarify the results of other experimental investigations (Strome and Tuccio 1971; Smolskaya and Rubinov 1971).

Whilst the above methods of triplet quenching rely on intermolecular processes, intramolecular triplet quenching proved to be most efficient, albeit more difficult to implement. Here again two chromophores are chemically linked by a short saturated hydrocarbon chain to form a bifluorophoric laser dye, as described above (p. 53). But now one of the chromophores acts as a

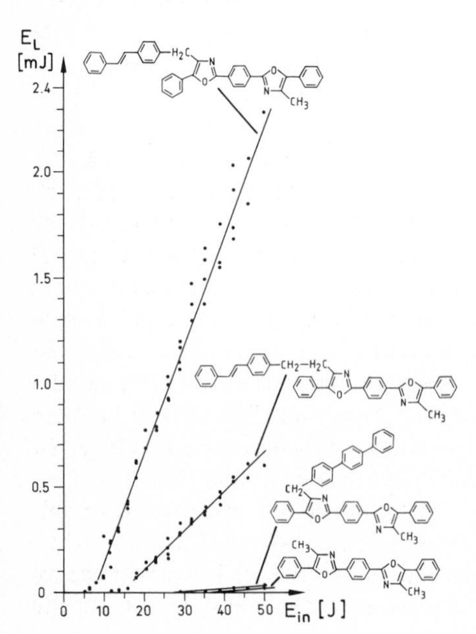


**Fig. 1.38.** Comparison of input–output curves for different dyes with intramolecular energy transfer with that of dimethyl-POPOP. (From Liphardt et al. 1981)

triplet quencher for the lasing chromophore. The best example is again dimethyl-POPOP as the lasing chromophore, but now linked to *trans*-stilbene, which acts as triplet quencher, since its $T_1$ state is lower than the $T_1$ state of dimethyl-POPOP (Liphardt et al. 1981). At the same time, *trans*-stilbene absorbs at a shorter wavelength than dimethyl-POPOP and its fluorescence band has a good overlap with the absorption band of dimethyl-POPOP. This means, *trans*-stilbene plays the double role of donor for singlet–singlet energy transfer to dimethyl-POPOP and acceptor for triplet–triplet energy transfer from it. The triplet lifetime of the dimethyl-POPOP moiety of this molecule was found to be only $7 \pm 3$ ns when it was linked to the quenching moiety via one $CH_2$ group, which rose to $70 \pm 10$ ns for a link using $-CH_2-CH_2-$ (Schäfer et al. 1982). Consequently, the output energy of this new bifluorophoric dye with intramolecular triplet quenching is many times higher than that of dimethyl-POPOP or even the corresponding bifluorophoric dye with p-terphenyl, as shown in Fig. 1.38. Actually, using the same flashlamp-pumped dye laser at 50 J electrical input energy, the outputs of the new bifluorophoric dye with one methylene group as a link and the dye using two methylene groups as a link are respectively 110 and 30 times higher than the output energy of dimethyl-POPOP (Liphardt et al. 1981).

These bifluorophoric dyes with intramolecular triplet-triplet energy transfer are a lucid example of the power of preparative organic chemistry for solving long-standing problems in dye laser technology.

### 1.5.2 Practical Pumping Arrangements

Flashlamp-pumped dye laser heads consist in principle of the dye cuvette, the flashlamp and a pump light reflector or diffuser. The latter serves to concentrate the pump light emitted from the extended, uncollimated, broadband source, the flashlamp, onto the absorbing dye solution in the cuvette. The reflector can be of the imaging type, e.g. an elliptical cylinder whose focal lines determine the position of linear lamp and cuvette, or it can be of the close-coupling type, which is especially advisable where there are several flashlamps surrounding the cuvette. Instead of a specular reflector, a diffusely reflecting layer of MgO or $BaSO_4$ behind a glass tube surrounding flashlamp and cuvette is often used. The design of the pump cavity is thus similar to that of solid-state lasers, except that for dye lasers it is even more important to prevent nonuniform heating of the solution in order to avoid thermally induced schlieren. Furthermore, it is advisable to use some means of filtering out photochemically active wavelengths which might decompose the dye molecules. It it often sufficient to use an absorbing glass tubing for the cuvette, otherwise a double-walled cuvette or a filter solution surrounding the flashlamp can be used (Jethwa et al. 1982; Calkins et al. 1982).

For maximum utilization of the pumplight output the length of the cuvette will generally be about the same as that of the flashlamp. This in turn makes a flow system almost mandatory, because in a long cuvette even small thermal gradients can severely degrade the resonator characteristics. The dye flow in

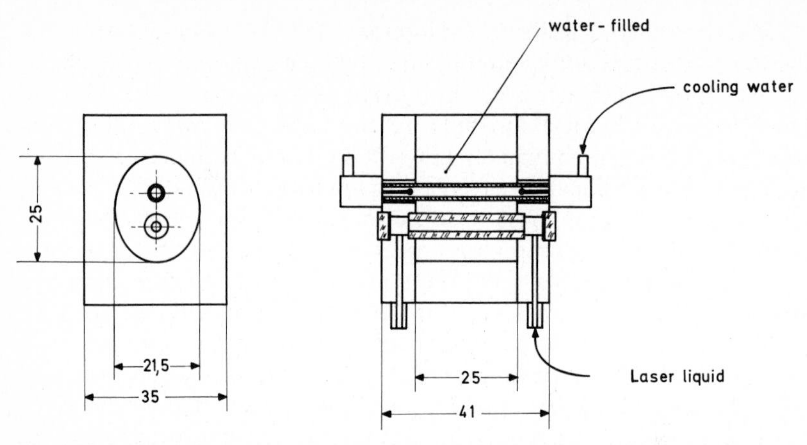


**Fig. 1.39.** Cross-sections of small dye-laser head for 100 Hz pulse repetition frequency (dimensions in mm). (From W. Schmidt 1970)

the cuvette may be longitudinal or transverse. In either case the flow should be high enough to be in the turbulent regime. This rapidly mixes the liquid and hence reduces thermal gradients due to nonuniform pump-light absorption in the cuvette. Both flow orientations are used in high-repetition-rate dye lasers (Boiteux and De Witte 1970; W. Schmidt 1970). If the cuvettes are not of all-glass construction, windows are generally pressed against the cuvette end and sealed by O rings. These O rings, and the hoses connecting the cuvette with the circulating pump and the reservoir, can be of silicone rubber for use with methanol or ethanol solvents. Commercial O rings, and even clear plastic hoses, usually give off absorbing or quenching filler material and hence should not be used. For other solvents, like dimethylformamide or dimethylsulfoxide, Teflon coatings are required on O rings and hoses. A similar choice is required with the circulating pump and the reservoir for the dye solution. Magnetically coupled centrifugal or toothed wheel pumps made of Teflon or stainless steel are well suited for this application, whereas membrane pumps give less reproducible results because of the pulsation in the flow rate. The metal end pieces carrying the nipples for the inflow and outflow of the dye solution, situated between the cuvette made of glass tubing and the windows, are also best made of stainless steel and should contain as little unpumped length of dye solution as possible so as to reduce the reabsorption of the dye-laser beam. As an example, the cross-sections of a small dye-laser head for 100 Hz pulse-repetition frequency and several watts average output power are given in Fig. 1.39. A great variety of flashlamps have been used in dye lasers. The simplest possibility is the use of commercial xenon flashlamps as in the laser head of Fig. 1.40. The risetime of linear and helical flashlamps can be reduced and the output power increased by the introduction of a spark gap in series with the lamp; this allows the lamp to be operated at a voltage that is much higher than the self-firing voltage. The excess voltage applied to the lamp ensures a rapid build-up of the plasma and a much higher peak power in the lamp. In this ar-

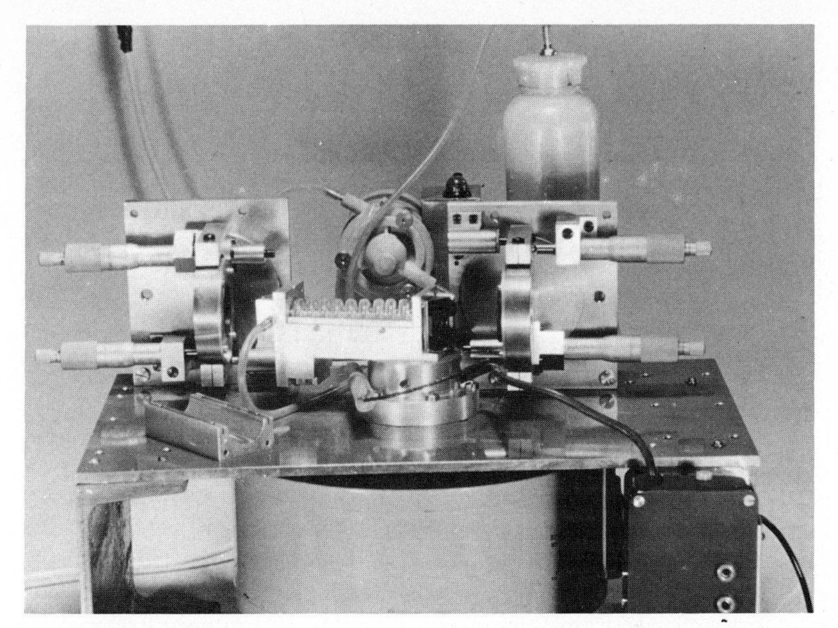

**Fig. 1.40.** Photograph of a dye laser with helical flashlamp

rangement with a linear, 8-cm-long xenon lamp of 5 mm bore diameter and a low-inductance 0.3 μF/20 kV capacitor, a risetime of 300 ns can be obtained. By comparison, a helical flashlamp of 8 cm helix length and 13 mm inner diameter of the helix gives a risetime of about 0.5 μs. For reasons connected with the accumulation of molecules in the triplet state, discussed in the last section, much effort has gone into the development of high-power lamps of short risetime. Very fast risetimes (70 ns) can be achieved by the use of small capillary air sparks fed by an energy storage capacitor in the form of a flat-plate transmission line (Aussenegg and Schubert 1969). Another type with a very fast risetime is a low-inductance coaxial lamp in which the cylindrical plasma sheet surrounds the cuvette. This type was first developed for flash photolysis work (Claesson and Lindquist 1958). Later it was used successfully for dye-laser pumping (Sorokin and Lankard 1967; W. Schmidt and Schäfer 1967). An improved version of this lamp was developed by Furumoto and Ceccon (1969a). With this lamp too a spark gap is used in series with the lamp, so that a voltage much higher than the breakdown voltage of the lamp can be used. At the same time the pressure can be adjusted so that the plasma fills the lamp uniformly. By comparison the original design shows constricted spark channels which move from shot to shot. The improved version offers nearly uniform illumination of the cuvette and the pulse height is reproducible from shot to shot. Risetimes of 150 ns using a 0.05 μF capacitor were achieved with this lamp. For smaller capacitors even shorter risetimes can be obtained.

This configuration is also amenable to up-scaling, and this has been done by Russian workers (Alekseev et al. 1972; Baltakov et al. 1973) who obtained

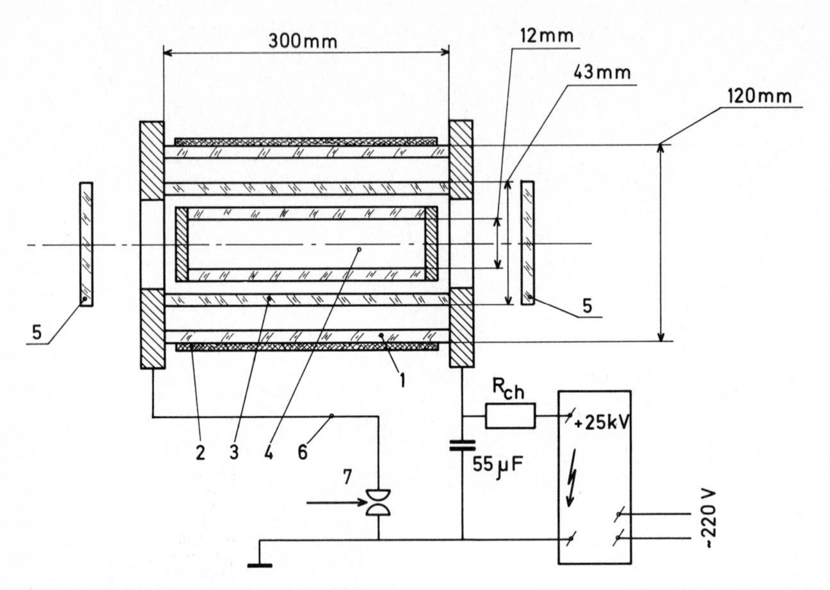


**Fig. 1.41.** Cross-section through a high energy dye laser with coaxial flashlamp. (*1*) external tube, (*2*) silicon dioxide coating, (*3*) internal tube, (*4*) dye cuvette, (*5*) mirrors, (*6*) external current lead, (*7*) triggered spark gap, $R_{ch}$ = charging resistor. (From Alekseev et al. 1972)

dye-laser pulses of up to 150 J output energy. Figure 1.41 shows a cross-section through a dye laser using such a lamp. A great disadvantage of this arrangement, however, is the enormous increase in beam divergence during the laser pulse from 1.8–3 mrad at the start of the pulse to 36–91 mrad near the end of the pulse (Baltakov et al. 1974).

A significant improvement in reliability and flashlamp lifetime, in particular for high repetition rate lasers, was achieved by the introduction of a simmer mode of operation of the flashlamps (Stephens and Hug 1972) and by solid-state switches (Jethwa and Schäfer 1974). This development resulted in dye lasers of over 100 W average power at 50 Hz repetition frequency (Jethwa et al. 1978). Other reports on flashlamp-pumped dye lasers of 100 W and 90 W average power were published by Morey and Glenn (1976) and Mack et al. (1976).

### 1.5.3 Time Behavior and Spectra

The time behavior of flashlamp-pumped dye lasers is more complex than that of laser-pumped dye lasers because of time-dependent triplet losses and thermally or acoustically induced gradients of the refractive index. Several authors have derived solutions of the rate equations for flashlamp-pumped dye lasers under various approximations (Bass et al. 1968; M. J. Weber and Bass 1969; Sorokin et al. 1968). In view of the many quantitative uncertainties, a calculation of the time behavior of flashlamp-pumped dye lasers is of rather limited value. Instead, experimental results are given here.

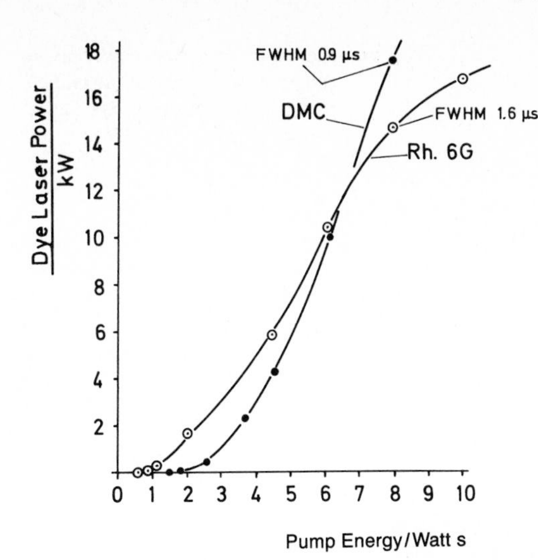


**Fig. 1.42.** Dye laser peak output power vs flashlamp pulse energy at 100 Hz pulse repetition frequency for rhodamine 6G (Rh. 6G) and 7-diethylamino-4-methylcoumarin (DMC). (From W. Schmidt 1970)

For small coaxial lamps having a fast risetime and short pulsewidths, the time behavior is similar to that of laser-pumped dye lasers. Thus, Furumoto and Ceccon (1970) obtained a pulse of 40 kW peak power and 100 ns duration in the ultraviolet from a solution of p-terphenyl in DMF using a lamp of 50 ns risetime and 20 J energy capacity. With similar lamps Hirth et al. (1972) and Maeda and Miyazoe (1972) were able to obtain laser emission from many cyanines in the visible and near-infrared.

Lasers equipped with commercial xenon flashlamps have slower risetimes. Consequently the number of dyes that will lase in these devices is restricted and triplet quenchers must be used if long pulse emission is wanted. Nevertheless, high average and peak powers and relatively high conversion efficiencies can be obtained in this way. Figure 1.42 shows a plot of dye-laser peak power versus flashlamp energy for the small laser head shown in Fig. 1.39 (W. Schmidt 1970). An optimized version allowed operation at 100 Hz pulse repetition frequency and an average dye-laser output power of 3.5 W (W. Schmidt and Wittekindt 1972). With linear flashlamps in an elliptical pump cavity powers of up to 1 MW were obtained (Bradley 1970). Figure 1.43 shows the average power vs the repetition frequency of a 100 W dye laser, and Fig. 1.44 gives the corresponding oscillograms of flashlamp current and laser power per pulse.

With a lumped constant transmission line in place of a single capacitor, a long, flat-top pump-light pulse can be formed. Figure 1.45 is an oscillogram of such a pump-light pulse from a helical flashlamp and the dye-laser pulse excited by it (Ringwelski and Schäfer 1970). A 400 µs long dye-laser pulse is obtained from an air-saturated methanol solution of rhodamine 6G. Here, proper filtering of the pump-light through a copper sulfate solution was required to reduce unnecessary heating. Also, an optimal concentration of the dye was chosen, so that heating due to pump-light absorption was reasonably uniform throughout the cuvette volume. With the same pumping arrangement Snavely

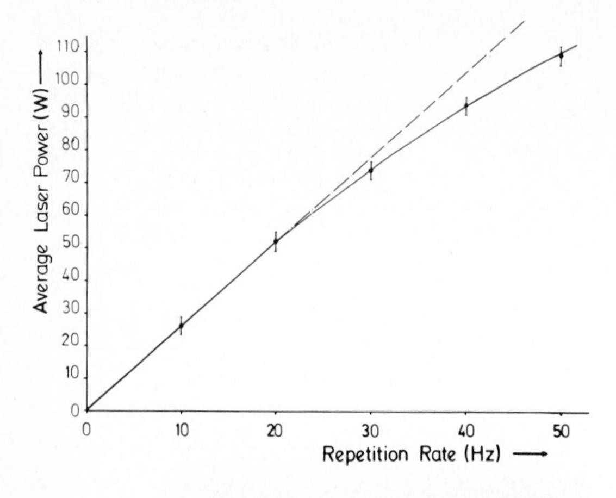


**Fig. 1.43.** Average power as a function of the repetition rate of a flashlamp-pumped dye laser. (From Jethwa et al. 1978)

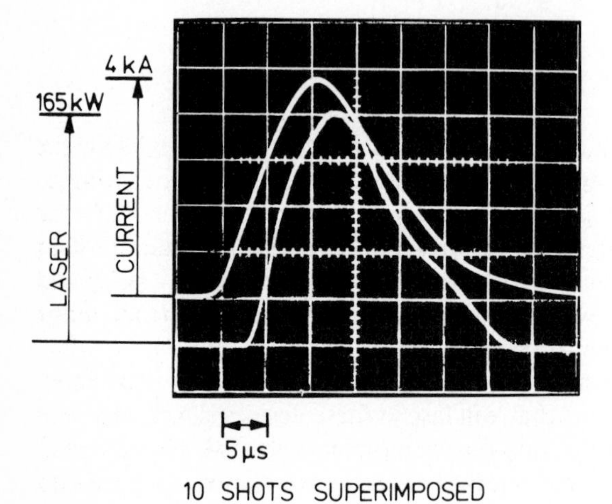


**Fig. 1.44.** Current and laser pulse forms for the laser of Fig. 1.43. (From Jethwa et al. 1978)

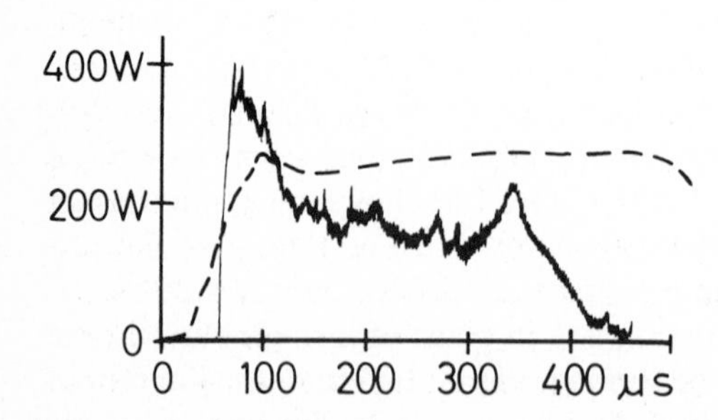


**Fig. 1.45.** Oscillogram of pump-light pulse (*broken line*) of a dye laser using a helical flashlamp and a lumped parameter transmission line and dye-laser pulse (*solid line*) from an air-saturated rhodamine 6G solution

and Schäfer (1969) had obtained 140 μs long pulses from rhodamine 6G and rhodamine B solutions; the duration clearly indicated that a steady state of the triplet population had been reached, provided the solution was saturated with oxygen or air. No laser emission was obtained, even at twice the original threshold, if nitrogen was bubbled through the solution for a time sufficient to purge the oxygen. Pappalardo et al. (1970b) using a laser with a 600-μs pump pulse obtained a 500 μs dye laser pulse from a $5 \times 10^{-5}$ molar rhodamine 6G solution containing $5 \times 10^{-3}$ mole/l of cyclooctatetraene as triplet quencher. Such results with long pulses first proved that triplet absorption cannot prevent cw dye-laser emission if the necessary triplet quenchers are added. In fact, it was concluded from these experiments that an absorbed pump-light power of less than 4 kW/cm$^3$ for a $5 \times 10^{-5}$ molar air-saturated methanol solution of rhodamine 6G should be sufficient to reach threshold with steady-state triplet population. The thermal problem, on the other hand, remains a serious one. To alleviate this problem one might employ a solvent with higher specific heat and use it at a temperature where the variation of the refractive index with temperature is at a minimum. In this respect water near freezing point or, even better, heavy water at 6 °C would be an ideal solvent.

A miniature long-pulse dye laser with 60 s pulse duration, 100 Hz repetition frequency, and a very low threshold of only 6 J was described by Hirth et al. (1977).

Extremely high dye-laser pulse energies have been obtained with large coaxial lamps. The laser shown in Fig. 1.41 gives an output energy of 32 J with an alcoholic rhodamine 6G solution at 17.3 kJ electrical energy input, i.e. 0.2% efficiency, and a specific output energy of 1 J/cm$^3$. The output power was reported as 10 MW, so that the half-width of the pulse must be about 3.2 μs. The lamp was filled with xenon at 1 Torr pressure in this case. The laser reported by Baltakov et al. (1973) generated 110 J in 20 μs, and 5.5 MW peak power.

The characteristics and limitations of coaxial and U-shaped flashlamps as pumping sources for dye lasers are described in (Anikiev et al. 1976; Marling et al. 1974; Hirth et al. 1973b; Drake and Morse 1974; Baltakov and Barikhin 1976; Strizhnev 1976; Ornstein and Derr 1974; Maeda et al. 1977). Guided sparks, discharges in a gas vortex, and laser-produced plasmas were tested in some investigations as pump sources: (Weysenfeld 1974; Brown 1975; Ferrar 1972; Silfvast and Wood II 1975a and 1975b; Ferrar 1973). Of the more exotic pumping sources the plasma focus (N.T. Kozlov and Protasov 1975/6), the high pressure mercury capillary lamp (Dal Pozzo et al. 1975), and semiconductor diodes (G. Wang and Webb 1974; G. Wang 1974) should be mentioned. Four papers discussed the prospects for dye lasers pumped by electrochemiluminescence: (Measures 1974; Keszthelyi 1975; Measures 1975; C.A. Heller and Jernigan 1977). Xenon-ion lasers proved useful because of their longer pulse width as replacements for nitrogen lasers in ultra-narrow bandwidth dye laser work (Hänsch et al. 1973; Schearer 1975; Levenson and Eesley 1976), while copper-vapor lasers as pump sources can improve the overall efficiencies of dye lasers (Decker et al. 1975; Pease and Pearson 1977), and excimer lasers can extend the short-wavelength limit of dye lasers (Sutton and Capelle 1976; Bücher and Chow 1977).

The spectral range of flashlamp-pumped dye lasers at present extends from 340 nm in p-terphenyl to 850 nm in DTTC (Maeda and Miyazoe 1972). Time-resolved spectra of flashlamp-pumped dye lasers show even greater variety than those of laser-pumped dye lasers (Ferrar 1969a). Some dyes show almost no wavelength sweep during an emission of 300 ns, while others have either monotonic or reversing sweeps. In these experiments triplet–triplet absorption and thermal effects due to nonuniform illumination of the cuvette may have been of importance. If the triplet–triplet spectrum is known, the sweep can be predicted. Thus in rhodamine 6G a sweep towards shorter wavelengths should be observed in a uniformly illuminated laser cuvette (Snavely 1969).

In addition to these fast sweeps there is a long-term drift in wavelength associated with increasing loss due to products of photochemical decomposition of the dye. Thus, in a 100 Hz flashlamp-pumped rhodamine 6G laser, the laser wavelength was observed to drift from 570–600 nm in a 10 min period (W. Schmidt 1970). The absorption spectrum of the solution taken after this experiment gave clear evidence of a photodecomposition product with absorption increasing towards shorter wavelengths.

## 1.6 Wavelength-Selective Resonators for Dye Lasers

A coarse selection of the dye-laser emission wavelength is possible by judicious choice of the dye, the solvent, and the resonator Q, as described in Sect. 1.4.1. Fine tuning and simultaneous attainment of small linewidths can only be achieved by using a wavelength-selective resonator.

Up to now the following four classes of wavelength-selective resonators seem to have been employed:

1) resonators including devices for spatial wavelength separation,
2) resonators including devices for interferometric wavelength discrimination,
3) resonators including devices with rotational dispersion,
4) resonators with wavelength-selective distributed feedback.

The various implementations of these classes of wavelength-selective resonators and their relative merits will be discussed in the above order.

The first wavelength-selective resonator was constructed by Soffer and McFarland (1967). They replaced one of the broad-band dielectric mirrors by a plane optical grating in Littrow mounting. This arrangement is shown in Fig. 1.46 together with a diagram of laser output vs wavelength obtained with a grating of 610 lines per mm in the first and second order for a rhodamine 6G solution. Consider the grating equation $m\lambda = \delta(\sin\alpha + \sin\beta)$ which for autocollimation ($\alpha = \beta$) reduces to $m\lambda = 2d\sin\alpha$. Here $m$ is the order, $\lambda$ is the wavelength, $\beta$ is the angle of incidence, $\alpha$ is the angle of diffraction from the normal to the grating, and $d$ is the grating constant. Then the angular dispersion is $d\alpha/d\lambda = m/2d\cos\alpha$.

If the dye laser has a beam divergence angle $\Delta\alpha$, the passive spectral width of this arrangement would be

$$\Delta\lambda_\alpha = \frac{2d\cos\alpha}{m}\,\Delta\alpha\ . \tag{1.13}$$

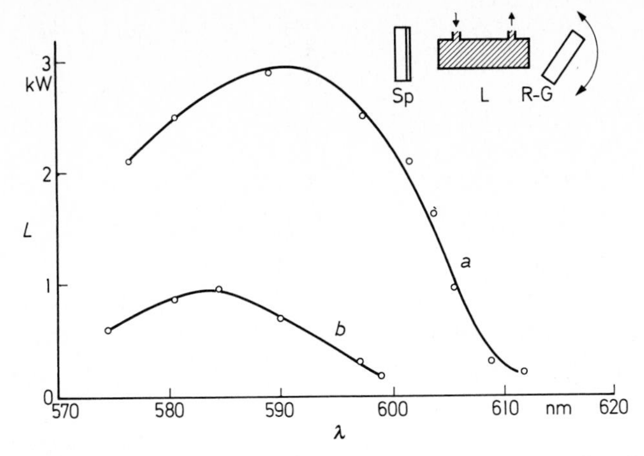


**Fig. 1.46.** Tuning of dye-laser wavelength with an optical grating. Output powers vs wavelength for a $10^{-4}$ M solution of rhodamine 6G in methanol, grating 610 lines/mm, a first order, b second order. (Insert: optical arrangement; Sp = 98% mirror, L = laser cuvette, R-G = optical grating in rotatable mount)

On the other hand, if the laser has diffraction-limited beam divergence, $\Delta\alpha = 1.22\lambda/D$, where $D$ is the inner diameter of the cuvette, then the passive spectral width is

$$\Delta\lambda_D = 2.44 d\lambda \cos\alpha/mD \ . \tag{1.14}$$

For a typical case, $\Delta\alpha = 5$ mrad, $D = 2.5$ mm, $d = (1/1200)$ mm, $m = 1$, and $\lambda = 600$ nm, one obtains $\Delta\lambda_\alpha = 7.8$ nm, and $\Delta\lambda_D = 0.37$ nm. If beam expanding optics are used within the cavity to reduce the beam divergence by a factor of ten, this would give $\Delta\lambda_{\alpha\mathrm{r}} = 0.78$ nm and $\Delta\lambda_{D\mathrm{r}} = 0.037$ nm. The active spectral width at threshold is smaller than the passive width depending on the available gain. Passive bandwidths are quoted here since these give an upper limit for the spectral bandwidth. The experimental value of the spectral width for a rhodamine 6G laser with 5 mrad beam divergence and a 2160 lines/mm grating was 0.06 nm, as compared with a passive width of 4.6 nm (Soffer and McFarland 1967).

For another laser with a 600 lines/mm grating the spectral width was 2 nm in the first and 0.4 nm in the second order, as compared with the passive width of 16 nm and 8 nm, respectively (Marth 1967). Thus the active bandwidth is smaller by a large factor (between 8 and 80 in these cases) than the passive bandwidth. The peak power is reduced by a factor of only two to five if the grating is blazed for this wavelength. In a laser with a Brewster angle cuvette and resultant polarized emission, this ratio can be even more favorable for the polarization which gives higher grating efficiency.

While high-quality gratings can have efficiencies of up to 95% at the blaze wavelength, most gratings have lower efficiencies, 65% being a realistic value. Thus, the insertion loss due to the grating is substantial.

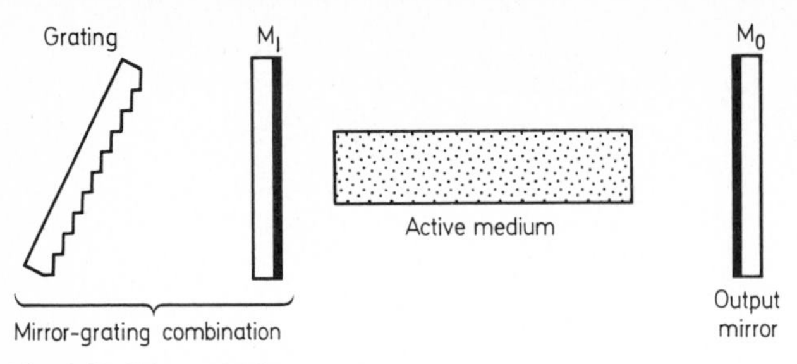


**Fig. 1.47.** Schematic diagram showing the use of a mirror-grating combination in a laser. $M_1$ is the intermediate mirror. (From Bjorkholm et al. 1971)

Another disadvantage of the grating is the reflecting metal film which may be damaged by high power and energy pulses. An improvement may be expected from holographically produced bleached transmission gratings (Kogelnik et al. 1970). This problem can also be circumvented by the use of a high-power beam-expanding telescope. The use of such a beam-expanding device is especially indicated in the case of laser-pumped dye lasers, as shown in Fig. 1.21. Most suitable for this purpose are prism beam expanders, of which many different designs have been described. Simple ones use a single prism (Myers 1971; Hanna et al. 1975; Wyatt 1978); more complicated designs, using several prisms in series, allow higher magnification (Klauminzer 1978; Duarte and Piper 1980, 1982). A comparison of the various designs can be found in (Rácz et al. 1981). An unexpanded beam of a fraction of one mm diameter would cover only a few lines on the grating and thus also seriously impair the spectral resolution. Another way to prevent the burning of a grating is by the use of an additional semitransparent mirror in front of the grating, as shown in Fig. 1.47 (Bjorkholm et al. 1971). This scheme not only reduces the power incident on the grating to a few percent of what it would have been without the mirror, but also significantly reduces the laser threshold, since the mirror-grating combination acts as a high-reflectivity resonant reflector for the tuned wavelength. The authors report a reduction in threshold by a factor of two and in bandwidth by a factor of 3.3 over the use of a grating only.

An important improvement was developed by several groups in 1977 when they used a grating in grazing incidence in combination with a maximum reflectivity mirror as shown in Fig. 1.48. As is immediately obvious, this arrangement obviates the need for intracavity beam expanders. Outcoupling is via the zeroth order (Shoshan et al. 1977; Littman 1978; Saikan 1978). A further improvement was the use of a second grating instead of the mirror, as shown in Fig. 1.49 (Littman and Metcalf 1978). Single-mode operation could easily be achieved in such an arrangement.

Alternatively, tuning and spectral narrowing may be achieved by one or more prisms in the laser cavity (Yamaguchi et al. 1968). The relatively small angular dispersion of a single prism is sufficient to isolate one of several sharp lines in gas lasers, for example, where this method has long been used. But it

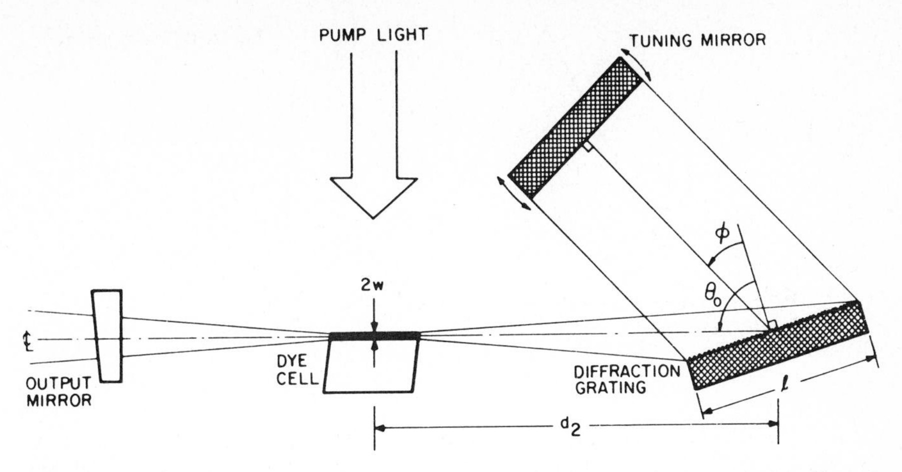


**Fig. 1.48.** Grazing incidence laser-pumped tunable dye laser. (From Littman and Metcalf 1978)

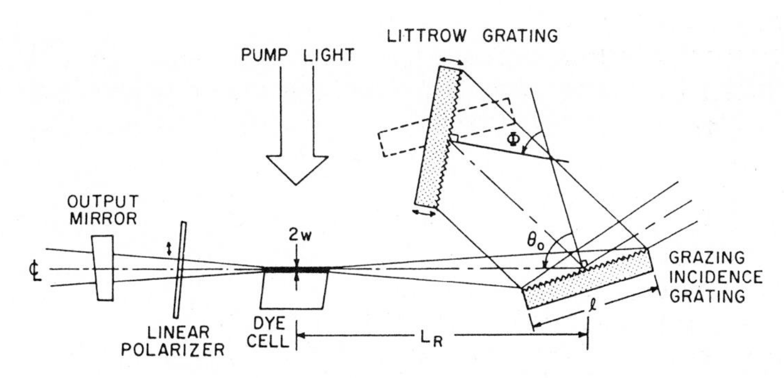


**Fig. 1.49.** Grazing incidence laser-pumped tunable dye laser with a second grating. (From Littman 1978)

gives hardly any reduction in spectral bandwidth of a flashlamp-pumped dye laser, so that multiple-prism arrangements have to be used. With the notation of Fig. 1.50 one has $\alpha = 2i - \beta$ and $r = \frac{1}{2}\beta$ so that

$$\frac{d\mu}{d\alpha} = \frac{\cos\frac{1}{2}(\alpha+\beta)}{2\sin\frac{1}{2}\beta} . \tag{1.15}$$

Since it is better to work near the Brewster angle where $d\alpha/d\mu = 2$, the angular dispersion of a prism is

$$d\alpha/d\lambda = 2\,d\mu/d\lambda . \tag{1.16}$$

Using $z$ prisms in autocollimation with a dye laser of beam divergence $\Delta\alpha$, the passive spectral width is

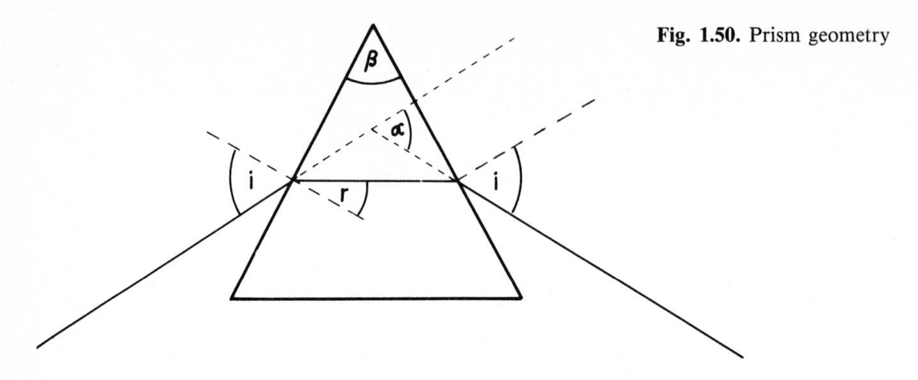


**Fig. 1.50.** Prism geometry

$$\Delta\lambda_\alpha = \frac{\Delta\alpha}{4z\,d\mu/d\lambda} \,. \qquad (1.17)$$

If $\Delta\alpha = 1.22\lambda/D$, in the diffraction-limited case one has

$$\Delta\lambda_D = \frac{1.22\lambda/D}{4z\,d\mu/d\lambda} \,. \qquad (1.18)$$

Consider 60°-prisms of Schott-glass SF10 for which $\mu_D = 1.72802$ and $d\mu/d\lambda = 1.35 \times 10^{-4}\,\mathrm{nm}^{-1}$. Then the following values are obtained for 1 or 6 prisms in autocollimation:

Values of $\Delta\lambda$ [nm] for a number of prisms

| No. of prisms | $\Delta\lambda_\alpha$ | $\Delta\lambda_{\alpha r}$ | $\Delta\lambda_D$ | $\Delta\lambda_{Dr}$ |
|---|---|---|---|---|
| 1 | 9.3 | 0.93 | 0.54 | 0.05 |
| 6 | 1.5 | 0.15 | 0.09 | 0.01 |
| grating 1200 lines/mm | 7.8 | 0.78 | 0.37 | 0.04 |

As above, it is assumed that $\Delta\alpha = 5$ mrad and $D = 2.5$ mm, with and without tenfold beam expanding optics.

A comparison of passive spectral widths, as given in the above table, shows that it should be better to use two or more prisms rather than one grating. In addition, the cumulative insertion loss of even 6 prisms near the Brewster angle is much smaller than that of one grating.

A five-prism arrangement has been reported by Strome and Webb (1971) and a six-prism arrangement by Schäfer and Müller (1971). Another method for obtaining increased angular dispersion uses a prism at angles of incidence of slightly less than 90 degrees. This results in a very high dispersion at only slightly reduced resolution, as in early spectrographs. The high reflection at the prism face that was detrimental in those spectrographs is an advantage here,

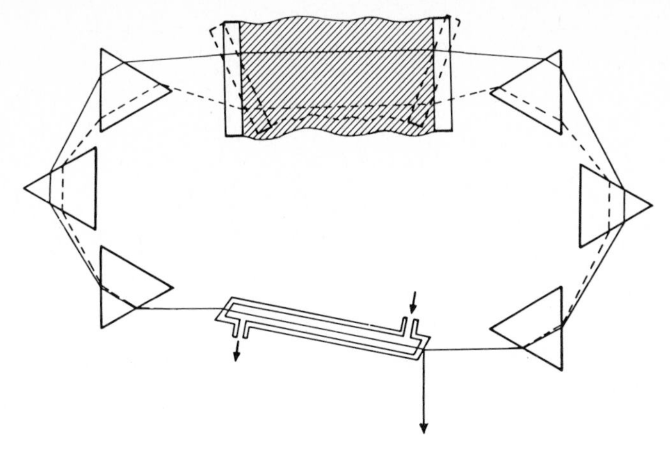

**Fig. 1.51.** 6-prism ring laser with variable wedge. (From Schäfer and Müller 1971)

since it serves as an outcoupling device, variable between 20% reflection at 80° incidence to 100% reflection at 90° (Myers 1971). In the prism there is a considerable increase in beam width and a concomitant reduction in beam divergence, which makes the additional insertion of a grating attractive, again particularly in nitrogen-laser-pumped dye lasers.

The prism method lends itself to use in ring lasers. Schäfer and Müller (1971) incorporated six 60°-prisms made of the Schott-glass SF10 as dispersive elements in a ring laser whose optical path is completed by two glass plates connected by a rubber bellows which is filled with an index-matching fluid so that a variable wedge is formed. The laser wavelength is selected by the setting of the refracting angle of the wedge (Fig. 1.51). A linewidth of less than 0.05 nm was obtained with 40 J electrical input energy pumping a $10^{-4}$ molar rhodamine 6G solution flowing in a 7.5-cm-long dye cuvette of 2.5 mm inner diameter, placed in the center of a 2″ helical flashlamp. The Fresnel reflection of one of the dye cuvette windows, intentionally set a few degrees off the Brewster angle, generated the output beam. It is noteworthy that the spectrum showed none of the satellite lines that are usually found in linear lasers using prisms or gratings for spectral narrowing. These satellite lines near the selected wavelength are probably associated with rays passing through inhomogeneities in the dye solution. Hence they form an angle with the optical axis which may be greater than the beam divergence of the main part of the beam. Insertion of an aperture into the cavity for reducing the beam divergence eliminates these lines. In a ring laser the cuvette evidently acts in this way, resulting in much improved discrimination against satellites. Another ring laser using four Abbé or Pellin-Broca prisms was described by Marowsky et al. (1972). The four 90° constant-deviation prisms made of SF10 glass are so arranged at the corners of a square that a closed path exists for the selected wavelength. The wavelength tuning was achieved by simultaneous counter rotation of the four mechanically coupled prisms. The spectral range covered reached from 430 nm

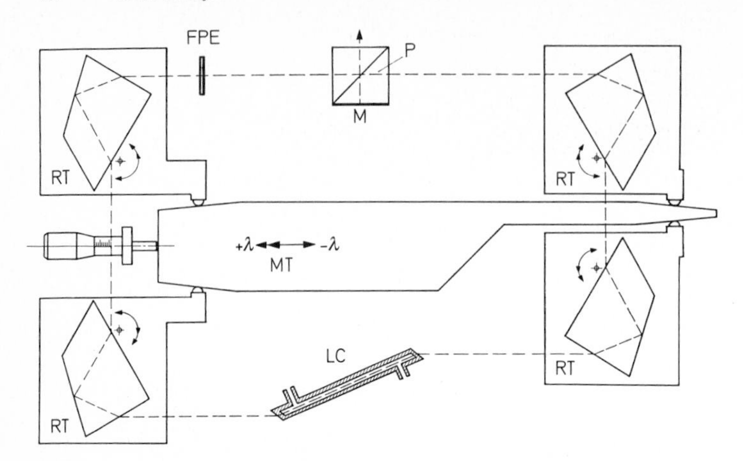


**Fig. 1.52.** Ring laser. (LC: laser cuvette, RT: rotating prism tables with Abbé prisms, axis and sense of rotation indicated, MT: movable table, FPE: Fabry-Perot etalon, P: beam-splitting output prism with high reflectivity mirror M). (From Marowsky et al. 1972)

to beyond 700 nm and the spectral width was 0.8 nm at 600 nm and decreased towards shorter wavelengths (Marowsky 1973a). This type of ring laser is shown in Fig. 1.52.

A noteworthy advantage of these multiprism ring lasers is that they obviate the need for mirrors with broadband dielectric reflective coatings. Additional aspects of prism tuning are discussed in (Marowsky 1973a, c, 1975; Marowsky et al. 1975; Yamagishi and Inaba 1976).

If only fixed wavelengths are needed, the simplest way of utilizing interference effects is to have narrowband reflective coatings on the resonator mirrors. Often supposedly broadband reflectors show a certain selectivity. At wavelengths where the reflectivity drops slightly, e.g. a quarter of one percent, holes are produced in the broadband spectrum of the dye laser at this wavelength; these holes have sometimes been misinterpreted as due to some property of the dye molecules. Resonant reflectors and reflectors of the Fox-Smith type are useful only in conjunction with suitable preselectors because of their narrow free spectral range.

In a method which is especially well suited for achieving a small spectral bandwidth, one or more Fabry-Perot etalons or interference filters are inserted into the cavity (Bradley et al. 1968a, b). The wavelength $\lambda$ of maximum transmission in $k$th order for a Fabry-Perot of thickness $d$, refractive index $\mu$, and with an angle $\alpha$ between its normal and the optical axis, is given by $k\lambda = 2\mu d \cos\alpha'$. Here $\alpha'$ is the refracted angle $\mu \sin\alpha' = \sin\alpha$. Thus, for air ($\mu = 1$) the angular dispersion is

$$d\lambda/d\alpha = \lambda \tan\alpha \quad . \tag{1.19}$$

Hence the spectral bandwidth for beam divergence $\Delta\alpha$ is (independent of the finesse)

$$\Delta\lambda_\alpha = \lambda \Delta\alpha \tan\alpha \ . \tag{1.20}$$

The wavelength shift $\Delta\lambda_s$ for turning the Fabry-Perot from a position normal to the optical axis ($\alpha = 0$, corresponding to a wavelength $\lambda_0$) through an angle $\alpha$ is

$$\Delta\lambda_s = (1 - \cos\alpha)\lambda_0 \ . \tag{1.21}$$

The free spectral range $\Delta\lambda_F$ between adjacent orders is

$$\Delta\lambda_F = \lambda/k \ . \tag{1.22}$$

The spectral width near $\lambda_0$ is determined by the reflection coefficient $R$ of the Fabry-Perot mirrors,

$$\delta\lambda = \Delta\lambda_F/F \ , \tag{1.23}$$

where

$$F = \frac{\pi\sqrt{R}}{1-R} \tag{1.24}$$

is the so-called finesse factor.

From these relations it is easy to determine the required properties of the laser and the Fabry-Perot for narrow band emission and wide band tunability. The attainable minimum bandwidth is determined by the minimum angle which avoids reflection from the first mirror of the Fabry-Perot back into the cuvette. If $q$ is the ratio of the diameter of the cuvette to the distance between the Fabry-Perot and the nearest cuvette window, the minimum bandwidth is

$$\Delta\lambda_{\alpha\,\mathrm{min}} = \lambda \Delta\alpha \tan\tfrac{1}{2}q \ . \tag{1.25}$$

Assume a prism preselector in the cavity which gives $\Delta\lambda_\alpha = 3$ nm for $\Delta\alpha = 5$ mrad at 600 nm. Then the optimum free spectral range of the Fabry-Perot should be $\Delta\lambda_F = \lambda/k = 3$ nm so that $k = 200$ and $d = 60$ µm. With a typical value of $q = 0.05$ this yields a beam-divergence-limited bandwidth of $\Delta\lambda_{\alpha\,\mathrm{min}} =$ 0.075 nm. As the angle is increased to tune over the free spectral range, the bandwidth increases according to (1.20) to $\Delta\lambda_{\alpha\,\mathrm{min}} = 0.3$ nm. Thus, a large tuning range is possible only at the expense of a relatively large increase in bandwidth. In addition to an increasing bandwidth, the use of Fabry-Perot etalons at high angles also introduces serious walk-off losses, which become more serious the larger the ratio of etalon thickness to beam diameter and the higher the angle. In order to realize a specified narrow bandwidth, one would have to reduce the tuning range and/or the beam divergence of the laser. Wavelength selection and simultaneous spectral narrowing were achieved in this way with laser-pumped and flashlamp-pumped lasers. An emission bandwidth of less than 50 pm was first achieved in practice (Bowman et al. 1969a), later less than 1 pm (W. Schmidt 1970). Hänsch (1972) reported that the insertion of a Fabry-Perot etalon into the dye laser cavity shown in Fig. 1.21 c reduced the bandwidth from 3 pm, obtained with the grating only, to 0.4 pm. At the same time the output dropped from 20 kW to about 3 kW, thought to be primarily due to high losses in the available broadband coatings. Probably at least some of these losses are due to other sources, like the walk-off in the etalon. This is substantiated by the small reduction of output powers after the successive insertion of an interference filter (low-order Fabry-Perot etalon) and two Fabry-Perot etalons used near the minimum useful angle. In his rhodamine 6G laser at 600 nm wavelength and 20 J pump energy Marowsky (1973 b) obtained 2.4 kW at 50 pm bandwidth with only an interference filter in the cavity. The insertion of the first Fabry-Perot decreased the peak power to 2.2 kW and the bandwidth to 7 pm. The insertion of the second etalon decreased the peak power to 2.0 kW and the bandwidth to 0.1 pm.

As Fabry-Perot etalons one usually uses either plane-parallel quartz plates coated with dielectric multilayer broadband reflective coatings, or optically contacted air Fabry-Perots (Bates et al. 1968).

For high resolution spectroscopy a very convenient method is the pressure scanning of a Fabry-Perot etalon in the resonator of a dye laser, as described by J. M. Green et al. (1973 a); Flach et al. (1974) and Wallenstein and Hänsch (1975).

If the length of the dye laser resonator is very small, only a few well-separated longitudinal modes can oscillate, of which one can easily be singled out by some simple filter. When the length is decreased to only a few meters, only one longitudinal mode lies within the fluorescence band, and small changes in resonator length allow an easy scanning of the laser wavelength. Figure 1.53 gives an example of this (Schäfer 1970). Later, this tuning method for short cavity dye lasers was further refined by Cox and Scott (1979) and Cox et al. (1982).

Several methods of wavelength selection make use of the rotation of polarization. One method utilizes birefringent filters in the cavity (Soep 1970). A simple arrangement consists of a quartz plate cut parallel to the optic axis, which has a retardation of several half-wavelengths at the center of the tuning range, and a set of Brewster plate polarizers as shown in Fig. 1.54. In this case there are transmission maxima for retardations of multiple half-wavelengths,

$$k\lambda/2 = \Delta\mu x_0 \cos\alpha \ . \qquad (1.26)$$

Here $k$ is the order number, $\Delta\mu$ the birefringence and $x_0$ the crystal thickness, both for normal incidence. Thus the wavelength spread for beam divergence $\Delta\alpha$ is

$$\Delta\lambda = -\lambda \tan\alpha\,\Delta\alpha = -\lambda \frac{\cos\alpha \sin\alpha}{\mu^2 - \sin^2\alpha}\,\Delta\alpha \ . \qquad (1.27)$$

This expression is very small near $\alpha = 0$.

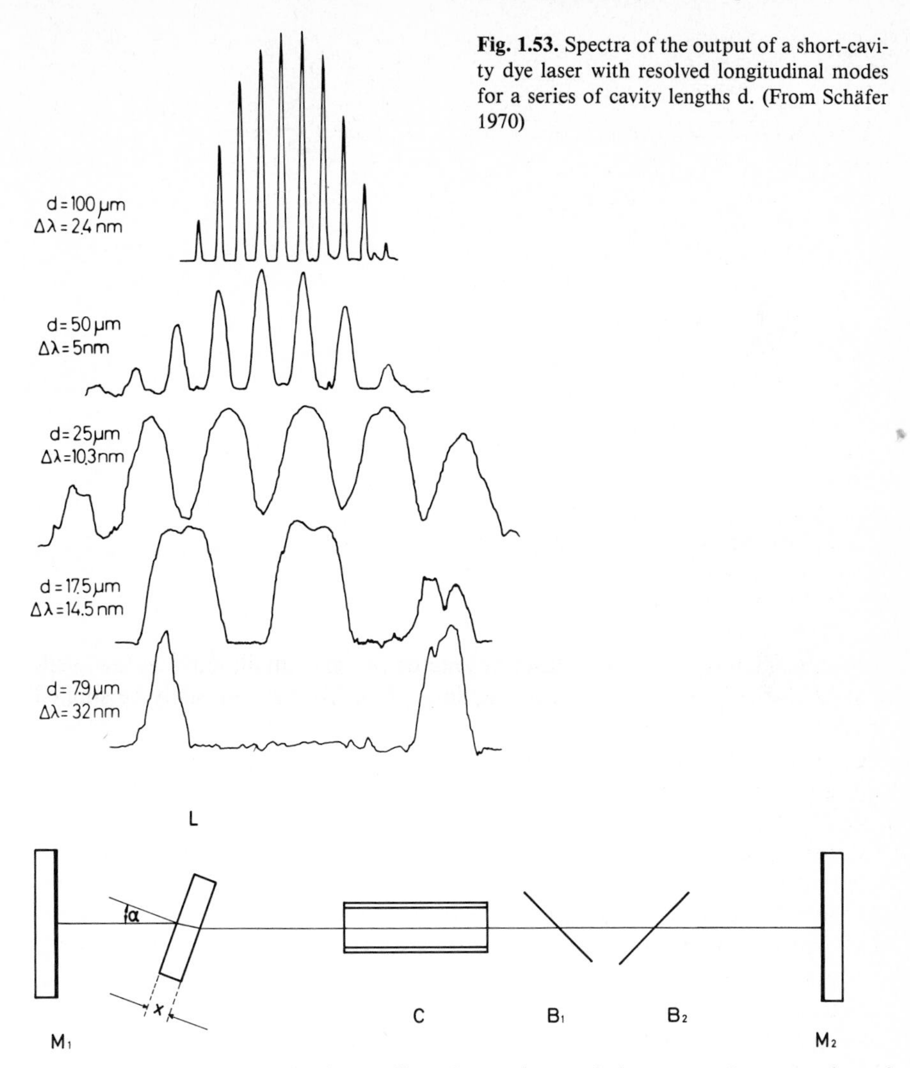


**Fig. 1.53.** Spectra of the output of a short-cavity dye laser with resolved longitudinal modes for a series of cavity lengths d. (From Schäfer 1970)

**Fig. 1.54.** Dye laser with birefringent filter. ($M_{1,2}$ mirrors, C dye cuvette, L quartz plate of thickness $x$, $B_{1,2}$ Brewster angle polarizers). (From Soep 1970)

Now, however, the bandwidth is not determined by the beam divergence of the laser, as in the methods discussed above, but rather by the transmission $T$ of the birefringent filter,

$$T = \cos^2(\pi \Delta\mu d/\lambda) \ . \tag{1.28}$$

If a reduction of 10% in transmission compared to maximum transmission brings the laser below threshold, one would expect an active bandwidth of

$$\Delta\lambda_a = \tfrac{1}{6}\lambda^2/\Delta\mu d \ . \tag{1.29}$$

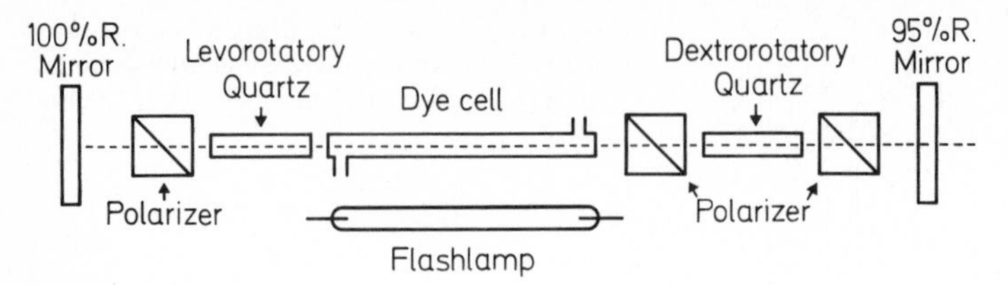


**Fig. 1.55.** Dye laser with double-state rotatory dispersive filter. (From D. Kato and Sato 1972)

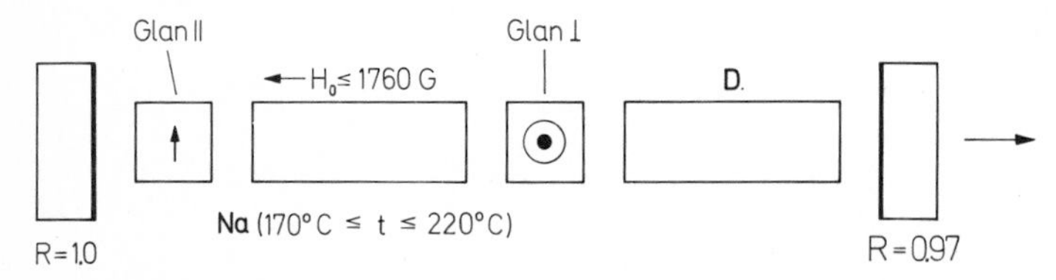


**Fig. 1.56.** Experimental arrangement to lock the dye laser wavelength to a spectral line. (D dye cuvette, Na sodium vapor cell, $H_0$ indicating magnetic field of solenoid surrounding cell). (From Sorokin et al. 1969)

This spectral width can be reduced further by the introduction of one or more additional quartz plates of greater thickness, as in Lyot or similar birefringent filters. Using KDP crystals of suitable orientation instead of the quartz plates, one can vary the transmission wavelength by applying a voltage (Walther and Hall 1970). The actual tuning range in a flashlamp-pumped rhodamine 6G laser with a quartz plate of 0.36 nm thickness and an angle of incidence between 35° and 50° was from 570–600 nm with a spectral bandwidth of 1 nm (Soep 1970). Walther and Hall (1970) obtained a spectral bandwidth of less than 1 pm and an electrical tuning range of 0.4 mm. The birefringent Fabry-Perot etalon is treated in (Holtom and Teschke 1974; Okada et al. 1975, 1976).

Another method makes use of the rotatory dispersion of *z*-cut quartz crystals. D. Kato and Sato (1972) used one dextro- and one levorotatory quartz crystal of 45 mm length each between 3 polarizers (Fig. 1.55). The tuning rate of this arrangement is 0.24 nm/degree if the central polarizer is rotated. The spectral half-width was rather wide, about 2 nm, since the rotatory dispersion of quartz is rather weak.

In an ingenious method, the large Faraday rotation in the vicinity of an atomic absorption line is used to lock the laser wavelength to the line (Sorokin et al. 1969). In the experimental arrangement of Fig. 1.56, the dye laser emission obtained consisted of two doublets locked to the sodium D lines. The components of both doublets are displaced symmetrically above and below the atomic line and each has a spectral width of less than 0.1 $cm^{-1}$. The splitting of the laser doublets may be adjusted by varying sodium vapor pressure and magnetic field.

Another important aspect of dye laser tuning methods is dual-wavelength operation. Many different designs have been devised, most of which make use

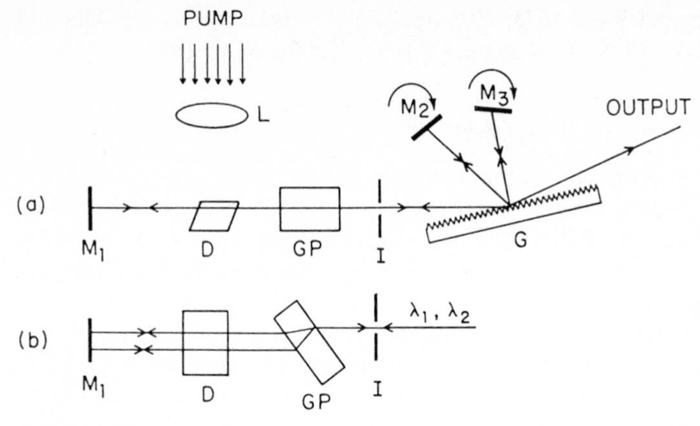


**Fig. 1.57.** Dual-wavelength grazing incidence dye laser, (**a**) top view, (**b**) side view. Dye cell D, glass plate GP, iris I, grating G, mirrors $M_{1,2,3}$, cylindrical lens L. (From Prior 1979)

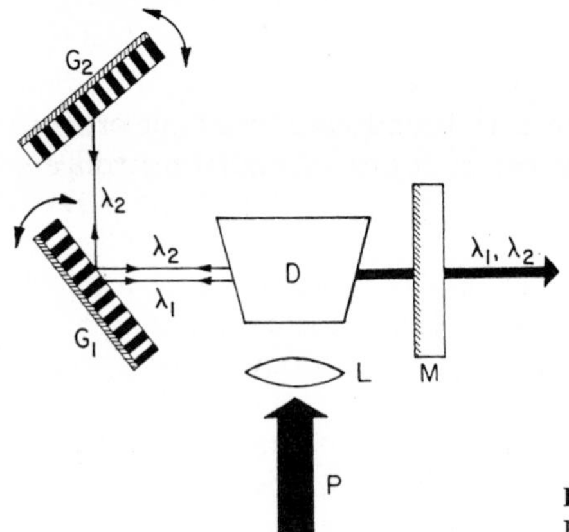


**Fig. 1.58.** Dual-wavelength dye laser. (From Friesem et al. 1973a)

of a grating in combination with two mirrors for feedback of the desired wavelengths, as shown in the example of Fig. 1.57, taken from (Prior 1979). Other papers describing similar designs are (Dinev et al. 1980; Inomata and Carswell 1977; Kittrell and Bernheim 1976; Dorsinville 1978).

Some other designs make use of two gratings, as shown in the example of Fig. 1.58 taken from (Friesem et al. 1973a). Further papers on this type of dual-wavelength dye lasers are (Pilloff 1972; Wu and Lombardi 1973; Dorsinville and Denariez-Roberge 1978; Nair 1979; Nair and Dasgupta 1980; Kong and Lee 1981). Still other variants can be found in (A. J. Schmidt 1975b; Lotem and Lynch 1975; Matsuzawa et al. 1976; Winter et al. 1978; Chou and Aartsma 1986).

Rapid tuning is possible with electro-optic elements in the resonator as described above (p. 76) or using an acousto-optic deflector in combination with a grating (Streifer and Saltz 1973; Hutcheson and Hughes 1974a, b; Telle and Tang 1974a, 1974b, 1975; K. Kopainsky 1975; Turner et al. 1975).

### 1.6.1 Distributed-Feedback Dye Lasers

Dye lasers with distributed feedback might become very important active elements for integrated optics. Their time behavior, which makes them very suitable for the production of single picosecond pulses, is described in Chap. 4. The first distributed-feedback dye laser was described by Kogelnik and Shank (1971). They produced a distributed-feedback structure by inducing a periodic spatial variation of the refractive index $\mu$ according to $\mu(z) = \mu + \mu_1 \cos Kz$, where $z$ is measured along the optic axis and $K = 2\pi/\Lambda$, $\Lambda$ being the period or fringe spacing of the spatial modulation and $\mu_1$ its amplitude. They calculated that a threshold could be reached in a dye laser with a gain of 100 and a length of 10 mm, if $\mu \geqq 10^{-5}$. This was easily obtained by exposing a dichromated gelatin film to the interference pattern of two coherent uv beams from a He-Cd laser. After exposure the gelatin film was developed in the usual manner and soaked in a solution of rhodamine 6G to make the dye penetrate into the porous gelatin layer. After drying, the film was transversely pumped with the uv radiation from a nitrogen laser. Threshold was reached at pump densities of 1 MW/cm$^2$ and dye-laser emission less than 0.05 nm wide was observed at about 630 nm. In uniform gelatin under the same pumping conditions the emission was 5 nm wide and centered at about 590 nm.

Shank et al. (1971) showed that a distributed-feedback amplifier can also be operated with the feedback produced by a periodic spatial variation of the gain of the dye solution. They used the experimental arrangement shown in Fig. 1.59, pumping a rhodamine 6G solution with the fringes of two coherent beams from a frequency-doubled ruby laser, meeting at an angle $2\theta$. Then the wavelength of the dye laser is given by $\lambda_L = \mu_s \lambda_p / \sin\theta$, where $\mu_s$ is the index of refraction of the dye solution at the lasing wavelength $\lambda_L$, and $\lambda_p$ is the pump wavelength. Thus, tuning is possible by varying either $\theta$ or $\mu_s$. The result of angle tuning is given in Fig. 1.60, while the results of tuning by variations of the refractive index of a solvent mixture of methanol and benzyl alcohol are given in Fig. 1.61. At about 180 kW peak pump power, the peak output power of the distributed-feedback dye laser was 36 kW. The spectral width with reduced pump power was less than 1 pm and apparently due to a single mode.

Narrowband laser oscillation from such a system was also observed when the period of the distributed-feedback structure was two or three times larger than the oscillation wavelength (Bjorkholm and Shank 1972b), probably due to harmonic frequencies in the spatial gain modulation.

The pumping arrangement described above needs as pump light source a laser of sufficiently small spectral bandwidth and good spatial coherence over the beam cross-section in order to create an interference pattern of high visibility. A nitrogen laser for example would not give satisfactory results.

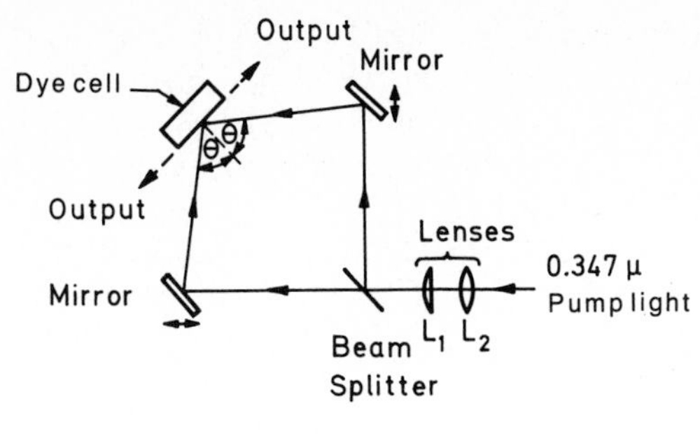


**Fig. 1.59.** Experimental arrangement of distributed feedback dye laser. (From Shank et al. 1971)

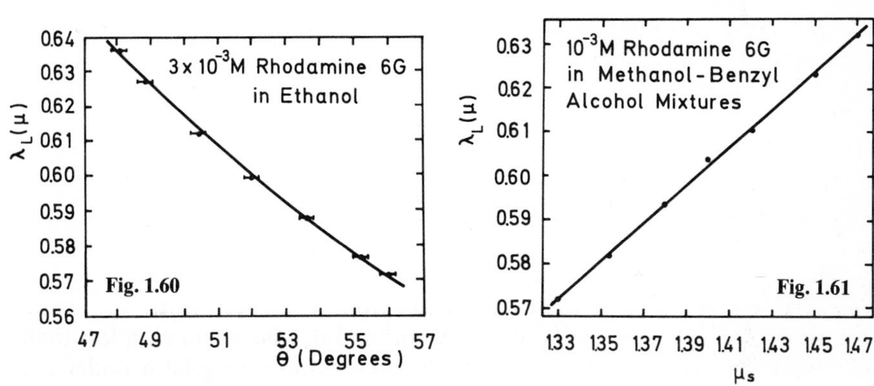


**Fig. 1.60.** Lasing wavelength $\lambda_L$ as a function of the angle $\theta$ for $3 \times 10^{-3}$ M rhodamine 6G in ethanol. The points are experimental, the curve is theoretical. (From Shank et al. 1971)

**Fig. 1.61.** Lasing wavelength $\lambda_L$ as a function of the solvent index of refraction $\mu_s$ for $\theta = 53.8°$. The dye solution was $10^{-3}$ M rhodamine 6G in a methanol-benzyl alcohol mixture. The points are experimental and the curve is theoretical. (From Shank et al. 1971)

A greatly improved pumping arrangement avoiding these restrictions was described by Bor (1979). It is shown in Fig. 1.62. The use of a holographic grating as a beamsplitter and a judicious choice of the geometry results in a high visibility interference pattern even with broadband lasers of low spatial coherence as pump lasers, e.g. nitrogen, excimer, and dye lasers.

Another very recent version of a distributed-feedback laser that allows broadband pumping is shown in Fig. 1.63. It uses a coarse grating of about 50 lines/mm that is imaged by a microscope objective into a dye cell in contact with the objective. A beam stop blocks the zeroth order. This arrangement is widely tunable by changing the distance between grating and microscope objective (Szatmári and Schäfer 1988).

Another possibility for distributed feedback was described by Kaminow et al. (1971). They used a sample of polymethylmethacrylate with the dimensions $4 \times 10 \times 38$ mm, doped with $8 \times 10^{-6}$ mole/l of rhodamine 6G. In this sample they produced two three-dimensional phase gratings of about $2 \times 2 \times 2$ mm by a 2-minute exposure of two intersecting uv beams of 0.7 mW each from a He-Cd laser as shown in Fig. 1.64. The sample was pumped by the second har-

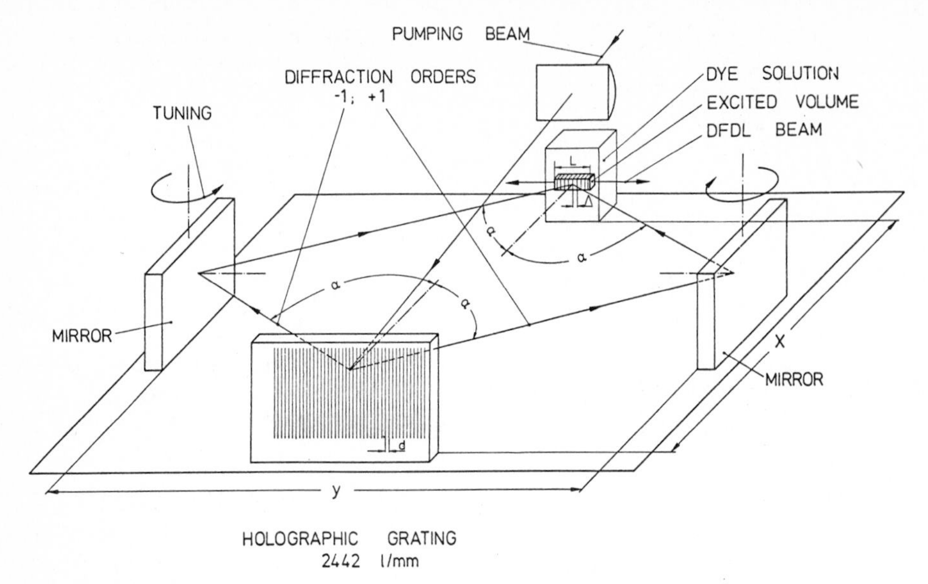


**Fig. 1.62.** Experimental arrangement of an achromatic distributed-feedback dye laser. (From Bor 1979)

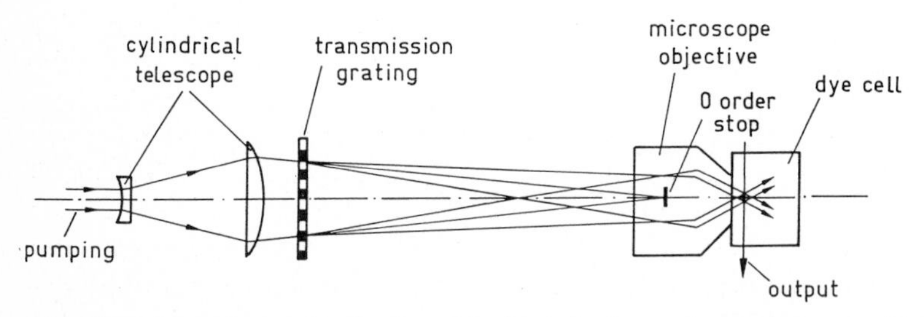


**Fig. 1.63.** Achromatic, tunable distributed-feedback dye laser using a microscope objective and a coarse grating. (From Szatmári and Schäfer 1988)

monic from a Nd laser. Again, the dye-laser wavelength was determined by the Bragg condition $\lambda_L = 2\mu_s\Lambda$ that must be fulfilled with $2\Lambda = \lambda_{uv}/\sin\theta$. The output consisted of one strong line of 0.5 pm near threshold. At higher pumping power, several modes separated by $c_0/2\mu_s L$ were observed, where $L = 20$ mm is the distance between the two gratings. The application of mechanical stress to the region of the gratings allowed a tuning of 1.1 nm.

Yet another possibility of producing a phase grating in polymethylmethacrylate was reported by Fork et al. (1971). They dissolved 2 mmole/l rhodamine 6G tosylate and 0.1 mole/l acridizinium ethylhexanesulfonate photodimers in methylmethacrylate and acrylic acid and polymerized the resulting solution to form a hard, transparent plastic. The sample was then cut and polished into 1 cm × 1 cm × 1 mm chips. The photodimers were first broken

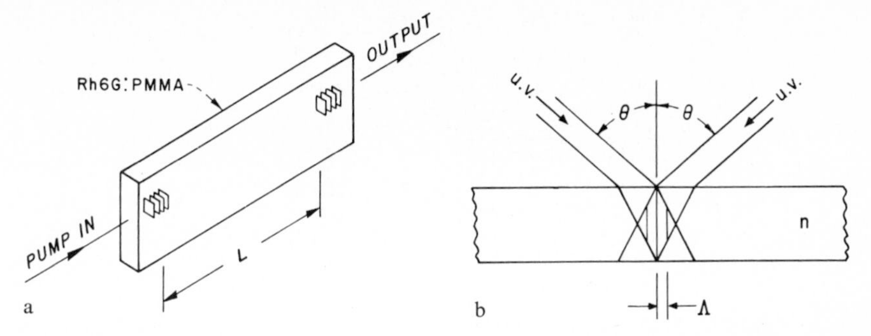


**Fig. 1.64.** (**a**) Dye-doped polymethylmethacrylate laser with internal grating resonator. (Dimensions: $4 \times 10 \times 38$ mm; grating spacing $L = 20$ mm). (**b**) Preparation of gratings by intersecting uv beams through the broad face of the plastic sample. (From Kaminow et al. 1971)

to a depth of 80 µm by illumination with an erase beam of 313 nm light from a mercury arc and then selectively remade in a grating pattern by two intersecting writing beams from an argon laser, in the manner described above for the other examples. The sample was then pumped by the 10-kW pulse from a neon laser focused on the sample by a cylindrical lens. The output showed a few narrow lines with a spacing determined by the $c_0/2L$ separation for the length $L = 1$ cm of the laser.

Fork and Kaplan (1972) rpeorted on a distributed-feedback dye laser with a variable phase grating which can be optically written into photodimer optical memory material with the interference fringes of a 364 nm $Ar^+$ laser, or erased with the 313 nm light of a Hg arc lamp. The laser material is a solid PMMA host, doped with rhodamine 6G and photodimers of acridizinium ethylhexanesulfonate. The photodimer can be broken with the erasing light and remade with the writing laser light of longer wavelength. Writing times as short as 3 ns appear possible with high-power pulsed lasers. A gain of 8 $cm^{-1}$, twice above threshold, was obtained, when the small active volume (1 cm$\times$80 µm$\times$10 µm) was pumped with a pulsed neon laser at 540 nm with 10 kW peak power. The narrowband dye-laser output consists of several peaks, some less than 1.2 GHz wide.

Instead of having the distributed feedback within the laser beam, one can also provide feedback for the evanescent wave and gain within the main laser beam in the dye solution adjacent to the distributed-feedback structure (Hill and Watanabe 1972). A schematic cross-section of such a laser is shown in Fig. 1.65. In the experimental implementation of the device the cover plate was a quartz optical flat, the organic dye solution a $3 \times 10^{-2}$ molar solution of rhodamine 6G in benzyl alcohol, or a mixture of benzyl and enthyl alcohols, and the feedback structure was a gelatin film grating on a glass substrate. The dichromated gelatin film had been exposed to two coherent intersecting laser beams and developed as described above. By varying the relative refractive indices of dye solution and gelatin film one could either have normal-wave gain and evanescent-wave feedback, or evanescent-wave gain and normal-wave feed-

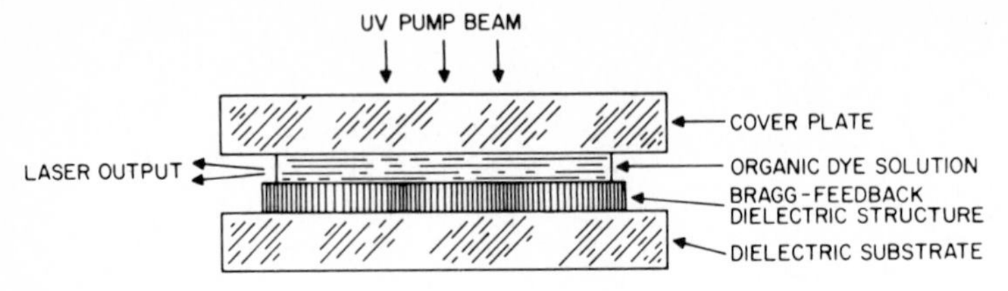


**Fig. 1.65.** Cross-section of the distributed feedback side-coupled laser. (From Hill and Watanabe 1972)

back, or gelatin film and dye solution acting together as a waveguide for higher-order modes for the case of nearly equal refractive indices.

Distributed feedback in a thin film dye laser can also be produced by providing a periodic perturbation of the film thickness, which can be achieved using photoresist and ion-milling techniques (Schinke et al. 1972; Kotani et al. 1976; Deryugin et al. 1976b; Kolbin et al. 1976).

Distributed feedback dye laser action in an optical fiber by evanescent field coupling was described by Periasamy and Bor (1981).

### 1.6.2 Thin-Film and Waveguide Dye Lasers

It is easy to prepare a dye laser structure having transverse dimensions of a few micrometers which supports only a small number of low-order waveguide modes. Modes of low losses are known in straight dielectric waveguides, even if the embedding medium has a higher index of refraction than the guiding core, as in hollow waveguide gas lasers. Burlamacchi and Pratesi (1973a) have utilized this phenomenon in a flashlamp-pumped superradiant dye laser, contained in a small-bore glass capillary. In general, it is more desirable to use a surrounding dielectric medium of lower refractive index to permit waveguiding by total internal reflection, and the wide choice of available liquid and solid dye-laser host materials makes it easy to meet this condition. Such waveguiding liquid dye lasers have been constructed by using benzyl alcohol ($n = 1.538$) as a solvent and filling the liquid dye solution into thin glass capillaries (Ippen et al. 1971; Zeidler 1971) or sandwiching a liquid dye film between flat glass substrates (Zeidler 1971).

H. P. Weber and Ulrich (1971) have reported the successful operation of a solid thin-film dye laser. They produced the active waveguiding structure by coating glass substrates with a thin film of polyurethane ($n = 1.55$) doped with $8 \times 10^{-3}$ M/l rhodamine 6G. At a typical thickness of 0.8 μm, such a film can support only the fundamental $TE_0$ and $TM_0$ modes. A gain as high as 100 dB/cm was obtained when the film was pumped by a pulsed $N_2$ laser. In order to provide feedback for laser oscillation, the doped light-guiding film was applied on the surface of a cylindrical glass rod of 5 mm diameter, as shown in Fig. 1.66. In this way a closed optical path is established along any circumference of the rod. A narrow circumferential strip of the beam is il-

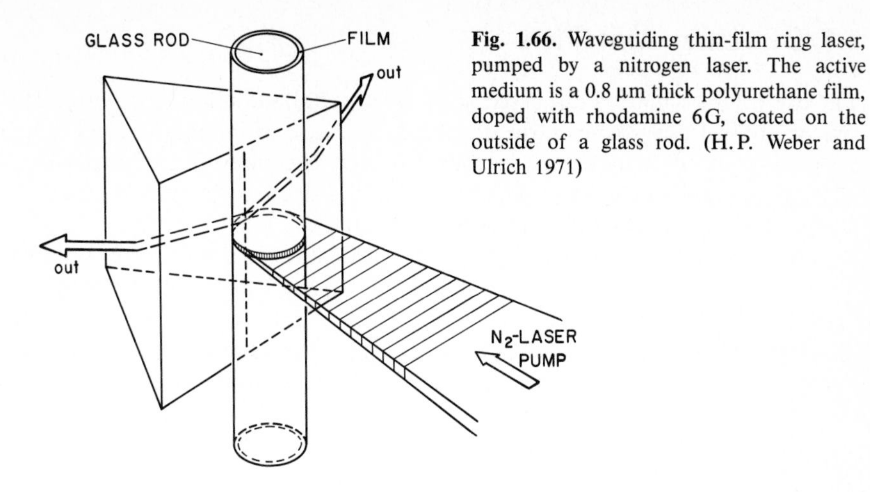


**Fig. 1.66.** Waveguiding thin-film ring laser, pumped by a nitrogen laser. The active medium is a 0.8 μm thick polyurethane film, doped with rhodamine 6G, coated on the outside of a glass rod. (H.P. Weber and Ulrich 1971)

luminated by a sheetlike beam of the $N_2$ laser. The dye-laser light is coupled out via its evanescent wave by a closely spaced prism. Two beams are obtained, corresponding to the two opposite directions of rotation. A peak power of 100 W was measured in each beam when the film absorbed about 1 kW of the incident pump light. The laser operated near 620 nm with a bandwidth of 11 nm. Individual axial modes of the ring resonator could be resolved with a Fabry-Perot interferometer, confirming the feedback around the rod. A similar arrangement was described by Deryugin et al. (1976a).

An entirely different approach to waveguiding dye lasers is realized in the evanescent-field-pumped dye laser, as reported by Ippen and Shank (1972a). Here the dye molecules are not incorporated in the waveguide, but are located in the surrounding (liquid) medium. They are pumped by the evanescent wave of the pump-laser light travelling through the waveguide, and they radiate by stimulated emission into a waveguide mode. In the experiment the pump light of a frequency-doubled Nd:glass laser was coupled via a prism into a thin waveguiding glass film, covered by a solution of rhodamine 6G in a benzyl alcohol glycerol mixture. The superradiant dye laser output was measured for different liquid refractive indices and different pump powers. The reported scheme appears particularly attractive because it is easy to replenish photobleached dye molecules. A theoretical estimate indicates that it should be possible to construct a rhodamine 6G dye laser with a waveguide cross-section of $1 \times 3\ \mu m^2$, which could exhibit a threshold of only 5 mW at a round-trip loss of 10%.

In a similar way a rhodamine-B-doped thin polyurethane film was pumped by a nitrogen laser and light of a He-Ne laser coupled in and out through two prisms. A gain of $13\ cm^{-1}$ was observed (Chang et al. 1972).

A special class of waveguide dye lasers is the one using thermally produced waveguiding structures, as mainly developed by Burlamacchi and coworkers (Burlamacchi and Pratesi 1973b, c; Burlamacchi et al. 1974, 1975a, b; Burlamacchi and Salimbeni 1976; Burlamacchi et al. 1976; Fowler and Glenn 1976).

## 1.7 Dye-Laser Amplifiers

The dye-laser oscillators discussed above are broadband amplifiers with selective or non-selective regenerative feedback. Because of its high inherent gain, a dye laser needs very little feedback to reach the threshold of oscillation. Thus, it is usually somewhat difficult to build a dye-laser amplifier, carefully avoiding all possibilities of regenerative feedback.

The first report on broadband light amplification in organic dyes pumped by a ruby laser was by Bass and Deutsch (1967). They set a Raman cell containing toluene in the ruby-laser beam and behind it a dye cell containing DTTC dissolved in DMSO. The ruby beam and the first stimulated Raman line pumped the dye solution, and broadband laser emission was obtained with the four-percent Fresnel reflection from the cuvette windows. However, when the concentration of the dye was set to a value such that the dye would lase near or at the wavelength of the second Stokes line at 806.75 nm, the broadband oscillation of the dye was quenched and the sharp Stokes line strongly amplified instead. The Raman signal being present from the beginning of the pump process used up all available inversion in the dye so that no free oscillation could start. This, too, is an experimental proof of the homogeneous broadening of the fluorescence band of the dye, at least on a nanosecond scale. Since there was four-percent feedback here, the amplification process was a multipass amplification. Similar results were obtained with $CS_2$ as Raman liquid and with cryptocyanine as amplifying dye. These results were confirmed by a similar investigation by Derkacheva and Sokolovskaya (1968).

In the ingenious experimental arrangement shown in Fig. 1.67 Hänsch et al. (1971b) obtained broadband, wide-angle light amplification in several dye solutions pumped by a nitrogen laser. The amplifier calls with an active length of 1.3 mm were made from 10-mm-diameter Pyrex tube sections. The antireflection-coated windows were sealed to the ends under a wedge angle of about 10° to avoid multiple reflections. In a typical experiment two cells were used, one acting as an amplifier and the other as an oscillator. Both were filled with the same dye solution and excited simultaneously by the same nitrogen laser. Part of the dim fluorescent light of the oscillator cell, showing a noticeable preference for near-axial propagation (superradiance), was collected by a field lens and focused by an additional multielement photographic lens into the active volume of the amplifier cell. Here it was amplified and emerged as a bright light cone at the other end, illuminating a circular area on a projection screen. If now any object, such as a photographic transparency, was put into the object plane of the photographic lens, it gave rise to a bright projected image on the screen despite its own faint illumination. The gain was determined by comparing the output with and without excitation of the amplifier, absorption in the amplifier cell and background, i.e. stimulated emission of the amplifier alone, being taken into account as corrections. For small signals (input energies of up to 8 μJ), a single-pass gain of 1000 or 23 dB/mm was obtained when a rhodamine 6G solution was excited by 100-μJ pulses of the nitrogen laser. At input signals of 25 μJ the gain dropped to 14 dB/mm, indi-

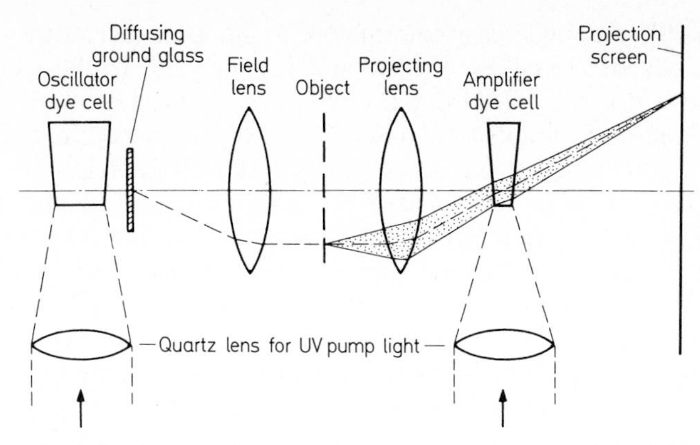


**Fig. 1.67.** Test setup for image amplification using a dye laser. (From Hänsch et al. 1971 b)

**Fig. 1.68.** Prototype of dye-laser image amplifier. (Hänsch et al. 1971 b)

cating saturation. With rhodamine B and fluorescein, the gain coefficients were about 4 dB/mm lower. Figure 1.68 shows a photograph of the experimental setup.

Erickson and Szabo (1971) also used a dye cell pumped by a nitrogen laser as an amplifier. The 1-cm cell was placed in a resonator consisting of a 99% and a 40% reflectivity mirror spaced 4.2 cm apart. When the acid form of 4-methylumbelliferone was pumped 20% above threshold, the spectral width of the dye-laser emission was 40 nm. Injecting the 514.5 nm line of an argon laser into the resonator caused practically the same energy to be emitted in the region around 514.5 nm in a bandwidth of only 0.16 pm, or about 4 times the width of the injected argon line. This is equivalent to a multi-pass amplification in the dye cell of $10^5$ or a single-pass gain of 100. A similar regenerative amplifier experiment was described by Vrehen and Breimer (1972). They longitudinally pumped a dye cell filled with a mixture of cresyl violet and rhodamine B with a frequency-doubled Nd laser. The cell was placed in a resonator consisting of two mirrors, one of which had 10% reflectivity at 530 nm to pass the pumping laser beam and 95% reflectivity at 632.8 nm, while the other had 20% transmission at this wavelength for the injection of 1 mW from a He-Ne laser. Here, too, the total output energy from the dye laser with and without injection was found to be constant, the effective total gain for the injected radiation being $10^6$. The output of the regenerative amplifier consisted of one or several longitudinal modes centered around the injected line, the number and linewidth depending on the length of the cavity that could be tuned piezoelectrically.

The gain obtainable in flashlamp-pumped dye-laser amplifiers is much smaller than that in laser-pumped amplifiers because of the high triplet losses. Huth (1970) measured the wavelength and time dependence of a flashlamp-pumped amplifier. The amplifier cell had a 5.3 mm inner diameter and was 3.8 cm long. It had antireflection-coated windows with 30-minute wedges. Pumping was by a flashlamp of 3-μs half-width in an elliptical cylinder and with typically 20 J energy. The dye solution was a $2 \times 10^{-4}$ molar solution of rhodamine 6G in ethanol. The signal to be amplified was derived from a dye-laser oscillator, had a bandwidth of 0.1 nm tunable by a grating over the range from 570–630 nm, and a peak power of typically 20 W/cm$^2$. It had a duration of about 400 ns and could be shifted in time over the flashlamp pulse length. The maximum gain thus found was 2.3, or 95 dB/m. Even less gain was found in a six-stage amplifier chain by Flamant and Meyer (1971) who measured an energy gain of only 6.0 in the whole chain. This low value was attributable to the very high transmission losses, indicated by the fact that the ratio of the amplifier outputs when pumped and not pumped, termed "apparent gain" by the authors, was 700. These investigations suggest that great improvements in the operation of amplifiers are possible when all parameters are carefully optimized. Injection of a strong monochromatic signal into a regenerative dye-laser amplifier (termed "forced oscillator" in this work) was used by Magyar and Schneider-Muntau (1972). The amplifier cell had an inner diameter of 9 mm and a length of 160 mm and was pumped in a close-coupled configuration by six linear air-filled flashlamps, enclosed by a cylindrical reflector of aluminum foil. The energy of the pump pulse was 4.2 kJ, its risetime 2 μs and its half-width 15 μs. The dye was rhodamine 6G in water of somewhat less than $7 \times 10^{-5}$ molar concentration, with the addition of 1.5% Ammonyx and 0.2% cyclooctatetraene. One resonator mirror was 99% reflecting, the outcoupling mirror 50%. The output without injection was 1.6 J in a 12 nm wide spectral band. A cell could be placed into the cavity under the Brewster angle that had the twofold purpose of containing an absorbing dye solution and injecting the signal by reflection from its front window. The injected signal was derived from a dye laser oscillator that was spectrally narrowed and tuned by two tilted Fabry-Perot interferometers and had a maximum output of 55 mJ in 400 ns and a bandwidth of less than 10 pm. Careful timing of the signal resulted in most of the energy of the amplifier appearing in the amplified injected line. Complete frequency locking, however, could only be achieved by adding a few drops of a solution of a suitable absorbing dye, e.g. 1,1′-diethyl-4,4′-cyanine iodide, to the absorber cell which previously contained only solvent. The output then contained only the injected line and had a total energy of 600 mJ. This frequency locking by an absorbing dye was first demonstrated with ruby lasers (Opower and Kaiser 1966). Since the injected signal had a duration of only 400 ns while the amplified signal was 5 μs long, only the front part was true amplification and the forced oscillator continued its emission at the same frequency in the later part of the pulse. Because of the complexity of the operation of this device, it is difficult to make a meaningful statement concerning power or energy gain. Nevertheless, the practical interest of this type of regenerative amplifier or forced oscillator is considerable.

The maximum useful single-pass gain is also determined by the amplified spontaneous emission, because a high noise level at the output reduces the gain by saturation. Typical saturation parameters are on the order of 100 $kW/cm^2$ to 1 $MW/cm^2$. The intensity of the amplified spontaneous emission must be kept well below this saturation intensity if "superradiant" emission is to be avoided. Amplification factors of up to 1000 have been realized in practice without violating this condition. Much higher gains should be possible if the dye-laser amplifier is subdivided into several stages and the number of modes of the transmitted noise radiation reduced by spatially limiting apertures and spectral filters.

Unlike solid-state lasers, dye lasers cannot store pump energy for longer than a few nanoseconds, i.e. the lifetime of the excited state. Hence dye-laser amplifiers are only suitable for the amplification of relatively short pulses to extremely high powers. An example of this type of multistage amplifier for the amplification of picosecond or femtosecond pulses to the gigawatt level is to be found at the end of Chap. 4.

## 1.8 Outlook

To end this chapter, the reader might be interested in a few speculative remarks about possible trends for future developments of dye lasers. Regarding the chemical aspects, the reader is referred to the concluding remarks in Chap. 5. The most important topics with regard to the physical aspects of dye lasers are new pumping methods and pump-light sources, followed closely by the physical dimensions of dye lasers. Some properties can be extrapolated to foreseeable physical limits.

The usual pumping methods using lasers and flashlamps will certainly be improved with respect to efficiency, power, control over pulse shape, and several other parameters. The long-standing problem of an incoherently pumped cw dye laser has just been resolved by Drexhage and coworkers (Thiel et al. 1987) using a specially constructed arc lamp and dye jet technology. Further developments of this first realization might scale it to very high output powers at reasonable efficiencies.

For very small pulsed or continuous dye lasers direct electrical pumping might be feasible. Some dyes are reported to form relatively stable anions and cations in certain aprotic solvents, e.g. the ions of diphenylanthracene (DPA) in dimethylformamide, which can be formed electrochemically and which, at recombination, leave one molecule in the excited state so that the fluorescence of the neutral molecule is emitted (Chandross and Visco 1964; Hercules 1964; Measures 1974, 1975; Keszthelyi 1975):

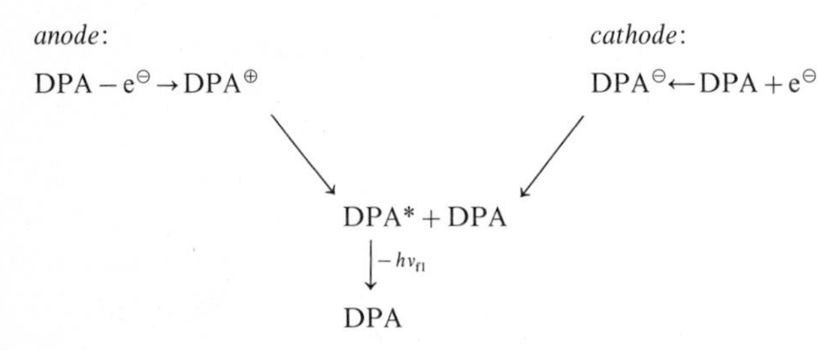


This method would have definite advantages for applications with integrated optics.

For very large volumes of dye solution, flashlamp-pumping would be very cumbersome and expensive; in this case chemical energy storage is more economical, one kilogram of explosive storing about 5 MJ of mechanical energy. This high energy content can be used to excite a shock wave in a gas (so-called argon bombs) (Held 1968) which thus becomes brightly luminescent. A cross-section through a possible structure utilizing such an argon bomb is shown in Fig. 1.69 (Schäfer 1969). The cylindrical mantle of explosive is ignited simultaneously at many lines of the circumference and excites a compressive shock wave in the argon layer that is traveling towards the symmetry axis of the structure. The dye solution is excited by the luminescent output from the shock wave and the concomitant superradiant output can be focused, e.g. by a parabolic ring mirror to the center point of the axis. One can expect outputs of several kJ focused on this spot, which would be useful for plasma experiments.

An extrapolation of present dye-laser properties into the next few years would give an estimate of extended wavelength coverage of 300 to 2000 nm for pulsed lasers and 380 to 950 nm for cw lasers. Together with frequency mixing and multiplication and stimulated Raman emission, this would in effect give complete coverage from the vacuum ultraviolet to the far-infrared.

The maximum power output of a laser is reached when a pulse containing the saturation energy (in photons per square cm) $E_s = 1/\sigma_{fl}$ is being amplified. For a dye $\sigma_{fl} \approx 10^{16}$ cm$^2$, and thus $E_s = 10^{16}$ photons/cm$^2$, equivalent (at 600 nm) to a power of 3.3 GW/cm$^2$ for a pulse of 1 ps duration. Since the cross-section of a dye solution is practically unlimited, pulses of terawatt peak power could be generated with dye lasers.

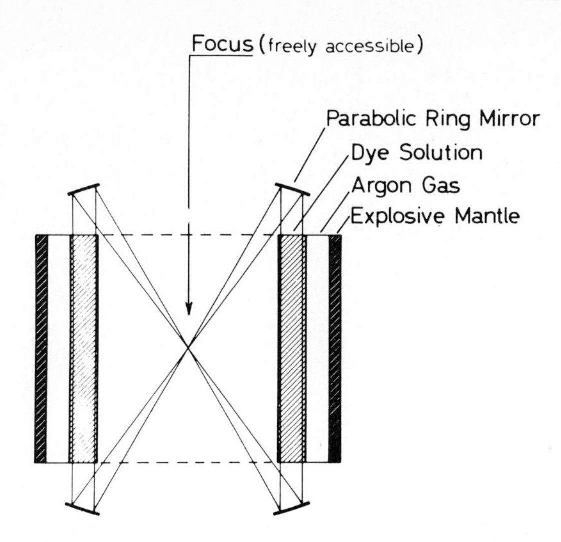


**Fig. 1.69.** Cross-section of a dye laser pumped by an argon bomb. (From Schäfer 1969)

The pulse energies of more than 1 J/cm$^3$ of dye solution that have been reported are higher than the stored energy and can only be obtained by multiple pumping of the dye molecules by a strong pump-light pulse. The stored energy at 10% inversion, 600 nm laser wavelength and a concentration of $10^{-3}$ mole/l is only 20 mJ/cm$^3$. Since this energy is stored for typically a few ns, the molecules can be pumped many times during a pulse of some μs duration. Thus, multikilojoule pulses from a few liters of dye solution appear feasible provided the problem of sufficiently strong pumping of such large volumes is resolved.

For a discussion of ultimate frequency stability of cw dye lasers, the reader is referred to the review by Salomon et al. (1988) while the question of the shortest possible pulses is discussed in Chap. 4.

# 2. Continuous-Wave Dye Lasers I

Benjamin B. Snavely[1]

With 18 Figures

The operation of the continuous-wave (cw) dye laser is based upon the same molecular states as that of pulsed lasers. However, some loss mechanisms that are relatively unimportant in the pulsed laser tend to dominate the performance of the cw laser. For example, the accumulation of molecules in the triplet state plays a major role in determining the efficiency of the cw laser, whereas the triplet state is relatively unimportant in pulsed lasers. Also, optical inhomogeneities produced in the active medium by heating resulting from excitation must be carefully controlled for the laser to operate continuously. Excitation sources for pulsed lasers are generally capable of producing intensities greatly in excess of that required to reach laser threshold. Pump sources for cw lasers, on the other hand, are often marginal. For these reasons, the efficient operation of a cw dye laser requires careful design and construction to minimize extraneous optical losses. The mechanical and optical tolerances are generally much more severe than those for pulsed dye lasers. In this chapter the analysis and design of cw dye lasers will be considered. Tuning systems will be discussed and the characteristics of some experimental cw laser systems reviewed.

As with pulsed dye lasers, the cw laser consists of an optically excited fluorescent dye solution within a suitable resonator. The mechanism for the production of stimulated emission in the dye solution is reviewed in Fig. 2.1.

Optical gain is associated with stimulated transitions $a \leftarrow B$ between the states labeled $S_1$ and $S_0$ of the singlet-state manifold. The population of $S_1$ is obtained by optical excitation at a wavelength corresponding to the transition $A \rightarrow b$. In some cases the excitation process may correspond to $A \rightarrow b'$, shown by the dashed extension of the $A \rightarrow b$ transition, though this is generally unfavorable owing to the large amount of energy that must be dissipated as heat in the subsequent relaxation to the upper laser level, $S_1 \leftarrow S_2$.

Molecules in the upper laser level, B, may decay by competing processes $S_0 \leftarrow S_1$ or $T_1 \leftarrow S_1$. $T_1$ is the lowest of a manifold of triplet states. The transition $S_0 \leftarrow S_1$ may occur either by spontaneous decay, which represents a loss, or by stimulated emission, the desired result. The relative transition rates for these processes are one factor that determines the operating efficiency of the laser. Any nonradiative $S_0 \leftarrow S_1$ decay also lowers the efficiency of the laser.

---

[1] This work was carried out when the author was with the Research Laboratories of Eastman Kodak Company, Rochester, N.Y. 14650/USA.

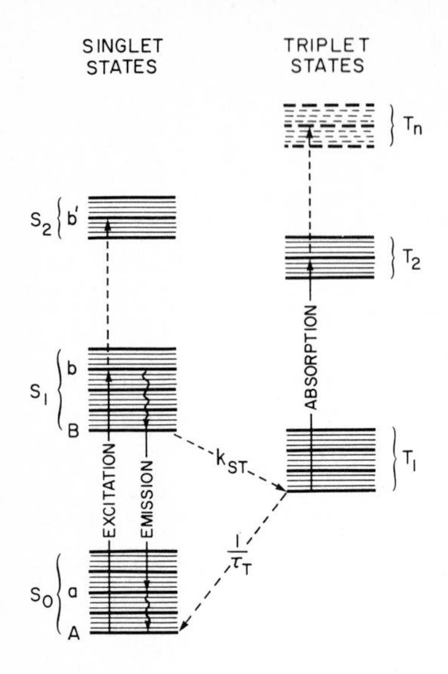


**Fig. 2.1.** Schematic energy level diagram for a dye molecule. The processes of importance in the operation of the cw dye laser are indicated by arrows. The heavy horizontal lines represent vibrational sublevels of the electronic states and the lighter lines represent levels due to interaction of the dye molecule with a solvent

This process, termed *internal conversion*, is generally negligible for dyes capable of lasing continuously.

An essential difference between the operation of the cw laser and that of pulsed lasers is the importance of the $T_1 \leftarrow S_1$ process. Although the probability for this process is relatively small, in good laser dyes, several percent of the excited molecules may make the transition. Molecules which arrive in $T_1$ remain there for a time that is long compared to the characteristic times involved in the lasing process. The transition $S_0 \leftarrow T_1$ is forbidden and the rate constant for this process ranges from $10^3$ to $10^7\,s^{-1}$, depending upon the environment of the dye molecule. The lifetime may be shortened if molecules, such as $O_2$, which tend to enhance the $S_0 \leftarrow T_1$ (Snavely and Schäfer 1969; Pappalardo et al. 1970a, b) process are nearby and, conversely, it may be quite long if there are no triplet quenching molecules in the dye solution.

The transition $S_0 \leftarrow T_1$ is, in all cases, slow with respect to $S_0 \leftarrow S_1$ which has a rate constant for spontaneous decay of about $2 \times 10^8\,s^{-1}$. The state $T_1$ thus acts as a trap and, for high excitation rates, a non-negligible fraction of the total number of dye molecules may reside in $T_1$. Of course, if $N_T$, the population of $T_1$, approaches $N$, the total dye concentration, lasing is not possible because there are no molecules available in the $S_0 \leftarrow S_1$ lasing path.

Lasing may also be quenched when $N_T \ll N$. $T_1$ is the lowest of a manifold of states of which the next higher state $T_2$ is shown. Transitions between the states of this manifold, $T_1 \rightarrow T_2$ for example, are allowed since no change in electron spin is required. The $T_1$ molecules have a large absorption cross section for the $T_1 \rightarrow T_2$ and $T_1 \rightarrow T_n$ processes shown in Fig. 2.1. Unfortunately, the absorption spectrum associated with $T_1$ generally overlaps the $S_0 \leftarrow S_1$ fluorescence of a dye molecule. Thus, it produces an optical loss at wavelengths for which optical gain is to be produced. This loss must be over-

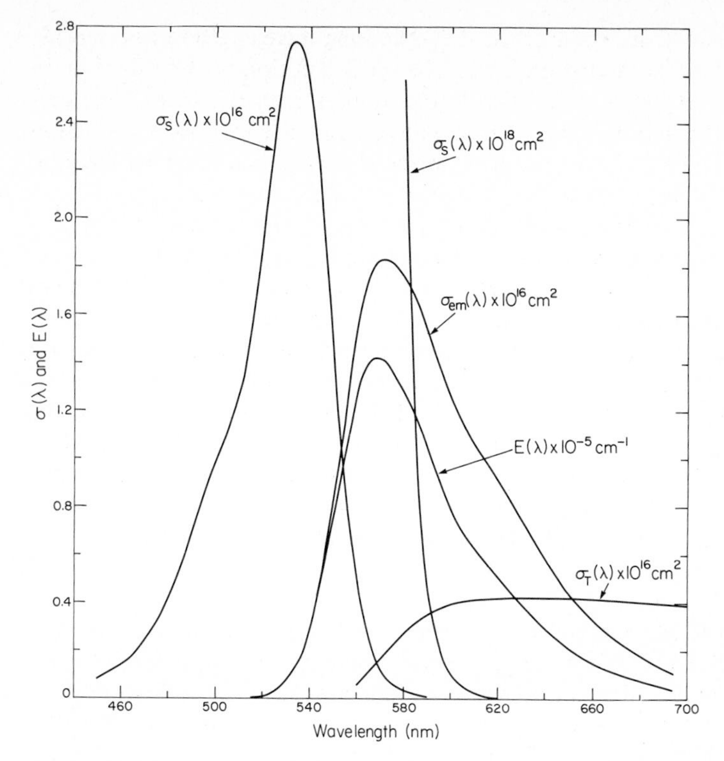


**Fig. 2.2.** Spectrophotometric data for the dye rhodamine 6G. The singlet absorption, $\sigma_s(\lambda)$, and emission, $E(\lambda)$, were obtained from a solution of the dye in water plus 2% Ammonyx LO, a commercial surfactant. The triplet-state spectrum, $\sigma_T(\lambda)$, was obtained by Morrow and Quinn (1973) with an ethanol solution. The dye concentration in all cases was $10^{-4}$ M. $E(\lambda)$ has been normalized so that $\int E(\lambda)d\lambda = 0.92$, the measured quantum yield for fluorescence

come by an increase in excitation power with respect to the power required if the loss were not present. At the least, the efficiency of the lasing process suffers as a result. If the absorption loss is too large, lasing may be prevented altogether.

The strength of an absorption process may be expressed in terms of the *molecular absorption cross section* $\sigma(\lambda)$ defined by the relation

$$I(\lambda, d) = I(\lambda, 0)\,\mathrm{e}^{-N\sigma(\lambda)d} .$$

Here $I(\lambda, 0)$ is the intensity of a light beam incident upon an absorbing sample of thickness $d$ with a concentration of $N$ absorbing molecules per cm$^3$. $I(\lambda, d)$ is the intensity of the transmitted beam. The absorption cross section has the dimensions of cm$^2$ and is effectively the area of an absorbing molecule at the wavelength $\lambda$.

Absorption cross sections for various processes in the laser dye rhodamine 6G are shown in Fig. 2.2. The cross section for $S_0 \rightarrow S_1$, labeled $\sigma_s(\lambda)$, is presented for a water solution of the dye. The long wavelength tail of $\sigma_s(\lambda)$ is also shown with an expanded ordinate. The water solution is often used in cw dye lasers (O. G. Peterson et al. 1970). The fluorescence spectrum $E(\lambda)$ corresponding to the spontaneous $S_0 \leftarrow S_1$ process and the stimulated emission cross section $\sigma_{em}(\lambda)$ derived from $E(\lambda)$ are shown for comparison with $\sigma_s(\lambda)$. The derivation of $\sigma_{em}(\lambda)$ will be discussed in the following section.

The absorption spectrum for $T_1 \rightarrow T_n$, $\sigma_T(\lambda)$, is also shown in Fig. 2.2. This spectrum was measured by Morrow and Quinn (1973) for an ethanol solution of rhodamine 6G. Unfortunately, the $\sigma_T(\lambda)$ spectrum has not been measured for a water solution. From the comparison of triplet spectra for different solvents with other dye molecules it is not expected that the triplet spectrum for rhodamine 6G in water would differ greatly from that shown.

The data of Fig. 2.2, along with a knowledge of the rate constants for the processes shown in Fig. 2.1, may be used to deduce the longest laser pulse which can be obtained from this solution if the triplet state population is uncontrolled. That is, the longest laser pulse duration $t_{max}$ consistent with a triplet lifetime $\tau_T$ which is very large with respect to $\tau$, the lifetime for the spontaneous $S_0 \leftarrow S_1$ process.

If the rate constant for the $T_1 \leftarrow S_1$ process is $k_{ST}[s^{-1}]$, the rate at which molecules enter $T_1$ is given by $N_{1C}k_{ST}$ where, $N_{1C}$ is the *critical inversion*, the population of $S_1$ required for the gain of the active medium to balance the intrinsic optical losses. Above laser threshold the population $S_1$ is fixed at $N_{1C}$; it does not increase significantly with excitation (W. V. Smith and Sorokin 1966). The population of $T_1$, $N_T$, will then be governed by the rate equation

$$\frac{dN_T}{dt} = N_{1C}k_{ST} - \frac{N_T}{\tau_T} . \qquad (2.1)$$

If it is assumed that the critical inversion is produced at time $t = 0$, and if $\tau_T \rightarrow \infty$,

$$N_T(t) = N_{1C}k_{ST}t . \qquad (2.2)$$

It is necessary at this point to anticipate a result from the analysis in the following section. If the molecular cross section for stimulated emission is $\sigma_{em}(\lambda)$ and the cross section for triplet-triplet absorption is $\sigma_T(\lambda)$ the intrinsic gain will just balance the loss due to the triplet state when

$$N_{1C}\sigma_{em}(\lambda) = N_T\sigma_T(\lambda) ,$$

when $N_1$ is the molecular concentration in the state $S_1$. From Fig. 2.2 it is seen that $\sigma_T(\lambda) \sim \sigma_{em}(\lambda)/10$ at the peak of $\sigma_{em}(\lambda)$. For the corresponding wavelength the net gain vanishes when $N_T \approx 10N_{1C}$. From (2.2), therefore,

$$t_{max} \approx 10/k_{ST} .$$

The quantity $k_{ST}$ can be estimated from a knowledge of $\phi$, the fluorescence quantum efficiency, and $\tau$ by means of the relation

$$k_{ST} = \tau^{-1}(1-\varphi) \ .$$

For $\tau = 6\times10^{-9}$ s and $\phi = 0.92$[2], as is approximately true for rhodamine 6G, $k_{ST} = 1.6\times10^{7}\ \mathrm{s}^{-1}$ from which $t_{max} \sim 6\times10^{-7}$ s. This is consistent with the observation of the laser pulse duration for systems in which no control is exercised over the population of $T_1$ (Snavely 1969; Sorokin et al. 1968).

From the foregoing discussion it is apparent that the population of $T_1$ must be limited if the cw operation of a dye laser is to be achieved. This can be accomplished by the control of $k_{ST}$ or $\tau_T$. The maximum allowable value for $k_{ST}$ is readily estimated by an extension of the reasoning presented above. In the steady state, $dN_T/dt = 0$ and (2.1) yields for the equilibrium population of $T_1$

$$N_T = N_{1C} k_{ST} \tau_T \ . \tag{2.3}$$

Again assuming that the intrinsic gain vanishes when $N_T/N_{1C} = \sigma_{em}/\sigma_T = k_{ST}\tau_T$, a maximum value of $k_{ST}\tau_T$ of $\approx 10$ is obtained from the data of Fig. 2.2. If cw operation is to be achieved, some means of quenching the triplet state concentration rapidly enough that $k_{ST}\tau_T < 10$ must be provided. This may be achieved by chemical additives or by rapid flow of the dye through the excited region.

## 2.1 Gain Analysis of the cw Dye Laser

### 2.1.1 Analysis at Laser Threshold

Analysis of the triplet state kinetics on the basis of the triplet state rate equation has provided a useful insight into the importance of the control of the triplet state population. Many other aspects of the performance of the dye laser can be understood from an analysis in terms of rate equations. In the treatment which follows the rate equation approach developed by Snavely (1969) and by O.G. Peterson et al. (1971) will be followed for the description of the laser at threshold. The effects of system parameters upon tuning and threshold will be examined.

A system of the form shown in Fig. 2.3, consisting of a longitudinally excited region within a two-element optical resonator will be considered. The resonator is hemispherical. It is assumed that the concentration of molecules in $S_1$ is $N_1\ \mathrm{cm}^{-3}$ and is uniform throughout the active volume. This assumption will be modified when the behavior of the laser above threshold is considered. For the moment, however, it is not restrictive.

If $I(\lambda, z)$ is the intensity of the lasing mode the net rate at which $I(\lambda, z)$ increases along the axis of the laser cavity, labeled as the $z$ direction, will be given by

---

[2] Measured by F. Grum, Eastman Kodak Research Laboratories.

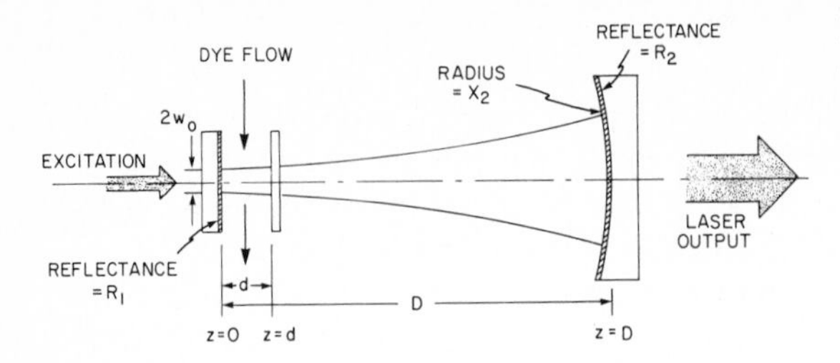


**Fig. 2.3.** Geometry of the cw laser considered in the analysis

$$\left(\frac{dI(\lambda,z)}{dz}\right)_{\text{total}}=\left(\frac{dI(\lambda,z)}{dz}\right)_{\text{stim}}-\left(\frac{dI(\lambda,z)}{dz}\right)_{\text{sing.}}-\left(\frac{dI(\lambda,z)}{dz}\right)_{\text{trip}} \tag{2.4}$$

where $(dI(\lambda,z)/dz)_{\text{stim}}$ is the rate at which the intensity increases due to stimulated emission, and $-(dI(\lambda,z)/dz)_{\text{sing.}}$ and $-(dI(\lambda,z)/dz)_{\text{trip.}}$ are the rates at which it decreases due to $S_1 \rightarrow S_2$ and $T_1 \rightarrow T_n$ processes, respectively. In the discussion which follows, the explicit functional dependence of quantities, such as $I(\lambda,z)$, will not be indicated when ambiguity is unlikely to result. For instance, "$I(\lambda,z)$" will be written simply as "$I$".

Expressions for the terms of (2.4) have been derived by many authors and will be taken from the literature. To obtain $(dI/dz)_{\text{stim}}$ use is made of the identity $(dI/dz)=(dI/dt)(dz/dt)^{-1}=(n/c)(dI/dt)$ in which case (Yariv 1967) $(dI/dz)_{\text{stim}}=N_1\lambda^4E(\lambda)I/(8\pi\tau cn^2)=\sigma_{\text{em}}(\lambda)I$. The quantities $n$ and $c$ are the refractive index and the velocity of light, respectively. The quantity $\sigma_{\text{em}}(\lambda)$ as defined by this equation is plotted in Fig. 2.2. $E(\lambda)$ is the spontaneous emission lineshape function normalized so that $\int_0^\infty E(\lambda)d\lambda=\phi$, the fluorescence quantum yield, and $\tau$ is the observed fluorescence decay time for spontaneous emission. The quantities $(dI/dz)_{\text{sing.}}$ and $(dI/dz)_{\text{trip.}}$ have the form $(dI/dz)=-IN_\alpha\sigma_\alpha(\lambda)$ where $\sigma_\alpha(\lambda)$ is an absorption cross section. The absorption cross section for the $S_0 \rightarrow S_1$ process has previously been denoted as $\sigma_s(\lambda)$ and that for $T_1 \rightarrow T_2$ processes as $\sigma_T(\lambda)$. $N_\alpha$ is the appropriate molecular density, the concentration of molecules in $S_0$, which is $N_0$, or $N_T$.

Using these relationships the rate equation for photon production becomes

$$\left(\frac{dI}{dz}\right)_{\text{total}}=I\{N_1\sigma_{\text{em}}(\lambda)-N_0\sigma_s(\lambda)-N_T\sigma_T(\lambda)\}\ . \tag{2.5}$$

Defining the gain of the active medium as

$$g(\lambda)=\frac{1}{I}\left(\frac{dI}{dz}\right)_{\text{total}}$$

and integrating (2.5) over a round trip through the active medium, accounting for the finite reflectances $R_1(\lambda)$ and $R_2(\lambda)$ of the mirrors, yields $G(\lambda)$, the round trip gain,

$$G(\lambda) = \int_{\text{round trip}} g(\lambda) dz$$

$$= N_1 \sigma_{\text{em}}(\lambda) 2d - N_0 \sigma_s(\lambda) 2d - N_T \sigma_T(\lambda) 2d - \ln [R_1(\lambda) R_2(\lambda)] \ . \quad (2.6)$$

At laser threshold $G(\lambda) = 0$ so that

$$N_1 \sigma_{\text{em}}(\lambda) - N_0 \sigma_s(\lambda) - N_T \sigma_T(\lambda) - \frac{1}{2d} \ln [R_1(\lambda) R_2(\lambda)] = 0 \ .$$

In the steady state the triplet and singlet state concentrations are related by (2.3) which expresses the *equilibrium triplet approximation* utilized by O. G. Peterson et al. (1971). Using this relationship and the condition $N = N_0 + N_1 + N_T$, the threshold condition can be written as

$$\{\sigma_{\text{em}}(\lambda) + \sigma_s(\lambda) + k_{ST} \tau_T [\sigma_s(\lambda) - \sigma_T(\lambda)]\} N_{1C} - \sigma_s(\lambda) N + r(\lambda) = 0 \quad (2.7)$$

where $r(\lambda) = -(1/2d) \ln [R_1(\lambda) R_2(\lambda)]$. The term in brackets is a growth coefficient relating the net rate at which photons are produced to the population of the upper laser level. It depends only upon the spectrophotometric parameters of the dye solution shown in Fig. 2.2 and is independent of laser system parameters which are contained in $r$.

If the growth coefficient in brackets is written as $\gamma(\lambda)$, (2.7) can be recast to solve for the critical inversion $N_{1C}$ as a function of wavelength and system parameters, i.e.

$$\frac{N_{1C}}{N} = \frac{1}{\gamma(\lambda)} \left[ \sigma_s(\lambda) + \frac{r(\lambda)}{N} \right] \ . \quad (2.8)$$

Equation (2.8) describes a critical inversion surface. As demonstrated by O. G. Peterson et al. (1971), examination of this surface provides insight into the operation of the cw dye laser, especially its tuning behavior.

The critical inversion surface has its *intrinsic* or *self-tuned* form when $r(\lambda)$ is nondispersive, i.e. the mirror reflectance is independent of wave-length. The self-tuned critical inversion surface for the dye rhodamine 6G is shown in Fig. 2.4. In plotting this surface a value for $k_{ST} \tau_T = 0.9$, as found by Siegman et al. (1972) has been used. The value for $\tau$ of $6 \times 10^{-9}$ s has been taken from the work of Cirkel et al. (1972). Other parameters have been taken from the data of Fig. 2.2.

Without excitation $N_1/N = 0$. As the excitation intensity is increased, $N_1/N$ increases and at some time intersects the critical inversion surface. When this occurs oscillation begins and the population of $N_1$ is clamped at $N_{1C}$. Thus, the critical inversion is constrained to lie along the valley of the surface as indicated by the dashed line in Fig. 2.4. The range over which the wavelength can be tuned by adjustment of mirror reflectance and dye concentration is found from the extremes of the dashed curve.

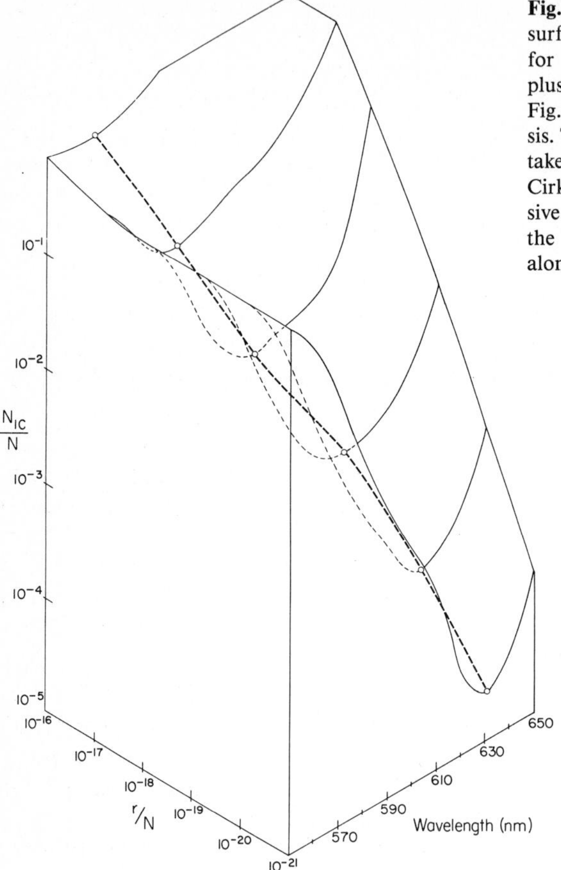


**Fig. 2.4.** Intrinsic critical inversion surface resulting from a plot of (2.8) for the dye rhodamine 6G in water plus 2% Ammonyx LO. The data of Fig. 2.2 have been used in the analysis. The fluorescence decay time $\tau$ was taken as $6.1\times10^{-9}$ s, as measured by Cirkel et al. (1972). With nondispersive mirrors forming the resonator, the laser is constrained to operate along the dashed curve

The self-tuning range for the rhodamine 6G laser is seen to extend from approximately 560–650 nm. At the short wavelength end of the tuning range $N_{1C}/N$ approaches unity while at the long wavelength limit the cross section for stimulated emission becomes too small to overcome the loss due to singlet state and triplet state absorption.

Tuning of a system over the self-tuned range is accomplished by adjusting the ratio $r/N$. From Fig. 2.4 it is seen that increasing $r$, while keeping $N$ fixed, causes the laser to oscillate at shorter wavelengths, while increasing $N$, keeping $r$ fixed, increases the laser wavelength. This behavior is consistent with experimental results (Sorokin et al. 1968; Schäfer et al. 1966).

In practice, the useful tuning range is less than that indicated above. It is generally not possible to provide an excitation intensity sufficient to produce an inversion greater than $N_{1C}/N\approx0.1$, which places a limitation upon the shortest wavelengths that may be attained. The minimum value of cavity loss due to scattering in the active medium and diffraction loss of the resonator determines the longest wavelength for oscillation. These losses have not been included in the rate equation analysis. For a carefully designed system they are generally much smaller than the loss due to triplet state absorption.

Of course, the laser may be operated at points other than those along the dashed line by using dispersive mirrors to form the optical resonator. A laser operated in this way is termed *extrinsically* tuned. The inclusion of dispersive optical elements within the cavity actually modifies the shape of the critical inversion surface. The modified surface can never lie lower than the intrinsic surface, however. Discussion of the surface shape under these conditions will be deferred to the section on tuning of the cw dye laser.

From the critical inversion surface it is possible to estimate the excitation power required to reach laser threshold for a cw device. The dependence of the excitation power upon some of the laser system parameters can also be determined.

Projection of the dotted curve of Fig. 2.4 onto the $N_{1C}/N$ vs. $r/N$ plane gives a representation of the critical inversion as a function of the extrinsic loss. Since the dotted path is the locus of points for which $d/d\lambda\,(N_{1C}/N)=0$ the condition

$$\frac{\gamma(\lambda)}{\gamma'(\lambda)}=\frac{[\sigma_s'(\lambda)+r'(\lambda)/N]}{[\sigma_s(\lambda)+r(\lambda)/N]}=\frac{\sigma_s'(\lambda)}{\sigma_s(\lambda)+r(\lambda)/N}\,,$$

where prime denotes the wavelength derivative, holds. This condition is satisfied at a particular wavelength for each value of $r/N$. Substitution of the wavelength found for a given $r/N$ into (2.8) then yields $N_{1C}/N$ for that value. Values of $N_{1C}/N$ vs. $r/N$ found in this way are plotted in Fig. 2.5 for rhodamine 6G. In these curves $\mu=k_{ST}\tau_T$ is chosen as a parameter.

Above the low $r$ transition region the value of $N_{1C}/N$ is relatively independent of $\mu$. O.G. Peterson et al. (1971) have shown that in this region the critical inversion is given approximately by

$$\frac{N_{1C}}{N}=5.8\times10^{11}\left(\frac{r}{N}\right)^{3/4}. \tag{2.9}$$

The total excitation power required is proportional to the product $N_{1C}d$. From (2.9) it is found that

$$P_p\propto N_{1C}d=5.8\times10^{11}\left(\frac{\ln R_2}{2}\right)d^{1/4}\,,$$

where it has been assumed that $R_1=1.0$. Thus, to minimize the excitation power the active length should be kept small.

From Fig. 2.5 it is seen that for the higher values of $\mu$, $N_{1C}/N$ approaches a constant, independent of $r/N$, for small values of $r/N$. For any given value of $\mu$, there is a practical minimum for $r/N$. Reduction of the extrinsic losses below this value does not significantly reduce the amount of excitation power

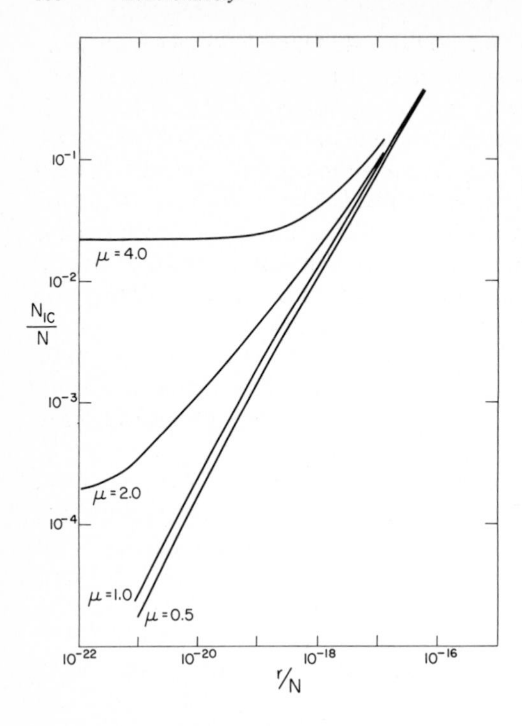


**Fig. 2.5.** Dependence of the critical inversion upon the ratio $r/N$ for non-dispersive resonator mirrors. Curves for various values of the parameter $\mu = k_{ST}\tau_T$ are shown

required to reach threshold. There is also a limit imposed upon $r/N$ by practical considerations. To reach the saturation region of the $\mu = 0.5$ curve for rhodamine 6G requires $r/N = 10^{-22}$. Values of $N$ much greater than $6 \times 10^{17}\ \mathrm{cm}^{-3}$ ($10^{-3}$ M) are not useful because the fluorescence quantum yield of the dye solution begins to drop at such high concentrations. The minimum value required for $r$ is, therefore, approximately $6 \times 10^{-5}$, requiring $R_2$ to be about 99.999%, an unrealistically high value. With mirrors of the highest quality 99.9% is attainable, yielding $r/N = 10^{-20}$. With this value of $r/N$ the laser is operating well out of the saturation region for small values of $\mu$. The critical inversion is then relatively independent of $\mu$.

The minimum excitation power density required to reach laser threshold, $P_p/A$, can be found directly from Fig. 2.5 since

$$\frac{P_p}{A} \cong \frac{N_{1C} h \nu_p d}{\tau}\ \mathrm{Watts/cm^2}\ .$$

This is the excitation power at the frequency $\nu_p$ per unit area of the lasing cross section in the longitudinally excited geometry of Fig. 2.3. For the dye rhodamine 6G with $\mu = 1$, an active length of 0.1 cm, and $R_2 = 0.995$, the calculated power density is approximately 3.5 kW/cm$^2$ at 530 nm.

The threshold excitation for practical systems is usually about 50 kW/cm$^2$. A minimum power density of 100 kW/cm$^2$ should be used in the excitation of a dye laser in order to provide a reasonable operating efficiency. With lower excitation power the dye laser operates close to the threshold and the overall efficiency is low. To obtain the necessary power density it is most attractive to use a cw gas laser for the excitation. The flux required to excite a yellow or red dye can be approached at the surface of a high pressure mercury arc lamp. However, the losses introduced by the optics required to transfer energy from the source to the active region are so great that available high pressure arcs are not attractive as excitation sources for cw dye lasers. No results have yet been published on the attempts to develop arc lamps suitable for the excitation of cw dye lasers. Of the available lasers operating in the visible region of the spectrum, the argon ion laser is capable of the highest output power. Commercial lasers with single or multiple line output powers of several watts are suitable for the excitation of a cw dye laser.

Given the excitation power of one to twenty watts, the beam diameter at the waist of the $TEM_{00}$ mode of the dye laser is fixed at $10-20\,\mu m$. The small mode diameter places restrictions upon the design of the laser resonator. To minimize diffraction losses a spherical or confocal cavity is chosen. The confocal parameter (Kogelnik and Li 1966), or optical length of the cavity, must be small to provide a beam waist small enough to match the excitation beam. For most cw dye lasers the confocal parameter is several millimeters.

The spherical, or hemispherical resonator is, in practice, more convenient than a confocal resonator in that the physical length of the spherical resonator is twice as long as that of the confocal resonator, for a given confocal parameter. This simplifies the fabrication of the laser since the dimensions of the optical elements are not small. Simple plane parallel resonators are not practical for the cw dye laser since the diffraction loss is large at the small Fresnel number required.

### 2.1.2 Gain Analysis Above Laser Threshold

In the preceding subsection it was assumed that the concentration of molecules in the upper laser level, $N_1$, was uniform throughout the active volume of the laser. Although the assumption simplified the derivation of the critical inversion and the threshold excitation power, as well as permitting the display of the self-tuning range, it does not correspond to an operating, longitudinally excited cw laser. For the practical device the population of the upper laser level is position dependent. It has the form $N_1(r,z)$ where $r$ is the radial distance from the cavity axis. When the position dependence of $N_1$ is incorporated the gain equation is not as simple as (2.6).

To develop a gain equation which can be used in the design of cw systems the active region of the laser, as shown in Fig. 2.3, is assumed to have the geometry shown in Fig. 2.6. The hemispherical resonator has a confocal parameter $b_L$, which determines the waist radius of the laser beam, $w_0$. Laser and excitation beams are assumed to be of the $TEM_{00}$ mode. The confocal parameter for the excitation beam will be taken as $b_p$ with the corresponding waist diameter $w_{p0}$. As drawn, the excitation and laser beam waists coincide at $z=0$.

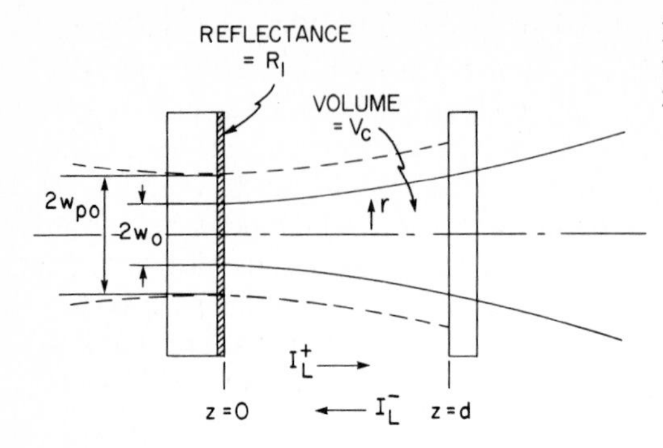


**Fig. 2.6.** Details of the active region of the two-element resonator shown in Fig. 2.3

Following the analysis developed by H. A. Pike (1971) the optical gain per unit length for a beam of intensity $I_L^+(r,z)$ traveling to the right and $I_L^-(r,z)$ traveling to the left can be expressed in terms of the growth coefficient $\gamma(\lambda)$, defined in connection with (2.6) as

$$dI_L^+(r,z) = I_L^+(r,z) N_1(r,z)\gamma(\lambda)dz \quad \text{and}$$

$$dI_L^-(r,z) = I_L^-(r,z) N_1(r,z)\gamma(\lambda)dz \ .$$

If the mirror at the right of Fig. 2.3 returns a fraction $R_2$ of the beam $I_L^+(r,d)$ to the active region, the on-axis round-trip gain of the system will be

$$G_0 = (1+R_2)\int_0^d N_1(0,z)\gamma(\lambda)dz \ . \tag{2.10}$$

The excitation intensity $I_p(r,z)$ determines the population of the upper laser level by $N_1(r,z) = I_p(r,z)N_0\sigma_p\tau$ where $\sigma_p$ denotes the absorption cross section $\sigma_S(\lambda_p)$ for excitation radiation. The excitation irradiance for a $TEM_{00}$ excitation beam will be

$$I_p(r,z) = \frac{4P_p(0)}{b_p hc}\left(\frac{w_{p0}}{w_p}\right)^2 \exp(-2r^2/w_p^2) \text{ photons cm}^{-2}\,\text{s}^{-1} \quad \text{with}$$

$$w_{p0}^2 = \frac{\lambda_p b_p}{2\pi} \quad \text{and} \quad w_p^2(z) = w_{p0}^2\left(1+\frac{4z^2}{b_p^2}\right) .$$

$P_p(0)$ is the incident power in watts.

From (2.10) it is clear that the round trip gain is maximized when the single pass gain is largest. Therefore, only the single pass gain needs to be considered to determine the effects of changes in parameters upon gain. Making use of the dependence of pump irradiance upon distance as given above, the on-axis single pass gain can be written as

$$G_0 = \frac{4\gamma\tau P_p(0)}{b_p hc}\int_0^d \frac{N_0\sigma_p e^{-N_0\sigma_p z}}{1+4z^2/b_p^2}\,dz \ .$$

H. A. Pike (1971) has shown that $G_0$ may be written in the useful form

$$G_0 = \frac{4\gamma\tau P_p(0)}{b_p hc} F(u, v) \tag{2.11}$$

where $u = N_0\sigma_p d$, $v = N_0\sigma_p b_p/2$ and $F(u,v) = v^2 \int_0^u e^{-x} dx/(v^2+x^2)$.

$F(u, v)$ has been expressed in terms of an exponential integral which attains its maximum value of unity when $u$ and $v$ are very large. This occurs when $1/(N_0\sigma_p)$, the absorption length in the dye is small in comparison with either the cell length, $d$, or the confocal parameter $b_p$. The primary conclusion to be drawn from (2.11) is that the front surface of the active medium should be placed at the waist of the excitation beam to obtain the maximum gain with a given excitation power. This is not too surprising when it is realized that the attenuation length in the dye, under the present assumptions, is independent of the excitation irradiance, whereas the gain is proportional to the irradiance. If the excitation beam is absorbed in a region of low irradiance, the gain of the system will be low as predicted by the analytical result.

The above analysis has been applied to a cw dye laser which is external to the cavity of the exciting laser. In some systems of interest the dye laser is inside the cavity of the exciting laser. In this case the excitation beam may be recirculated to provide a more uniform population of the active medium than in the case just considered. For this arrangement a low concentration of dye would be required, in order that the gain of the exciting laser would not be quenched. Although the dye concentration may be low, the upper laser level population must be comparable to that of the high concentration device. This is accomplished by the high excitation flux within the exciting laser cavity. The geometry of the system to be considered is shown in Fig. 2.7.

Assume that the intracavity power of the exciting laser, flowing in the $+z$ direction, $P_{1C}^+$, is the power incident upon the left side of the dye cell in Fig. 2.6. The excitation power within the dye cell will then be given by

$$\begin{aligned} P_p(z) &= P_{1C}^+(0)\,e^{-N_0\sigma_p z} + P_{1C}^+(0)\,e^{-N_0\sigma_p d}\,e^{-N_0\sigma_p(d-z)} \\ &= 2P_{1C}^+(0)\,e^{-N_0\sigma_p d}\cosh\left[N_0\sigma_p(d-z)\right] . \end{aligned}$$

This expression may be inserted into (2.11) to find the upper laser level population as a function of position along the axis. The on-axis gain is then calculated in the same manner as was done previously to yield

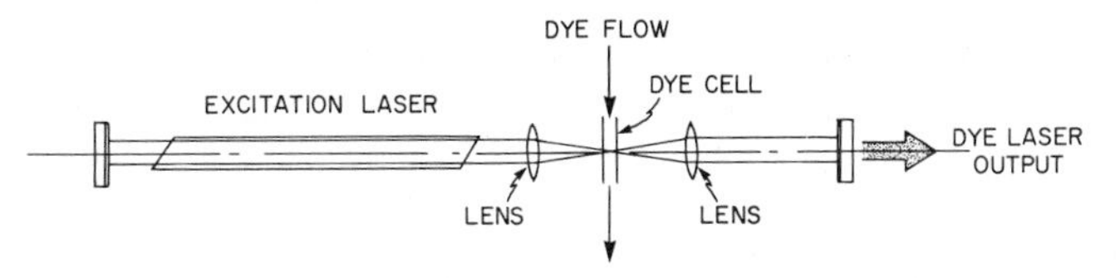


**Fig. 2.7.** Geometry of a cw dye laser internal to the cavity of the excitation laser

$$G_0 = \frac{8\gamma\tau P_{1C}^{+}(0)}{b_p hc} \, \mathrm{e}^{-N_0\sigma_p d} \sinh(N_0\sigma_p d) \; .$$

Since it has been assumed for this case that the absorption length $1/N_0\sigma_p \gg d$ the gain may be expressed as

$$G_0 \approx \frac{8\gamma\tau P_{1C}^{+}(0)}{b_p hc} N_0\sigma_p d \; . \tag{2.12}$$

To compare this result with the gain of the externally excited dye laser it is assumed that the exciting laser is operating under conditions which produce the maximum $N_{1C}$ in both cases. For the first case this occurs when $P_p(0)$ is a maximum. The maximum output power for a laser is

$$P_p(0) \cong 2\Gamma_1 P_{1C}^{+}(0) \; ,$$

where $\Gamma_1$ denotes the useful single pass gain of the excitation laser in excess of the nonuseful components such as scattering and absorption losses. For power extraction by means of a transmitting mirror, of reflectance $R$, $\Gamma_1$ represents the single pass loss associated with the mirror or $\Gamma_1 = -1/(2d \ln R)$. For steady-state operation the single pass loss and gain are equal.

For the optimum excitation of the intracavity laser

$$\Gamma_1 = N_0\sigma_p d$$

from which

$$P_p(0) = 2N_0\sigma_p d P_{1C}^{+}(0) \; .$$

Substituting this expression into (2.12) yields

$$G_0 = \frac{4\gamma\tau P_p(0)}{b_p hc} \; ,$$

which is identical to the maximum gain obtainable from the externally excited dye laser given by (2.11). On the basis of these results, obtained by H. A. Pike (1971), there appears to be no advantage in gain to be realized by placing the dye laser inside the exciting laser cavity. The intracavity geometry may offer some other advantages, however. One advantage with respect to the externally excited device is the uniform dissipation of heat produced by excitation over the active volume. This tends to alleviate hot spots, a serious problem in high-power cw dye lasers.

The on-axis gain expression derived above is valid when the diameter of the laser beam is much smaller than the diameter of the excitation beam. When the beam diameters become comparable it is necessary to integrate the intensity of the laser beam over the whole wave front, that is over all $r$, to determine the gain. The calculation yields a result which is somewhat less than $G_0$ since the active medium is not so highly excited for $r > 0$ as for $r = 0$. The off-axis gain is lower than the gain on the $r = 0$ axis.

A convenient way of expressing the relative sizes of excitation and laser beams is to take the ratio of their areas at the entrance face of the active medium,

$$\varrho = \frac{\pi w_{p0}^2}{\pi w_{L0}^2} = \frac{\lambda_p b_p}{\lambda_L b_L} .$$

H.A. Pike (1971) has shown that the net single pass gain depends upon $\varrho$ as

$$G(\varrho, G_0) \cong \frac{G_0}{1+\varrho} \tag{2.13}$$

to a good approximation. This approximation is derived for the case in which the excitation and laser beam radii do not change greatly over the active region, a situation which holds in most cases of practical interest.

With reference to (2.13) it should be pointed out that $G_0 \propto P_p(0)$, the incident excitation power even for $\varrho \neq 1$. The gain required to reach laser threshold is fixed by laser system parameters. Thus, (2.13) expresses the dependence of the pump power required to reach threshold upon the mode-match parameter $\varrho$. Note that the threshold pump power is a minimum when the laser beam is much smaller than the pump beam. The discussion that follows will show that the value of $\varrho$ giving minimum threshold is not consistent with maximum efficiency. In fact, the efficiency of the low threshold device is very low since the excitation beam is used very inefficiently.

H.A. Pike (1971) has also developed expressions for the dependence of laser output power and efficiency upon the mode matching parameter. In the derivation it is assumed that saturation of the ground state absorption could be neglected, that the excited state population follows a gaussian distribution across the active region, and that the active region is short in comparison with the pump and laser beam confocal parameters. These are good assumptions for a practical cw laser system.

Under these conditions the laser output power is given by

$$P_L = \frac{\Gamma_1}{\Gamma} (1 - \mathrm{e}^{-N_0 \sigma_p d}) \frac{\varrho}{\sqrt{1+\varrho^2}} (P_p - P_t) \tag{2.14}$$

where $\Gamma$ is the total single pass loss at the laser wavelength including useful and nonuseful components, such as scattering, singlet and triplet state absorption, extraneous reflections and so forth. $\Gamma_1$ is the useful loss associated with mirror transmission, as discussed previously, and

$$P_t = \frac{\Gamma(1+\varrho) b_p h c}{4 \sigma_L \tau [1 - \exp(-N_0 \sigma_p d)]}$$

is the threshold power.

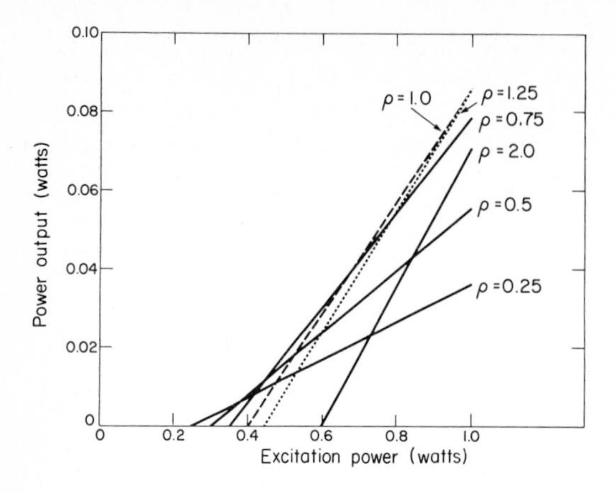


**Fig. 2.8.** Calculated dependence of dye-laser output upon excitation power. The curves illustrate the dependence of laser threshold and slope efficiency upon the choice of $\varrho$, the ratio of the areas of the laser and excitation beams at the beam waists (H. A. Pike 1971)

The slope efficiency is found from (2.13) to be

$$\eta_{\text{slope}} = \frac{\Gamma_1}{\Gamma}\,(1 - \mathrm{e}^{-N_0\sigma_p d})\,\frac{\varrho}{\sqrt{1+\varrho^2}}\;.$$

To illustrate the manner in which the slope efficiency and threshold power change with $\varrho$ (2.14) is plotted in Fig. 2.8 for a dye laser having a set of parameters corresponding to a practical, low power, cw dye laser.

Figure 2.8 illustrates graphically the trade-off between threshold and operating efficiency. Note that the slope efficiency becomes constant for large values of $\varrho$. The operating efficiency of the laser, defined as $\eta_{0p} = P_L/P_p$, is a maximum for some value of $\varrho$ which depends upon the available input power. It appears from the diagram that there is never an advantage to be gained from having $\varrho \gtrsim 1.25$, except perhaps for very high power devices. For low pump power, small $\varrho$ is favored, while for an input power of several watts the optimum value of $\varrho$ appears to be about 1.25.

The results displayed in Fig. 2.8 also give some indication of the dilemma faced by the designer wishing to maximize efficiency and tuning range of a tunable laser. For maximum tunable range, $P_t$ should be minimized. However, this results in an inefficient design. The desirability of maximum tunable range and maximum efficiency must be traded off against one another.

## 2.2 Tuning of the cw Dye Laser

The gain analysis of Sect. 2.1 centered upon a laser resonator geometry of the form shown in Fig. 2.3. It was assumed that the mirror spacing of the resonator could be adjusted to give the proper mode spot size in the active region. These assumptions are valid but a two-element resonator such as that of Fig. 2.3 possesses some practical disadvantages for tunable systems as Tuccio and Strome (1972) have pointed out. These are best illustrated by an example.

For a nearly hemispherical cavity, such as that shown in Fig. 2.3, the radius of the $TEM_{00}$ mode at the flat mirror will be given approximately by (Boyd and Gordon 1961)

$$w_0 = \left(\frac{\lambda}{\pi}\right)^{1/2} [D(X_2 - D)]^{1/4}$$

if $w_0 \gg \lambda$. The sensitivity of $w_0$ to changes in $D$ is displayed by differentiating $w_0$ with respect to $D$ to obtain

$$\frac{dw_0}{dD} = \left(\frac{\lambda}{\pi}\right)^2 \left(\frac{X_2 - 2D}{4w_0^3}\right) .$$

To a good approximation, therefore

$$\frac{\Delta w_0}{w_0} = \left(\frac{\lambda}{\pi}\right)^2 \left(\frac{X_2 - 2D}{4w_0^4}\right) \Delta D .$$

The desired spot size is fixed by the available excitation power, as discussed previously. For $w_0 \approx 5$ µm with a laser operating at 600 nm, the change in mirror spacing which will double the spot size may be easily calculated. For the minimum spot size, $D \approx X_2$, so that a doubling of $w_0$ occurs for

$$\Delta D = \frac{3.4 \times 10^{-4}}{D} \text{cm}^2 .$$

If $D$ is 0.3 cm, $\Delta D \approx 10$ µm. The tolerance on the position of the curved mirror will actually be somewhat less than this value. Although this tolerance is manageable, the cavity length of 0.3 cm does not allow space for the insertion of an intracavity tuning element. For this purpose $D$ must be extended to about 15 cm. For a cavity of this length $\Delta D$ drops to approximately 0.15 µm, an unreasonable tolerance on the location of the curved mirror. From these considerations it is seen that a two-element resonator is not suitable for tunable cw dye lasers.

The dimensional tolerances can be relaxed by the use of a three-element resonator. Two types of three-element resonators have been studied in some detail. The first of these systems, using transmitting optics, has been described by Tuccio and Strome (1972) and by Hercher and Pike (1971 a). A system using reflecting optics has been designed by Kohn et al. (1971) and has been discussed by Kogelnik et al. (1972). The design considerations for these two types of systems will be discussed in turn.

The basic form of the three-element resonator using transmitting optics is shown in Fig. 2.9a. In this system the curved mirror of the two-element

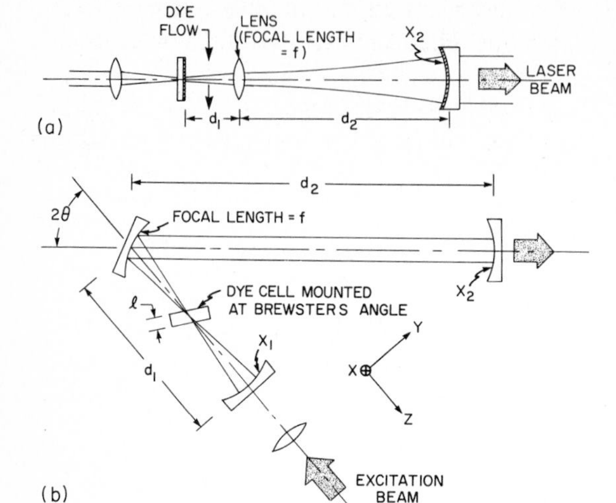


**Fig. 2.9.** **(a)** Three-element dye-laser resonator using transmitting optics. **(b)** Three-element dye-laser resonator using reflecting optics (Kohn et al. 1971)

resonator is replaced by a lens plus a mirror with long radius of curvature. Tuccio and Strome (1972) have shown that with this arrangement the position of the long-radius mirror is not nearly as critical as with the two-element resonator. The position of the lens is critical, however, though it may be rigidly fixed with respect to the first, flat mirror when the laser is constructed and does not need subsequent adjustment.

Tuccio and Strome (1972), following the treatment of Kogelnik (1965) utilized the complex beam parameter $q$ to describe the propagation of the Gaussian $TEM_{00}$ mode within the laser resonator. The beam parameter is defined by the relation

$$\frac{1}{q} = \frac{1}{X} - \frac{i\lambda}{\tau w^2}$$

where $X$ is the radius of curvature of the wavefront at any point along the beam, and $w$ is the radius of the mode at the same point.

The minimum radius of the beam, $w_0$, occurs at the flat mirror. The complex beam parameter $q_0$ at this surface is given by

$$q_0 = i\pi w_0^2/\lambda$$

since $X$ is infinite. Utilizing $q_0$ the beam can be traced back through the system and, from a relationship given by Kogelnik et al. (1972), the complex beam parameter can be found at any point as a function of $f$ and the spacing between the components, namely

$$\frac{1}{q} = \frac{-q_0/f + (1 - d_1/f)}{(1 - d_2/f)\,q_0 + (d_1 + d_2 - d_1 d_2/f)} \,. \tag{2.15}$$

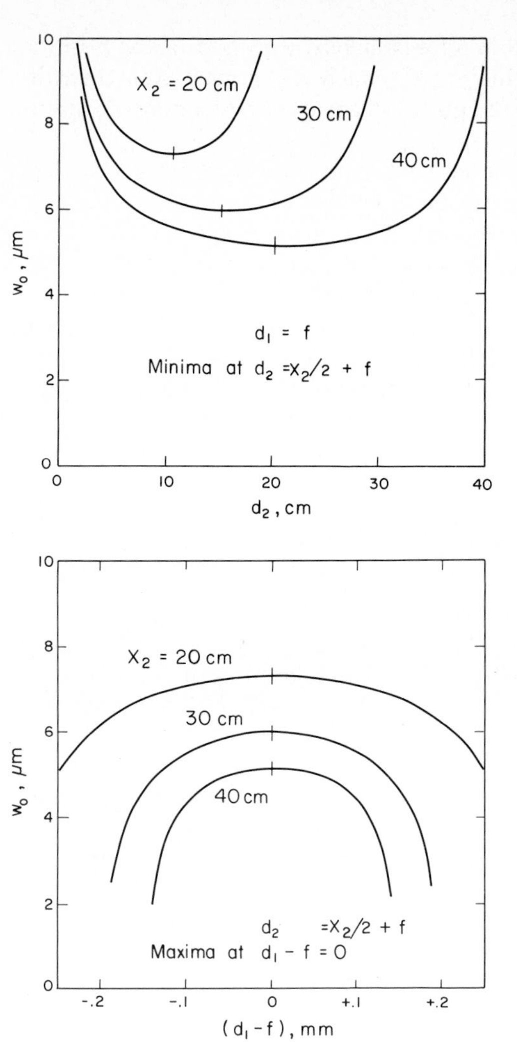


**Fig. 2.10.** Dependence of laser beam waist radius or "spot size" at the flat mirror of the resonator of Fig. 2.9a upon the spacing between the collimating lens and the curved mirror. The radius of curvature of the mirror, $x_2$, has been taken as a parameter. The curves are computed from (2.15) using a value of 0.53 cm for $f$ (Tuccio and Strome 1972)

**Fig. 2.11.** Dependence of the laser beam waist radius upon the spacing between the flat mirror and the collimating lens. The radius of curvature of the mirror at the end of the long arm of the cavity is taken as a parameter. The curves are computed from (2.15) using a value of 0.53 cm for $f$ (Tuccio and Strome 1972)

Tuccio and Strome demonstrated that the spot size at the flat mirror is stationary with respect to changes in $d_1$ and $d_2$ when $d_1 = f$, if the curvature of the mirror, $X_2$, is assumed to be the curvature of the wavefront and $d_2 = X_2/2 + f$. The sensitivity of the spot size to changes in $d_1$ and $d_2$ for several values of $X_2$, as obtained from (2.15) is shown in Figs. 2.10 and 2.11, respectively.

From Fig. 2.10 it is apparent that the allowable tolerance in the placement of the curved mirror increases as the radius of curvature of the mirror increases. The advantage with respect to the two-element cavity, in which the tolerance decreased with increasing mirror radius, is clear. From Fig. 2.11 an evaluation of the tolerance on the placement of the lens can be obtained. It appears that the longer cavity is again the less critical, though the location of the lens must be maintained within approximately $\pm 0.1$ mm of the position for minimum spot size. This value is very much less critical even than the tolerance on placement of the curved mirror in a two-element resonator several millimeters long.

In the preceding brief analysis it has been assumed that the intracavity lens is a thin lens. For a practical system the thickness of the lens will be comparable to the short focal length required. Thus, it will be necessary to use a thick-lens analysis. The modifications of the analysis resulting from this change were discussed by Tuccio and Strome (1972).

The principal disadvantage associated with the resonator of Fig. 2.9a is that the lens introduces two extra reflecting surfaces, with their attendant loss, into the cavity. This loss may be minimized by antireflection coating of the surfaces. However, even small reflections may be serious for some purposes, such as mode-locking (see Chap. 4). In these cases a three-element resonator using reflecting optics is advantageous.

A three-element reflecting cavity is shown schematically in Fig. 2.9b. The analysis of this system, as given by Kogelnik et al. (1972), is similar in many respects to that for the three-element cavity with transmitting optics. The principal difference in the analysis results from the folding of the system, which requires that the center mirror is operated off-axis. This introduces astigmatism. The rays in the plane of the drawing are focused at a different point than those perpendicular to the plane of the paper. The Brewster-angle dye cell also introduces astigmatism. Kogelnik et al. (1972) have shown that the thickness of the dye cell and the folding angle $\theta$ of the cavity can be adjusted so that the astigmatism of the mirror and dye cell cancel. Furthermore, the parameters of the compensated cavity may be such that the focal spot size is small and has a position which is relatively insensitive to the length of the short leg of the cavity. The treatment and results developed by Kogelnik et al. (1972) will only be outlined here. The reader is referred to the original paper for details of the analysis.

The three-element cavity will be stable for some range of variation about the spherical cavity separation $d_{1S} = S_1 + f$. The deviation of $d_1$ from this value is taken as $\delta$ where

$$d_1 = X_1 + f + \delta \ .$$

The limits upon $\delta$ for stability of the cavity are taken as $\delta_{\max}$ and $\delta_{\min}$ so that

$$d_{1\max} = X_1 + f + \delta_{\max}$$

$$d_{1\min} = X_1 + f + \delta_{\min}$$

with $\delta_{\max} = f^2/(d_2 - f)$ and $\delta_{\min} = f^2/(d_2 - R_2 - f)$. These stability limits are derived from the consideration of the equivalent, two-element, spherical resonator. Note that $\delta_{\min}$ is negative or zero, since $X_2 \geqq d_2 - R_2 - f$. Kogelnik et al. (1972) show that the focal spot varies, as $d_1$ is changed throughout its stability range approximately in accordance with

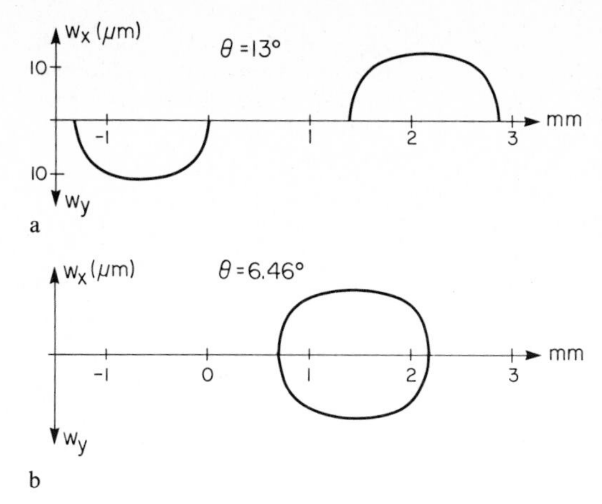


**Fig. 2.12.** (**a**) Focal-spot size and position for both sagittal ($xz$) and tangential ($yz$) planes of the resonator in Fig. 2.9b. The refractive index of the dye cell has been neglected in plotting the spot sizes according to (2.16). (**b**) Compensation of the astigmatism of (**a**) by the Brewster-angle-mounted dye cell. The dye cell has been taken to have a thickness $l = 0.15$ cm and refractive index $n = 1.45$. (In both plots the values $x_1 = 5$ cm, $x_2 = \infty$, $l = 5$ cm, and $d_2 = 190$ cm have been used) (Kogelnik et al. 1972)

$$\left(\frac{\pi w_0^2}{\lambda}\right) \approx (\delta_{\max} - \delta)(\delta - \delta_{\min}) \; . \tag{2.16}$$

The spot radius $w_0$ vanishes at the limits of the stability region and has its maximum size near the center of the range. As a result of astigmatism the focal length of the center mirror is different for rays in the $xz$ (sagittal) plane from that for rays in the $yz$ (tangential) plane. For rays in the $xz$ plane $f_x = f/\cos\theta$ and for the $yz$ plane $f_y = f\cdot\cos\theta$, where $f$ denotes the focal length of the mirror. The focal-spot size is dependent upon position and upon focal length.

Neglecting, for the moment, the presence of the dye cell, the spot size as a function of position in the sagittal and tangential planes is shown for one set of cavity parameters in Fig. 2.12a. The astigmatism resulting from $f_x \neq f_y$ is clearly seen. Such an optical cavity is unstable since the sagittal and tangential rays focus at different points. There is no value of $d_1$ for which both rays form stable modes.

The Brewster-angle cell introduces a compensating astigmatism which may be used to stabilize the resonator. The optical path length for the $xz$ and $yz$ rays is different in the Brewster-angle cell. The effective thickness for $xz$ plane rays is given by

$$d_x = \frac{t}{n^2}\sqrt{n^2+1}$$

and that for $yz$ plane rays by

$$d_y = \frac{t}{n^4}\sqrt{n^2+1} \; ,$$

where $n$ is the refractive index, and $t$ is the actual thickness of the cell. Kogelnik et al. (1972) demonstrated that the choice of the angle $\theta$ and cell thickness $t$ such that

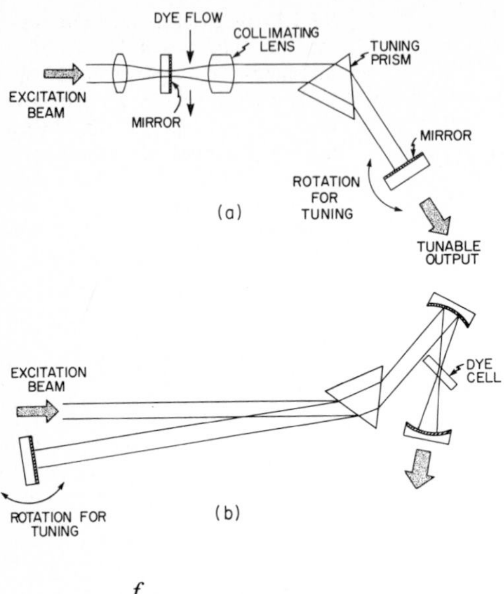


**Fig. 2.13 a, b.** Modification of the three-element resonators of Figs. 2.9 a and 2.9 b to include a tuning element (prism) in the long arm of the resonator

$$ZKt = \frac{f}{2} \sin\theta \tan\theta \ ,$$

where $K = [(n^2-1)/n^4]\sqrt{n^2+1}$, results in a complete overlap of the stability regions. The waist radii for the sagittal and tangential rays of a cavity compensated in this manner are plotted in Fig. 2.12 b for a cavity having the same parameters as that of Fig. 2.12 a.

In some cw dye-laser systems a spherical resonator has been used to separate the dye cell from the optical components of the resonator. In this system the flat mirror of Fig. 2.9 a is replaced with a mirror having a short focal length so that the beam waist falls within $d_1$ rather than at the mirror surface. This arrangement introduces extra reflections, associated with the dye cell windows, into the cavity. It is generally not possible to eliminate these by mounting the cell at Brewster's angle because of the astigmatism of the dye cell, as discussed above. In the cavity using transmitting optics there is no simple way to compensate for the astigmatism. Modification of the three-element resonators to provide for tuning is shown in Fig. 2.13 a and b. The space within the cavity between the collimating element and the resonator mirror is available for insertion of a tuning element. A prism is usually used.

The input-mirror requirements for the cavity with reflecting optics shown in Fig. 2.13 b are less stringent than for the cavity with transmitting optics. The input mirror needs only to have high reflectance, while in the arrangement of Fig. 2.13 a the input mirror must be highly transmitting at the excitation wavelength. This is avoided in the arrangement shown in Fig. 2.13 b by dispersing the excitation beam with the tuning prism (Dienes et al. 1972). The disadvantage of this system is that only a single output line of the exciting laser may be used. Thus the total power obtainable from a multiline source is not available. This is not a problem with the cavity design of Fig. 2.13 a.

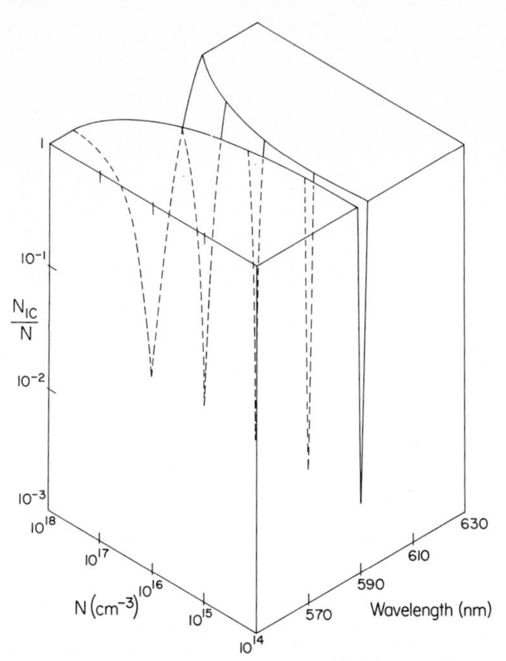


**Fig. 2.14.** Extrinsic critical inversion surface modified by the inclusion of a dispersive element within the resonator. A mirror with a reflectance spectrum $R = \exp[-(\lambda-\lambda_0)^2/\Delta\lambda^2]$ has been used, where $\Delta\lambda = 5$ nm and $\lambda_0 = 590$ nm

The effect of system parameters upon the laser with a dispersive resonator can be visualized by reference to the critical inversion surface. The surface for a laser with an active length of $d = 0.2$ cm, and a dispersive mirror having a reflectance $R_2 = R_0 \exp[-(\lambda-\lambda_0)^2/\Delta\lambda^2]$ where $R_0 = 1$, $\lambda_0 = 590$ nm and $\Delta\lambda = 5$ nm, is shown in Fig. 2.14. In this figure, one of the axes is dye concentration. It is seen that the width of the valley narrows as the dye concentration decreases. The bottom of the valley intersects the critical inversion surface with nondispersive mirrors along the $\lambda = 590$ plane. The plot suggests that the greater laser output line stability is associated with a low concentration of dye. This result is understood with reference to (2.7). As $N$ decreases the $r(\lambda)/N$ term becomes dominant with respect to the $\sigma_s(\lambda)$ term.

The $1/e$ linewidth, 10 nm, used for the tuning element in Fig. 2.13 is considerably broader than that associated with a practical tuning element, which should have a linewidth of less than 1 nm. The valley of Fig. 2.14 would be correspondingly narrower than that shown. This relatively large value was chosen only for illustrative purposes.

The linewidth limit of the cw dye laser will be determined in part by the linewidth of the tuning element and in part by statistical processes. Each of this contributions will now be considered.

The dispersive portion of a resonator tuned by a single prism is shown in Fig. 2.15a. For this arrangement, the linewidth $\Delta\lambda$ will be given by

$$\Delta\lambda = \frac{d\lambda}{dn}\frac{dn}{d\varphi}\Delta\varphi = \frac{1}{(dn/d\lambda)(d\varphi/dn)}\Delta\varphi\ ,$$

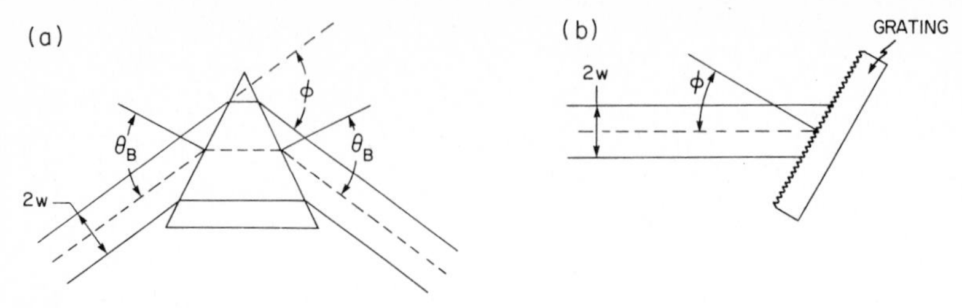


**Fig. 2.15 a, b.** Detail of the laser beam intersecting the tuning element of a tuned laser. These diagrams define the geometrical parameters used in estimating the linewidth of a laser tuned by a **(a)** prism or **(b)** grating

where $n$ denotes the refractive index of the prism, and $\varphi$ is defined by the figure.

If the laser is operating in its lowest-order ($TEM_{00}$) transverse mode, the divergence of the beam within the cavity, $\delta\varphi$ will be approximately $\lambda/(2w)$, where $w$ is the radius of the beam at the collimating element of the cavity. It was observed (Hercher and Pike 1971 a) that the laser ceases to oscillate when the mirror misalignment is about $\delta\varphi/20$. This figure may be taken as the angular deviation of the resonating beam from the resonance angle $\varphi$. The oscillating linewidth, therefore, will be approximately

$$\Delta\lambda = \frac{1}{(dn/d\lambda)(d\varphi/dn)} \frac{\lambda}{40\,w} .$$

Using the values for a phosphate flint prism (Tuccio and Strome 1972) ($n \cong 1.98$ at 598 nm) cut at Brewster's angle and operated at minimum deviation, with $2w = 0.2$ cm, the expected linewidth is about 0.1 nm. The linewidth may be narrowed further by incorporating etalons in the cavity in addition to the prism.

A similar argument may be applied to a system tuned with a grating. The grating is assumed to be mounted in the Littrow mounting, as shown in Fig. 2.15 b. In this case the linewidth will be given by

$$\Delta\lambda = \frac{\lambda}{\tan\varphi}\,\Delta\varphi = \frac{\lambda^2}{40\,w\tan\varphi} .$$

If $\varphi \approx 30°$, and $\lambda = 600$ nm, the expected linewidth is about 0.016 nm. It should be noted that the linewidth is independent of the ruling spacing of the grating. It depends only on the angle at which the grating is used. A coarse grating used at high order is completely equivalent to a fine grating used at low order. There may be an advantage in the use of the coarse grating in that it is not so susceptible to damage as the fine grating. The linewidth associated with the grating is seen to be much less than that associated with a single prism. For the cw laser, however, the prism is to be preferred since the losses are lower than those of a grating. Efficient pulsed dye laser systems have been made by using a grating in conjunction with a dielectric mirror (Bjorkholm et al. 1971). This technique should be applicable to cw lasers as well.

Of the statistical effects which influence the linewidth, three have been considered (Hercher and Snavely 1972): Spontaneous emission or phase diffusion, Brownian motion of the resonator, and statistical fluctuations in the density of the active medium. Only the last of these is large enough to be important in practical systems.

The density (fluctuations) in a liquid due to the statistical fluctuation in the number of molecules per unit volume have been treated by many authors. An early treatment by Einstein (1910) yields the rms fluctuation in density, $\varrho$, as

$$\left(\frac{\Delta\varrho}{\varrho}\right) \approx \sqrt{\frac{kT\beta}{V_c}},$$

where $\beta$ is the isothermal compressibility, $T$ is the absolute temperature, $k$ is Boltzmann's constant, and $V_c$ is the volume under consideration. The appropriate volume for the dye laser is the active volume of the system, as shown in Fig. 2.6. In determining the frequency shift associated with these density changes it is assumed that the polarizability is proportional to the number of molecules per unit volume. The refractive index is proportional to the square root of the polarizability so that

$$\left(\frac{\Delta n}{n}\right) = \frac{1}{2}\left(\frac{\Delta\varrho}{\varrho}\right).$$

Putting values appropriate to a water solution into this equation, it is found that $\Delta n/n \approx 2.3\times10^{-9}$ if a value of $10^{-7}\,\mathrm{cm}^3$ is chosen for $V_c$. To determine the frequency change $\Delta\nu$ produced by this fluctuation in refractive index, with reference to Fig. 2.3, it is noted that for an active length of 0.2 cm and a total length of 30 cm,

$$\frac{\Delta\nu}{\nu} = \frac{\Delta d}{D} = \frac{\Delta n d}{nD} \cong \frac{2.3\times10^{-9}\times0.2}{300} = 1.53\times10^{-12}\,.$$

For a laser operating in the yellow at $5\times10^{14}$ Hz, $\Delta\nu \approx 800$ Hz. This value is much smaller than the linewidth observed to date with any practical system. Reasons for this discrepancy will be discussed in the following section.

## 2.3 Performance of Experimental Systems

Reports on the operation of cw dye lasers have been published by several laboratories. Since the initial demonstration of cw operation (O. G. Peterson et al. 1970), power output, tuning range, linewidth, stability, and efficiency have im-

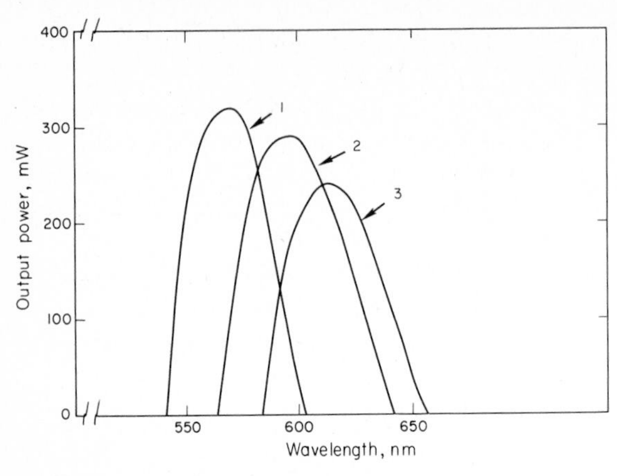


**Fig. 2.16.** Power output vs wavelength for three different dye solutions used in a tunable system having the geometry of Fig. 2.13a. Curve *1* was obtained with a $2 \times 10^{-4}$ M solution of rhodamine 6G in 3:1 water-hexafluoroisopropanol; curve *2* represents a $3 \times 10^{-4}$ M solution of rhodamine 6G in water plus 5% Ammonyx LO, and curve *3* was the output from a $3 \times 10^{-4}$ M solution of rhodamine B in 3:1 water-hexafluoroisopropanol. The excitation of the dye laser was 4 watts at 514.5 nm (Tuccio and Strome 1972)

proved steadily. The state of the art, as of early 1973, is summarized in Table 2.1. More up-to-date information can be found in Chap. 3.

The power output is limited at present (1973) by the power available from the gas lasers used for excitation. As more powerful lasers are developed, the output of the cw dye laser will increase.

The tuning range of the cw dye laser has recently been extended toward the blue region of the spectrum by Tuccio et al. (1973). The argon lines at 351 nm and 364 nm were used to excite coumarin derivative dyes. Operation at the short wavelength end of the range, 420 nm, is restricted to long pulses by the power available in these lines. Power output of the cw dye laser will certainly be increased and extended in the blue and near uv spectral regions. Extension of the output wavelength beyond 710 nm requires more powerful red and near infrared sources. It is possible that laser diodes will be useful for exciting dye lasers in this spectral region.

Tuccio and Strome (1972) have described the operation of a three-element dye laser. The device had an optical arrangement very similar to that of Fig. 2.11a. In this system a phosphate flint Brewster-angle prism was used for tuning. The device was excited with the 4-watt output of an argon laser at 514.5 nm. The laser mode and pump waist diameters were matched at 12 μm. The active length of the system was 0.3 cm.

The output power as a function of wavelength is shown for several dyes in Fig. 2.16. The tuning curves of Fig. 2.16 were obtained with water solutions of the dyes rhodamine 6G and rhodamine B using two different additives. The refractive index of water depends much less strongly upon temperature than that of methanol or ethanol. The use of water solutions of dyes tends to reduce the laser instability caused by nonuniform heating of the active region.

Many of the best laser dyes do not dissolve readily in water (see Chap. 5). Rhodamine 6G, for instance, tends to dimerize in water. To prevent this it is necessary to add a deaggregating agent to the dye solution. In Fig. 2.16, curve 1, hexafluoroisopropanol was used as the deaggregating agent. For curves 2 and 3 the surfactant Ammonyx LO was used. It was found that these additives also tend to quench the triplet state of the dye molecules. The reasons for this behavior are not yet understood.

Interestingly, one dye may produce quite different fluorescence spectra, as reflected by the range of the tuning curve, in different solvents. In Fig. 2.16, curves 1 and 2 are both rhodamine 6G in water. The different tuning curves result from the use of different deaggregating agents.

The total tuning range of about 70 nm for a given dye is typical of the performance of the cw dye laser. The efficiency of this laser however has not been optimized in this design, as pointed out by Tuccio and Strome (1972). The efficiency and power output of the system were found to be degraded by birefringence in the optical components of the dye cell. If the effect is not eliminated, spurious reflection losses will occur at the tuning prism since the polarization of the laser beam will not be determined completely by the Brewster-angle surfaces of the prism. This effect is very likely to occur when sapphire elements are used in the dye cell. Modification of the system to eliminate the effects of birefringence in the dye cell has yielded a slope efficiency of about 45% and over-all efficiency of approximately 30%. Similar results have been obtained by Dienes et al. (1972) with three-element reflecting cavities.

Tuccio and Strome (1972) developed an analysis similar to that of Pike (1971), which was reviewed in Sect. 2.1, and have compared the measured and calculated performance characteristics for tuned and untuned cw dye lasers. Figure 2.17 shows the comparison between calculated and observed power output vs power input. The output with the tuning prism in the cavity is seen to be slightly lower than that without the prism owing to the losses associated with the prism. The difference between the calculated and measured results is attributed to scattering losses which have not been included in the calculations. Nevertheless, the agreement is seen to be fairly good. Figure 2.17, and similar results obtained by H. A. Pike (1971), strongly support the validity of the rate-equation approach to the analysis of the cw dye laser.

Measurements were made with the same system to determine the effect of dye flow velocity upon the output characteristics. It was found that the amplitude noise decreases and the power output increases markedly as the flow rate increases from 2 m/s to 12 m/s. Experiment has also shown that the laser threshold decreases with increasing flow velocity, presumably because of the physical removal of triplet-state molecules from the active region. With transit times of less than $10^{-6}$ s across the active diameter, corresponding to a flow velocity of about 10 m/s, "mechanical quenching" becomes a useful technique for control of the triplet-state population. Flow velocities of up to 100 m/s do

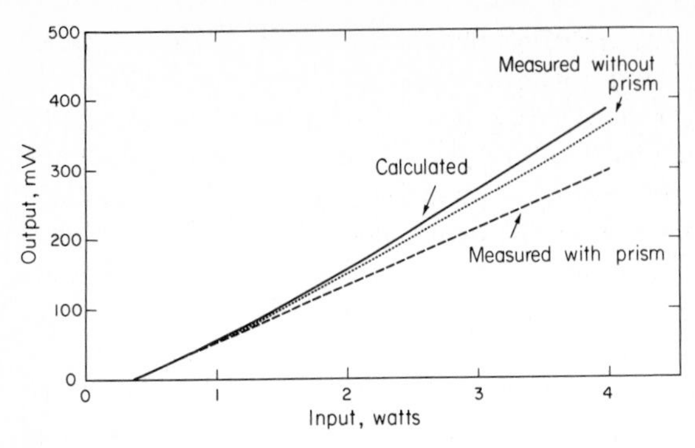


**Fig. 2.17.** Comparison of measured and calculated output power as a function of input power for a cw dye laser using the geometry of Fig. 2.13a. The data of Fig. 2.2 were used in the analysis with the following additional dye and resonator parameters: $N = 1.8 \times 10^{17}\ \mathrm{cm}^{-3}$, $W_0 = W_{p0} = 6\ \mu\mathrm{m}$, $d = 0.3$ cm, $R_1 = 0.992$, $R_2 = 910$, output mirror transmittance = 0.03. A single-wavelength excitation at 514.5 nm was assumed. Absorption loss of the intracavity lens was included in the analysis, though the absorption due to the prism was not. The slight lowering of efficiency in the prism-tuned system is due to the loss of the prism (Tuccio and Strome 1972)

not seem unreasonable, in which case the transit time across the 10 μm active diameter would be about $10^{-7}$ s. This is short enough to be used as the primary means for control of the triplet state population and permits consideration of dyes for which no chemical quenchers are known. With such large flow velocities, however, the flow geometry of the system must be carefully designed to avoid cavitation in the dye cell. In addition to possible mechanical damage to the system, cavitation would produce severe degradation of the optical quality of the resonator.

Several authors have reported on the single longitudinal mode operation of cw dye lasers (Hercher and Pike 1971 b; H. W. Schröder et al. 1973). In all cases an arrangement similar to that shown in Fig. 2.13a or b has been used with an etalon mounted in the long arm of the cavity in addition to the prism. In the system reported by Hercher and Pike a 2-mm uncoated glass etalon was used in conjunction with a single prism. The linewidth of the system without the etalon was approximately that given by geometrical considerations, as discussed in the preceding section. With the etalon in the cavity the short term (several seconds) linewidth was reduced to about 35 MHz. Figure 2.18a shows a scanned Fabry-Perot interferometer trace of the dye laser output. The scanned output of a multimode He-Ne laser is shown in Fig. 2.18b for comparison. Hercher and Pike (1971 a) observed that the line was stable to within 180 MHz over a period of minutes. Using a similar system, H. W. Schröder et al. (1973) have reported a short-term (one second) stability and linewidth of less than 10 MHz.

In more recent experiments Hartig and Walther (1973) have reported a somewhat narrower linewidth and greater stability. In their experiments a spherical three-element cavity with transmitting optics similar to that described by Hercher and Pike (1971 a) was used. A linewidth of less than

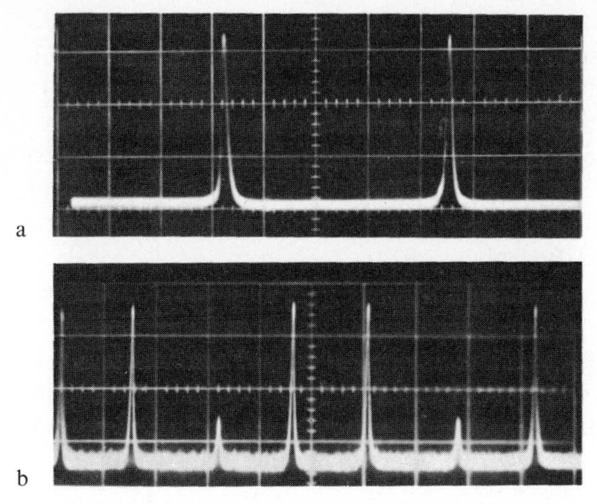

**Fig. 2.18.** (**a**) Single longitudinal mode output of a cw dye laser, as displayed by a scanning Fabry-Perot interferometer. The single mode output was obtained by use of an intracavity etalon (2 mm glass, uncoated) in addition to the prism in an arrangement similar to that shown in Fig. 2.13a. The free spectral range of the interferometer, as displayed, is 1.5 GHz. The laser linewidth is 35 MHz. (**b**) Interferometer trace of the multimode output of a He-Ne laser for comparison with (**a**) (Hercher and Pike 1971b)

20 MHz was deduced from measurement of the Na resonance line hyperfine structure. The laser output was locked to the Na $D_1$ line by conventional techniques in one experiment. Under these conditions a long term drift of the laser line center of less than $\pm 1.5$ MHz, with respect to the sodium line, was achieved. The laser produced a single mode output of 10 mW with an excitation power of 1.2 Watts at 514.5 nm.

The linewidths mentioned in the preceding paragraphs are considerably greater than the linewidth expected on the basis of statistical processes. In the preceding section arguments were presented for an ultimate linewidth of less than 1 kHz. The discrepancy does not seem to be due to the lack of dispersion in the resonator, but rather to problems associated with the flow of the dye solution through the active region. The dye cells utilized to date have not been designed to minimize cavitation or turbulence within the active region. The large observed linewidth suggests that there are gross instabilities in the dye stream as it flows through the active region.

Variation in the excitation power may also cause an artificially broad dye laser line. Most commerical argon ion lasers, of the sort used to excite dye lasers, have amplitude noise of several percent. This noise will produce temperature fluctuation in the active region which, in turn, may cause frequency fluctuations in the dye laser output. Narrow laser linewidth would seem to require stabilization of the excitation laser.

The excitation brightness necessary to excite a cw dye laser has led to operational problems in some cases. With an excitation power of several watts, damage may occur at the surface of the dye cell window at the beam waist.

The severity of this problem seems to increase with dye solutions containing triplet quenching and dye dispersing additives. Cell windows also seem to be more susceptible to damage when impure or unclean solvents and dyes are used.

One approach to a solution to the window burning problem has been the use of a windowless dye "cell" in the form of a free-flowing stream of dye from a nozzle. This technique was first described by Runge (1972) and has since been further developed (Runge and Rosenberg 1972).

Use of this technique places some restriction upon the choice of solvents and dyes. Unfortunately, the most favorable dye solvents, such as water and alcohols, do not seem to form stable streams from simple nozzles. To date experiments have been performed with ethylene glycol and glycerol-based dye solutions. Although the thermal properties of these solvents are undesirable, good performance is claimed for cw dye lasers using the free stream.

## 2.4 Conclusion

In this chapter the factors affecting power output, efficiency, tuning range, stability and linewidth of cw dye lasers have been discussed. Performance analysis has been approached from the rate-equation point of view. This approach has been found to give an acceptable description of the cw dye laser. At present it seems that no serious attempts to apply the more powerful semiclassical or quantum electrodynamic analyses to the dye laser have been made. Progress in this theoretical direction is to be anticipated.

**Table 2.1.** Summary of cw dye-laser characteristics

| | | |
|---|---|---|
| Power output | | 2 W |
| Tuning range | | 420 – 710 nm |
| Linewidth | ≈ | 2 MHz (tuned) |
| | ≈ | 3 nm (untuned) |
| Efficiency | | $0.35 \dfrac{\text{optical power out}}{\text{excitation power in}}$ |
| Mode-locked pulse-width | ≈ | $1.5 \times 10^{-12}$ s |

Commercial cw dye lasers with specifications comparable to those of Table 2.1 are becoming available. Powerful ultraviolet lasers as excitation sources for cw dye lasers operating in the blue spectral region are just becoming available at this time. As other new excitation sources become available extension of the tuning range into the near infrared is to be expected.

# 3. Continuous-Wave Dye Lasers II

Harald Gerhardt

With 10 Figures

In the foregoing chapter by Snavely on continuous-wave dye lasers, early results on cw dye lasers are discussed. Besides theoretical descriptions of dye properties and gain analysis, results on tuning of cw dye lasers are given. In this chapter the latest achievements in modern cw dye laser design will be described. In Sect. 3.1, cw dye lasers in linear configuration are discussed. Special emphasis is devoted to the free running jet technology and to tuning elements, e.g. birefringent filters. Section 3.2 treats single mode (SM) operation of linear laser systems. The most recent results on traveling-wave (TW) operation of ring dye lasers are described in Sect. 3.3. Special attention is given to SM operation and to some ring laser devices, e.g. undirectional devices. The chapter concludes with a short discussion of frequency doubling in cw dye lasers. Frequency stabilization techniques are not discussed, since the description of these techniques exceeds the scope of this book.

As discussed in Chap. 2, the threshold excitation power density of a practical cw dye laser system is about 50 kW/cm$^2$. This high power density is not easily obtained by arc discharge lamps. Therefore it was a widely held notion that cw dye lasers cannot be pumped with incoherent light sources. Thiel et al. (1987) have reported cw operation of a dye laser, pumped by a high pressure arc. With an electrical input power of 8.8 kW they were able to reach threshold. However, it remains to be seen whether the output power of an incoherently pumped dye laser can be raised into the milliwatt range.

## 3.1 Linear Dye Laser

### 3.1.1 Active Region

In early cw dye lasers the active region was formed by a dye cell. The dye cell contains the dye solution streaming in a laminar flow of about 1 mm thickness between two flat windows. Flow velocities of 10 m/s can be achieved with water solutions. Independent of the dye cell configuration, the cell windows become contaminated by decomposition products from the organic materials. As a consequence, the surface of the windows will be burned by the focused excitation beam even at moderate pump power densities of 300 kW/cm$^2$.

By using a free-flowing jet of dye solution with good optical surfaces, these problems can be avoided. Therefore, today free-flowing jets are generally used in cw dye lasers. The nozzle of the jet stream can be fabricated in different ways. However, the edge where the dye solution breaks away from the nozzle requires special attention. It must be polished to a minimum radius of curvature. A simple nozzle can be fabricated by flattening small-diameter (e.g. 3 mm) tubing, and carefully cutting and polishing a flat exit surface. A more advanced design is used by Wellegehausen et al. (1974). The nozzle is assembled from two polished metal plates. The thickness and height of the jet stream can be varied using two spacers. A completely new technical approach is described by A. Watanabe et al. (1989).

The problem of window contamination is totally eliminated by using a free-flowing jet. In exchange, problems with the surface quality of the jet require special attention. In general, a high optical quality of the active region can only be achieved by restricting the solution flow rate to the laminar flow regime. Therefore, the viscosity of the liquid solvent must be sufficiently large. The use of a highly viscous solvent has some additional advantages. Although the flow rate of the dye jet is kept in the laminar flow regime, the jet surface is deteriorated by surface waves, which are produced by irregularities in the nozzle and by pressure fluctuations. The damping of these surface waves is high for a high-viscosity solvent. In addition, a higher maximum laminar flow velocity can be reached, which leads to a more effective mechanical triplet quenching and reduces thermal problems. As high-viscosity solvents high alcohols such as ethylene glycol, benzyl alcohol, propylene glycol and glycerol or mixtures of these can be used. These solvents decrease the quantum efficiency of several dyes and do not always offer optimum thermal properties. Therefore, water-based dye solutions with viscosity-raising additives are used in addition. A viscosity-raising additive is a polymer or other large organic molecule that increases the viscosity of the solvent considerably (Leutwyler et al. 1976). Substantial improvement in laser performance was found especially for water-based viscosity-raising additives such as polyvinyl alcohol or polyvinylpyrrolidone. The viscosity of water of 1 cP can be raised by adding 3% polyvinyl alcohol to about 20 cP, which is comparable to the viscosity of ethylene glycol.

A water-based solvent is extremely useful in high power cw dye lasers. In these applications thermal problems play a major role (Wellegehausen et al. 1975). The focused pump beam of a high-power dye laser causes local thermal heating of the dye solution due to radiationless transitions. Since the pump intensity has a nonuniform distribution, an inhomogeneous temperature distribution is created which is related to refractive index variations $\Delta n$ by

$$|\Delta n| \propto \frac{1}{\varrho c}\left|\frac{dn}{dT}\right| , \tag{3.1}$$

where $\varrho$ is the density, $c$ is the heat capacity and $dn/dT$ is the temperature variation of the refractive index of the dye solution. The refractive index variations can be noticed as a thermal lens effect, which has some astigmatism (Teschke et al. 1976). Therefore, this thermal lens effect is hard to compensate and reduces the maximum pump power. The refractive index of water changes only slightly with temperature in comparison to ethylene glycol or ethanol (Leutwyler et al. 1976). By using viscosity-raising additives a water-based dye solution is best suited for high-power applications. Anliker et al. (1977) have reported on a 33 W cw dye laser. As solvent they used water with 1.3% polyvinyl alcohol. The dye laser was pumped by an all-line argon ion laser.

For better understanding, typical operating conditions in the active region of a medium-power cw dye jet stream laser are given in the following. If possible, ethylene glycol is used as solvent for ease of operation. A simple nozzle as described above creates a jet with a height of about 8 mm. The thickness is normally fixed at 0.2 mm. At a laminar flow velocity of about 10 m/s a Reynolds number of about 220, well in the laminar flow region, can be calculated. The concentration of the dye solution should not exceed $2\times10^{-3}$ M to avoid concentration quenching. Therefore, jet streams with a thickness of less than 0.1 mm are not feasible, due to the decreased pump light absorption. Considering a minimum excitation beam diameter of 10–20 μm in the jet stream, the transit time of dye molecules through the active region is about 1 μs. This transit time is short enough to establish sufficient mechanical triplet quenching for most cw dyes. Since jet streams have no antireflection coating, they are generally used in a folded cavity with the nozzle set at Brewster's angle.

### 3.1.2 cw Dye Laser Tuning

In Chap. 2, self-tuning of the cw dye laser was discussed. By adjusting $r$ and the concentration $N$, self-tuning can be achieved along the valley of the critical inversion surface. However, in practice tuning is achieved by means of a dispersive resonator. Prisms, tunable interference filters or birefringent filters can be used as tuning elements. The space within the cavity between the folding mirror and the output mirror is available for insertion of a tuning element, since in this region the dye-laser beam is almost parallel. A grating is not applicable in dye-laser technology due to the relatively high losses.

When a prism is used as the tuning element, several modifications of the folded resonator are possible as shown in Figs. 3.1 and 3.2. Tuning of the laser wavelength is accomplished by rotating the output mirror. Unfortunately, the direction of the output beam changes when the laser is tuned. A disadvantage of the collinear excitation arrangement illustrated in Fig. 3.1a is the need for a dichroic input mirror. This is avoided in the configuration shown in Fig. 3.1b. The excitation beam and dye-laser beam are combined by the tuning prism. However, in the case of argon-ion-laser pumping only one line can be used for excitation. A unique pumping arrangement is shown in Fig. 3.2. The noncollinear excitation avoids the use of a dichroic mirror or dispersive prism in the beam of the pump laser. The pump beam is focused by a highly reflective mirror into the active region and crosses the dye laser beam at a small angle without passing through any of the dye-laser optics. In this arrangement all-line pumping with an argon ion laser can be used and only small losses are produced by the reflecting mirror compared to the transmitting optics of the col-

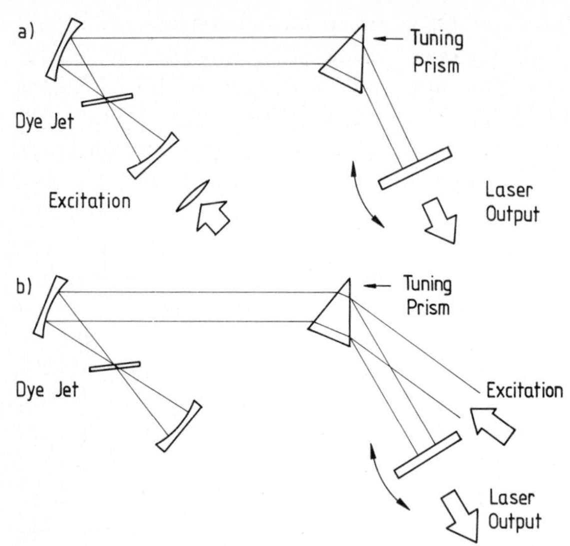


**Fig. 3.1 a, b.** Tunable three-element dye laser resonator using reflecting optics. (**a**) Collinear excitation through end mirror. (**b**) Collinear excitation through tuning prism

linear system. To avoid extra losses the tuning prism should be cut at Brewster's angle and operated at minimum deviation. With such a prism the laser linewidth of about 1 nm of a free-running dye laser can be reduced to several angstroms tunable linewidth. However, to cover a tuning range of a few hundred nanometers several prisms are needed.

By using a tunable interference filter the whole cw dye-laser tuning range from 380 – 1020 nm can be covered with one filter without increasing the tunable linewidth compared to a prism. In addition, the direction of the output beam is fixed. A tunable interference filter is comparable to a standard interference filter, except the spacing layer has the shape of a wedge. With a free spectral range of about 100 nm, the whole emission curve of one dye can be covered without producing sidebands. Tuning is accomplished by translating the wedged layer across the laser beam. A disadvantage of this filter is the relatively high losses common to interference filters.

One of the optical elements best suited for wavelength tuning of a cw dye laser is a birefringent filter. It depends on the interference of polarized light

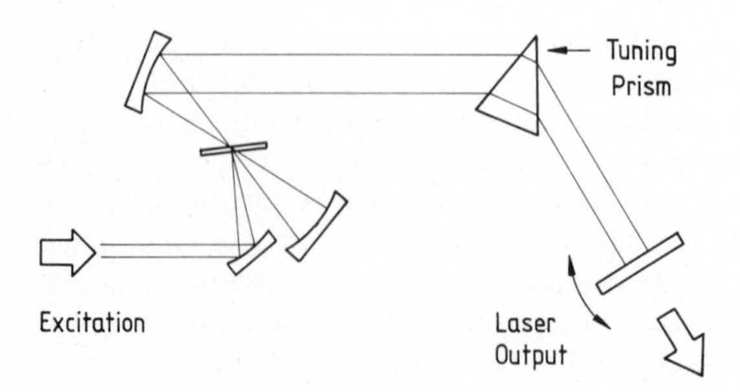


**Fig. 3.2.** Noncollinear excitation of tunable three-element dye laser resonator

transmitted through one or several birefringent crystals. The original birefringent filter was invented by Lyot (1933). It is constructed of a series of units, each consisting of a plane-parallel birefringent plate and a polarizer. For a uniaxial crystal, all plates have surfaces parallel to their optic axes and the slow axes are all oriented the same way. The direction of polarization as determined by the polarizers is oriented at an angle of 45° to the optic axes of the crystals. The thicknesses of the plates are such that the ratio between adjacent crystals is 1:2:4.... Light that enters a birefringent plate is divided into two components polarized parallel to the slow and fast axes. On emerging from the crystal, the two components have a relative retardation and interfere in a polarizer. When the wavelength corresponds to an integral number of full-wave retardations in the plates, the light is transmitted without losses. At any other wavelength the light suffers losses. Therefore, the spectrum of light transmitted by the filter consists of a series of widely spaced narrow bands. Their separation is equal to the separation of the transmission maxima of the thinnest plate, while their width is determined by the thickest plate (T. W. Evans 1949). Sidebands created by the Lyot filter are negligible. An electro-optically tuned Lyot filter has been used with pulsed dye lasers by Walther and Hall (1970). However, the use of a Lyot filter in cw dye lasers is not feasible due to the relatively high Fresnel losses.

For the wavelength tuning of cw dye lasers a variation of the Lyot filter has been developed by Yarborough and Hobart (1973). The birefringent plates of a Lyot filter are inserted within the laser cavity at Brewster's angle. In this case they can act as both retarding and polarizing elements. Since no polarizers are used and no surfaces normal to the laser beam exist, the birefringent filter can be regarded as a low-loss tuning element. Tuning is achieved by rotating the filter about an axis normal to the surface. This changes the polar angle between the optic and laser axes and, hence, the wavelength whose retardation is an integral number of wavelengths. However, compared to a Lyot filter, sidebands of a birefringent filter are not negligible. The sidebands are produced by incomplete polarizing action of the Brewster surfaces. For trouble-free laser action, the product of laser gain and transmission loss at the spurious peaks must be less than the gain of the laser at the transmission maximum. If this condition is fulfilled even at the edges of the dye gain curve, the laser can be tuned to any desired laser wavelength. In general, sideband maxima of less than 80% transmission are tolerable.

Practical filter configurations have been analyzed by Bloom (1974). In the following, calculated transmission curves for a stack of three plates with thicknesses in the ratio 1:4:16, with the thickest plate on the side adjacent to the end mirror, will be discussed. The transmission $T_{LR}$ is defined as the flux remaining in a linear resonator after one round trip. A dye jet stream oriented at Brewster's angle with the same refractive index as the plates is included in the calculations. In Fig. 3.3 $T_{LR}$ is displayed for two angles $\Theta$ as a function of the retardation $\delta$ for a maximum phase difference $\pi$ of the thick plate. The retardation between two transmission maxima of the birefringent filter amounts to $32\pi$ for the thick plate or $2\pi$ for the thin plate. The angle $\Theta$ is

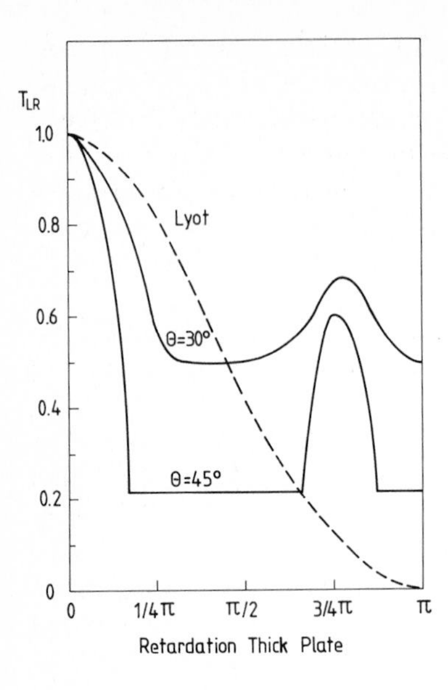


**Fig. 3.3.** Transmission $T_{LR}$ as a function of retardation of the thick plate of a 1:4:16 birefringent filter

between the plane of incidence and the plane containing the optic axis and the axis normal to the Brewster surface. Best blocking of the sidebands is achieved for $\Theta$ around 45°. Only one of the many spurious peaks between two transmission maxima is shown in Fig. 3.3. For the 1:4:16 filter, the transmission of the largest sideband peak is about 78%. Normally this is low enough to obtain stable laser operation. If a stronger blocking is desired, additional Brewster surfaces can be inserted or a different thickness ratio can be chosen, e.g. 1:2:15 (Holtom and Teschke 1974). However, this results in a slight increase of the bandwidth. A birefringent filter with a stack ratio of 1:2:4:8:16 creates no sideband peaks greater than 50%.

For comparison, the transmission curve of a five-element Lyot filter is also shown in Fig. 3.3. It is surprising that the width of the transmission maximum of the birefringent filter is smaller than that given by the Lyot filter. However, this narrowing can be explained by the polarization of the resonator modes. In a resonator containing a birefringent filter, modes in the region around $\delta = 0$ are generally not linearly polarized. Therefore, they suffer higher losses than in a resonator containing a Lyot filter. The result is a narrowing of the bandwidth of the transmission maxima.

For a practical design, low-loss crystalline quartz plates can be used. The retardation of a plate is given by (Born and Wolf 1970)

$$\delta = \frac{2\pi (n_e - n_o) d \sin^2 \gamma}{\lambda \sin \Theta_B}, \qquad (3.2)$$

where $n_e$ and $n_o$ are the index of refraction for the extraordinary and ordinary beams, respectively, $\Theta_B$ is Brewster's angle, $\gamma$ is the polar angle between the internal ray and the optic axis, $d$ is the plate thickness and $\lambda$ is the wavelength.

According to (3.2), a quartz plate of $d = 0.4$ mm cut parallel to the optic axis creates transmission maxima with a period of about 100 nm in the red part of the spectrum. Since $\delta \propto 1/\lambda$, the blue period is slightly smaller and the infrared period is slightly larger than 100 nm. The bandwidths of the dye gain curves are comparable to the periods of a 0.4 mm crystal plate. Therefore, this plate is well suited as the thinnest element in a 1:4:16 quartz filter that can be used to cover the whole wavelength range of cw dye lasers. A tunable laser linewidth of 0.05 nm is realistic with such a filter.

For tuning from one transmission maximum to the next a rotation of about $\pm 15°$ around $\Theta = 45°$ is required with the described filter. As can be seen from Fig. 3.3, a filter used at an angle of $\Theta = 30°$ has reduced blocking efficiency. Cutting the quartz plates at an angle of 25° to the optic axis results in a birefringent filter, which needs a rotation of only $\pm 5°$ for the same tuning range. With such a filter the entire tuning range can be covered with optimum blocking efficiency.

### 3.1.3 Dyes for cw Operation

Since the invention of the dye laser, many dyes suited for laser action have been found or synthesized, as can be learned from Chap. 5 of this book. But only a few that have minimal internal losses will operate cw. In Table 3.1 the

**Table 3.1.** Laser dyes for cw operation

| Dye | Short form | Solvent | Excitation | References |
|---|---|---|---|---|
| Polyphenyl 1 | PP1 | EG | Argon all lines UV | Hüffer et al. (1980) |
| Stilbene 1 | S1 | EG+ME (10:1) | Argon all lines UV | Hüffer et al. (1979) |
| Stilbene 3 | S3 | EG+ME (10:1) | Argon all lines UV | Kuhl et al. (1978) |
| Coumarin 102 | C102 | EG+BZ (10:1) | Krypton all lines violet | Yarborough (1974) |
| Coumarin 6 | C6 | GL+BZ (1:1) | Argon 488 nm | Yarborough (1974) |
| Rhodamine 110 | R110 | EG | Argon all lines | Yarborough (1974) |
| Rhodamine 6G | R6G | EG | Argon all lines | Yarborough (1974) |
| DCM | | EG+BZ (3:2) | Argon all lines | Marason (1981) |
| LD700 | | EG | Krypton all lines red | Johnston et al. (1982) |
| Styryl 9 | STY9 | EG+PGC (17:3) | Argon all lines | Hoffnagle et al. (1982) |
| HITC | | EG+DMSO (1:1) | Krypton all lines red | Romanek et al. (1977) |
| IR140 | | EG+DMSO (1:1) | Krypton all lines infrared | Leduc and Weisbuch (1978) |

Abbreviations: EG: ethylene glycol; ME: methyl alcohol; BZ: benzyl alcohol; GL: glycerol; PGC: propylene glycol carbonate; DMSO: dimethyl sulfoxide

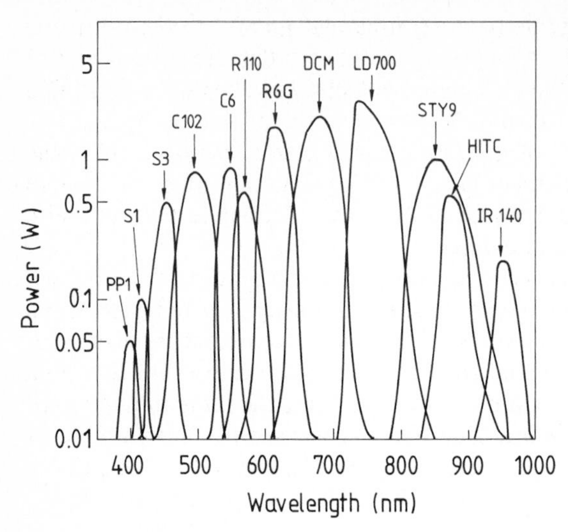


**Fig. 3.4.** Tuning curves of laser dyes used for cw operation as listed in Table 3.1

most effective and stable dyes covering the whole tuning range of cw operation are listed. In the references quoted in the last column, detailed descriptions of the cw performance of these dyes can be found. The references do not necessarily represent the first citation in the literature. The solvents are given as examples. Their use guarantees stable jet stream operation. As the pump source a cw argon or cw krypton ion laser can be used. The rhodamines and coumarins have been known since the early days of cw dye lasers, see Chap. 5. The stilbene dyes and the carbocyanine dyes were discovered recently for the blue wavelength region and the infrared wavelength region, respectively. The dye Styryl 9 has some interesting features. Besides a broad absorption spectrum covering about 200 nm, it has an unusually large Stokes shift of more than 200 nm. Although emitting around 850 nm, it can be pumped by the blue-green lines of an argon ion laser. Thus, by using Styryl 9 the wavelength range of dye lasers pumped by argon ion lasers can be extended into the infrared wavelength region.

In Fig. 3.4, tuning curves of the dyes listed in Table 3.1 are displayed. The tuning curves were obtained with typical linear jet stream dye lasers as described in this section. The whole wavelength range between 380 nm and 1020 nm is covered. The pump power was about 3 W in the UV, 4 W in the visible and 2 W in the infrared. The concentration of the dyes was adjusted so that the absorption of the pump beam was around 80% – 90%. When a 0.2 mm thick jet stream is used, a $10^{-3}$ M solution normally has the appropriate absorption. As can be seen from Fig. 3.4, the conversion efficiency of some dyes is as high as 40%.

## 3.2 Single-Mode Operation of Linear Dye Lasers

### 3.2.1 Spatial Hole Burning

The transversal mode pattern of a well-aligned three-element dye laser resonator is restricted to the fundamental $TEM_{00}$ mode due to an excitation diameter of only a few micrometers. The small excitation area acts as a perfect transverse mode selector. However, in order to obtain operation in a single longitudinal mode, special means have to be provided.

For a linear dye laser with a homogeneously broadened medium, one might expect spontaneous single-longitudinal-mode operation. However, due to the spatial hole burning effect (Tang et al. 1963) multimode operation will occur. The modes of a linear laser resonator are determined by standing waves. Therefore, spatial holes exist in the laser medium that are not saturated by a given standing-wave field. Thus, additional modes can reach laser threshold and will start oscillating unless mode selection techniques are utilized. The foregoing statement is often used in laser physics, but it is only partly true. A detailed theoretical description of the spatial hole burning effect based on third-order Lamb theory is given by Hambenne and Sargent (1975) and Sargent (1976). They describe the laser electric field by the superposition of two oppositely directed running waves $E_+$ and $E_-$. The equation of motion for the mode intensity $I_+ = E_+^2$ of a single mode laser is given by

$$\dot{I}_+ = 2I_+(\alpha_+ - \beta_+ I_+ - \theta_{+-} I_-) \; , \tag{3.3}$$

where $\alpha_+$ is the net linear gain for $I_+$, $\beta_+$ is the self-saturation coefficient and $\theta_{+-}$ is the cross-saturation coefficient. Self-saturation is caused only by a depletion of the population. This depletion results also in a cross-saturation because the oppositely directed waves interact with the same atoms. However, an additional cross-saturation exists. One can understand this by supposing that the spatial holes act like a Bragg grating. Some of the running waves are reflected by the grating. These scattered waves are out of phase with the running waves because the Bragg grating is out of phase with the standing-wave field. As a consequence, the intensity of the waves is reduced by destructive interference. The intermode coupling is described by a coupling parameter $C$:

$$C = \theta_{+-}\theta_{-+}/\beta_+\beta_- \; . \tag{3.4}$$

For a homogeneously broadened medium the Bragg-like contribution equals the population depletion contributions to both the cross-saturation and the self-saturation, yielding $C = 4$. For $C > 1$ strong coupling exists, which leads to an increased saturation of the laser medium and a corresponding decrease in the steady-state laser intensity. The equation of motion for the mode intensity (3.3) is also valid for the two-mode linear laser. In this case $I_+$ and $I_-$ represent the intensities $I_n$ and $I_m$ of the two modes. The scattering of one mode off the grating induced by the second mode gives a position-dependent

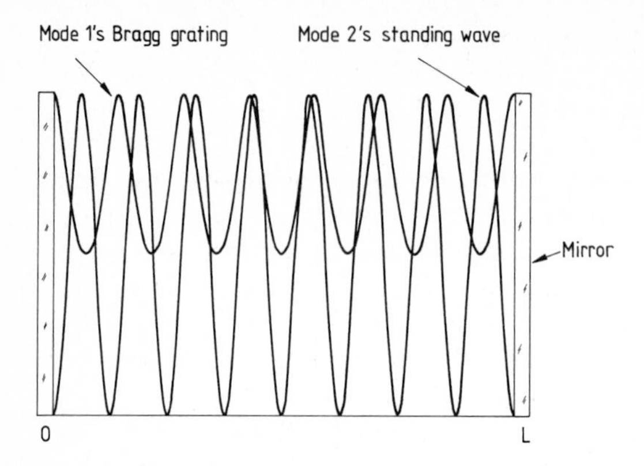


**Fig. 3.5.** Bragg grating induced by one mode relative to standing-wave intensity of next higher frequency mode. Scattering of one mode off the grating induced by the second mode gives position-dependent cross-saturation (Sargent 1976)

cross-saturation. In Fig. 3.5 the Bragg grating induced by one mode and the standing wave intensity of a second mode are displayed. The two modes differ by half a wavelength. In the middle of the resonator the scattering is constructive, resulting in a weak coupling. Alternatively, stronger coupling can be obtained when the laser medium is placed near the end mirrors of the laser cavity. The cross-saturation coefficient is given by (O'Bryan and Sargent 1973)

$$\theta_{nm} = \beta_n[1+K(n-m)](2+N_{2(n-m)}/N_{2(0)})/3 \ . \tag{3.5}$$

Here $\beta_n \approx \beta_m$, the second factor is given by population pulsations and the third factor is due to population depletion and Bragg-like contributions. In (3.5),

$$N_{2(n-m)} = \frac{1}{L}\int_0^L ds N(s)\cos\left[2(n-m)\pi s/L\right] \ , \tag{3.6}$$

where $N(s)$ is the population difference. The population depletion contribution is independent of position and gives the 2 in the parentheses of (3.5). The Bragg-like contribution is the only position-dependent contribution and gives $N_{2(n-m)}/N_{2(0)}$.

The population pulsations can lead to a stronger coupling if $K(n-m)$ is positive. For symmetric tuning $K(n-m)$ is given by

$$K(n-m) = \frac{\gamma_a\gamma_b}{\gamma_a+\gamma_b}\left[\frac{\gamma_a}{\gamma_a^2+\Delta^2}\left(1-\frac{\Delta^2}{2\gamma\gamma_a}\right)+\frac{\gamma_b}{\gamma_b^2+\Delta^2}\left(1-\frac{\Delta^2}{2\gamma\gamma_b}\right)\right] , \tag{3.7}$$

where $\gamma_a$ and $\gamma_b$ are the level decay constants of the upper and lower laser levels respectively, $\gamma$ is the dipole decay constant and $\Delta$ is the frequency separation between modes $n$ and $m$. For a dye laser, $K(n-m)$ is positive for a mode separation up to several 10 GHz. When the laser medium is placed near the end mirrors of the laser cavity, a coupling parameter of $C>1$ can be obtained for adjacent modes. However, for a mode separation of more than about 1 GHz a weak coupling is expected, due to the Bragg-like contribution, and two modes will oscillate. The cross-saturation has its lowest value when the argument of the cosine function in (3.6) equals $\pi$. In this case a frequency separation of $\Delta\nu = c/4s$ can be calculated. Therefore, when the laser medium is placed near the end mirrors at a distance $s$, two modes will oscillate at the preferred frequency difference of $c/4s$. This was experimentally verified by Pike (1974).

### 3.2.2 Single Mode Selection Techniques in a Linear Dye Laser

The axial mode spacing $c/(2\times$cavity length) of a common cw dye laser is on the order of several hundred megahertz. With a three-element birefringent filter, a linewidth of about 0.05 nm can be achieved. Thus, the emission spectrum of a broadband dye laser consists of several longitudinal modes.

Several techniques have been developed to select a single longitudinal dye laser mode. All techniques increase the resonator losses for the unwanted modes to a level such that lasing is not possible. A number of techniques can be called interferometric techniques. Grove et al. (1973) have used a Fox-Smith mode selector to obtain single-frequency dye-laser operation. The selector consists in principle of a Michelson-type interferometer and replaces the output mirror of the dye laser. The axial mode spacing is now determined by the entire length of the interferometer. By selecting a free spectral range that is larger than the linewidth given by the tuning element, a single axial mode can be obtained.

The most advanced technique for single-mode selection uses the insertion of tilted Fabry-Perot etalons in the space between the folding mirror and the tuning element. The free spectral range of the etalon has to be larger than the tunable laser linewidth of the dye laser, if only one etalon is used. For maximum efficiency the transmission peak of the etalon has to coincide with a laser resonator mode. This can be accomplished by tilting the etalon axis with respect to the axis of the laser resonator. The finesse of the etalon has to be sufficiently high to suppress modes with weak coupling. The desired finesse can be attained by using high-reflectivity coatings on the etalon surfaces. However, the insertion of tilted Fabry-Perot etalons into the laser cavity also creates losses for the lasing modes. Due to multiple internal reflections in the tilted etalon the laser beam is caused to walk off. This beam distortion results in walk-off losses which increase with the square of the incidence angle (Leeb 1975). Because this loss increases in addition with increasing reflectivity, it is often advisable to use two etalons with low reflectivities and with different thicknesses instead of one. Practical single-mode laser systems use either a single etalon [e.g. air-spaced etalon, 80% reflectivity, 0.8 mm thick (Gerhardt and Timmermann 1977)] or two etalons [e.g. solid etalons, 6 mm thick and 32% reflectivity plus 0.3 mm thick and 64% reflectivity (Wellegehausen et al. 1974)].

Continuous tuning of the single-mode laser requires synchronous tuning of the laser resonator length and of the transmission maxima of the selecting elements. The optical path length of the laser cavity can be controlled either

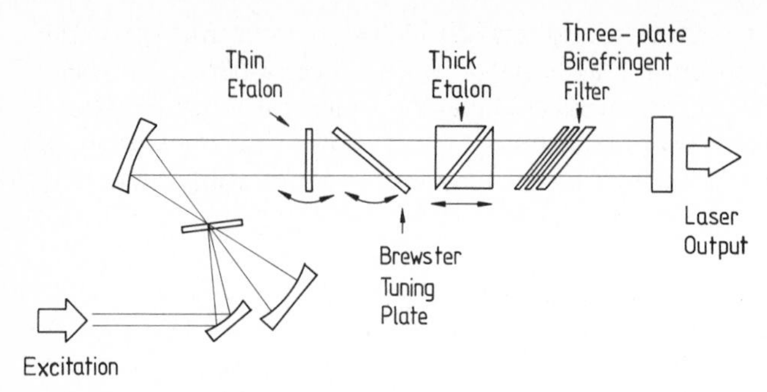


**Fig. 3.6.** Tunable linear single-mode dye laser using 1 thin etalon, 1 thick etalon, a Brewster tuning plate and a three-plate birefringent filter

by a linear motion of the output mirror, e.g. piezoelectric transducer (PZT) driven, or by tilting a plane-parallel glass plate inside the resonator. If the glass plate is inserted at Brewster's angle, reflection losses are negligible. Air-spaced etalons can also be tuned by a linear motion of one etalon mirror. Solid etalons need rotation to achieve wavelength tuning. However, when this technique is used the walk-off losses are changed as well. For thin etalons this change creates no problems. For thick etalons a rotation is not advisable. However, if a thick solid etalon is cut into two pieces (Johnston et al. 1982) at Brewster's angle, the tuning can be accomplished by a linear motion, eliminating additional walk-off losses. Figure 3.6 is a schematic diagram of a standard linear single-mode dye laser using a thin solid etalon (0.5 mm, 20% reflectivity), a thick etalon (10 mm, two pieces, 20% reflectivity), a Brewster tuning plate and a three-plate birefringent filter. The thick etalon is linear PZT scanned. The thin etalon and the Brewster plate are rotated by a galvanometer. This laser can be electronically tuned over more than 30 GHz without mode hopping.

### 3.2.3 Frequency Stability and Output Power of a Linear Dye Laser

The natural linewidth of a cw single-mode dye laser is below 1 Hz, as can be calculated by the Schawlow-Townes formula (Schawlow and Townes 1958). However, the emission linewidth of a cw single-mode dye laser is broadened by fluctuations of the optical path length of the laser cavity. The fluctuations are caused mainly by density and thickness fluctuations of the dye jet and by some insignificant mechanical instabilities of the laser cavity. The density and thickness fluctuations are created by thermal fluctuations of the dye solution, power fluctuations of the pump laser, pressure fluctuations, and surface waves of the dye jet as discussed in Sect. 3.1.1. The emission linewidth of a standard free-running single-mode jet stream dye laser is around 10–100 MHz. The smallest free-running linewidth of a jet stream dye laser reported in the literature is to our knowledge 2 MHz (Wellegehausen et al. 1974). By using a dye cell a similar emission linewidth can be obtained. With active stabilization techniques a reduction of the free-running linewidth is possible. These techniques will not be discussed in this book (see e.g. Johnston 1987).

The output power of a single-mode linear dye laser is less than that of a broadband dye laser. However, the tuning curves for the different dyes are similar to the broadband tuning curves displayed in Fig. 3.4. The maximum single-mode power in the visible is limited to about 300 mW at a pump power of around 3 W. In the infrared a maximum single-mode power of about 400 mW can be achieved at a pump power of around 6 W. At higher output powers several modes will appear due to the Bragg-like contribution to the cross-saturation, as explained in Sect. 3.2.1.

## 3.3 Single-Mode Operation of Ring Dye Lasers

### 3.3.1 Principles of Ring Dye Laser Operation

In Sect. 3.2.1 we discussed spatial hole burning and showed, that in a linear dye laser multifrequency operation is mainly caused by this effect. In addition, the same effect is responsible for a reduction of the single-mode output power. As can be calculated from (3.4, 5 and 7), without the population-depletion contribution to the cross-saturation coefficient, coupling parameters of $C > 1$ can be achieved for mode spacings up to several 10 GHz. As a consequence, without standing waves less selective elements are needed for single-mode operation and higher single-mode powers can be obtained. Therefore, for high-power single-mode dye lasers, traveling-wave (TW) operation should be preferred. This was extensively studied by J. M. Green et al. (1973a), Marowsky and Kaufmann (1976) and H. W. Schröder et al. (1977).

With several techniques TW operation has been demonstrated in lasers other than dye lasers. Mechanical motion of the laser medium relative to the resonator has been used by Danielmeyer and Nilsen (1970). Circularly polarized modes were used by Evtuhov and Siegman (1965). Tang et al. (1963) were the first to use a ring resonator for TW operation. For single-mode operation and a homogeneously broadened medium $C = 4$, as given by (3.4). This strong coupling leads to bistable operation of the single-mode ring resonator (Hambenne and Sargent 1975). Thus, the propagation direction of the mode is determined by a random process. Tang et al. (1963) used a Faraday rotator together with a birefringent polarization rotator to block one direction of propagation. The combination of the two elements is often called an optical diode. The Faraday rotator rotates the polarization of the TW mode by a certain angle. The birefringent rotator rotates the polarization back into the plane of incidence. For a mode running in the opposite direction the rotations will add and lead to high losses at the various Brewster-angle intracavity surfaces.

Tang et al. (1963) used a large angle of rotation, creating high insertion losses. However, a complete blocking of one TW direction is not needed. Clobes and Brienza (1972) have demonstrated that a differential loss obtained by less than 1° of rotation is sufficient to avoid bistable operation in a Nd:YAG laser and to select one single mode. However, a unidirectional ring dye laser will not run spontaneously in a single mode. Due to population pulsations, several weakly coupled modes will exist. Therefore, some selective elements for single-mode operation are needed in a ring dye laser.

### 3.3.2 High-Performance Ring Dye Lasers

Jarrett and Young (1979) were the first to demonstrate operation of a high performance single-mode ring dye laser. Figure 3.7 shows a ring dye laser with an optical diode. The cavity is formed by four mirrors. In one beam waist the dye jet is located. A second beam waist, which always exists in such a ring cavity, is available for intracavity second harmonic generation (SHG) or intracavity experiments where high power density is needed. The tuning elements are inserted in that part of the cavity where an almost collimated beam exists.

The elements of the optical diode can be made from material with either normal-incidence surfaces or Brewster-cut surfaces. However, in the case of normal-incidence surfaces a high-quality antireflection coating is required. Jarrett and Young (1979) have produced a rotation of about 2.5° at 653 nm using a 1.45-mm-long piece of Hoya FR-5 glass in an axial magnetic field of 0.42 T (4.2 kG) provided by a permanent magnet. The same rotation is obtained by a 0.125-mm-long piece of *c*-axis crystal quartz. Both elements have normal-incidence antireflection coated surfaces. This optical diode works over a wavelength range of 165 nm without adjustment. The total insertion losses are typically 1%.

Johnston and Proffitt (1980) have described optical diodes with Brewster-angled surfaces. In one example they use a 12-mm-long piece of SF-2 glass placed in a 0.36 T magnetic field in combination with a 0.18-mm-long piece of crystal quartz. This diode can be used in a wavelength range of 480–800 nm.

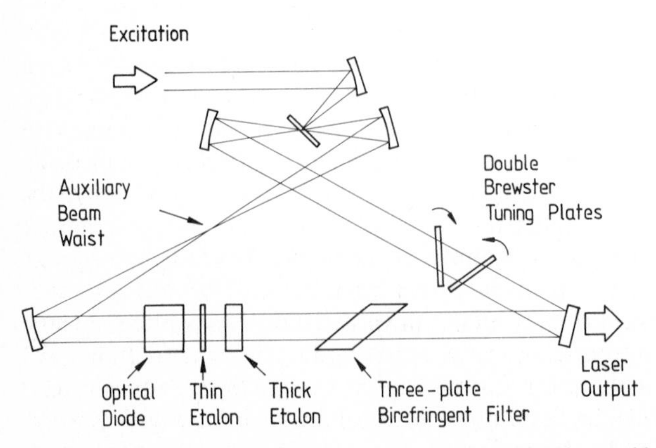


**Fig. 3.7.** Tunable single-mode ring dye laser using 1 thin etalon, 1 thick etalon, a double Brewster tuning plate, a three-plate birefringent filter and an optical diode

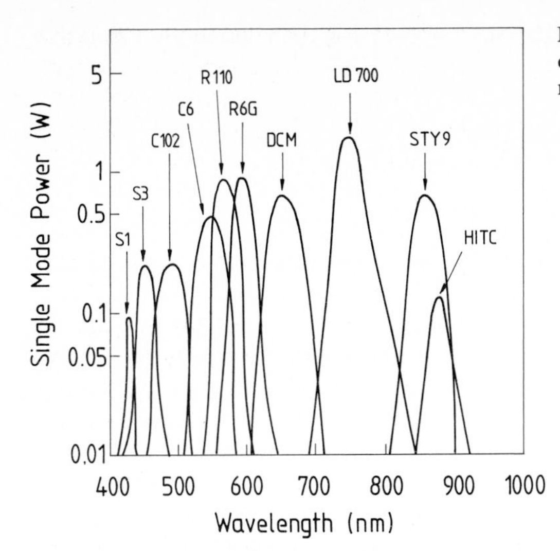


**Fig. 3.8.** Single-frequency tuning curves of ring dye lasers. For details refer to references listed in Table 3.1

Another example uses a unified optical diode. The Faraday material and the birefringent polarization rotator are combined in one piece. A left-handed optically active quartz crystal and a right-handed crystal of slightly different thickness were optically contacted to form the unified diode. This diode is useful in the blue spectral range, but is not practical in the red spectral range due to the small Verdet constant of quartz in this wavelength region.

When the optical diode produces an overall rotation of 3°–5°, stable TW operation can be achieved. Single-mode operation is obtained when in addition to a three-element birefringent filter the same two etalons are adopted as used for the linear laser displayed in Fig. 3.6 (thickness 0.5 mm, 20% reflectivity; thickness 10 mm, two pieces, 20% reflectivity). With this single-mode ring laser, output powers of several watts can be achieved. Single-mode conversion can be higher than 90%. In Fig. 3.8 single-frequency tuning curves of typical ring dye lasers are displayed. The pump power was about 3 W in the UV, 5 W in the visible and 3 W in the infrared. For higher pump powers solvents other than ethylene glycol should be used due to the thermal lens effect, as described in Sect. 3.1.1. Johnston et al. (1982) used a mixture of 25% ethylene glycol and 75% Ammonyx LO at a temperature of 10 °C. With a pump power of 24 W from an all-line argon laser a single-mode output power of 5 W could be achieved.

The frequency stability of a free-running single-mode ring dye laser is comparable to the stability of a linear dye laser. With active stabilization techniques linewidths in the kilohertz-range can be achieved. Single-mode tuning of the ring laser is accomplished with the same elements as used for tuning the linear dye laser shown in Fig. 3.6. To avoid beam displacements when tuning the laser frequency, two Brewster tuning plates instead of one should be adopted. No beam displacement will occur during laser tuning if the plates are rotated by equal amounts in opposite directions. A tuning range of more than

30 GHz without mode hopping can be achieved for the single-mode ring dye laser, as for the linear dye laser.

## 3.4 Frequency Doubling

Second harmonic generation (SHG) external to the dye laser cavity provides a tunable source of UV radiation with SHG powers in the microwatt range. However, by applying intracavity SHG much higher output powers can be achieved. As discussed in Sect. 3.3.2, the second beam waist of a ring cavity is a convenient location for a nonlinear crystal for efficient frequency doubling (Frölich et al. 1976). By using different nonlinear crystals like ammonium dihydrogen phosphate (ADP), lithium-formate monohydrate (LFM), potassium dihydrogen phosphate (KDP), beta barium borate (BBO) and $LiIO_3$ a wavelength range of about 240–400 nm is covered with frequency-doubled output powers of a few milliwatts. Thus, the tuning range of cw dye lasers is extended into the far UV spectral range. The shortest wavelength of 238 nm was obtained in an experiment by Bastow and Dunn (1980) where an LFM crystal was used. A drawback of the above-cited nonlinear crystals is that only angle tuning can be applied. This limits the frequency-doubled output powers to a few milliwatts (Johnston 1987).

An efficient way to produce high UV doubled output powers is to use temperature tuning instead of angle tuning. Couillaud et al. (1980) have reported intracavity frequency doubling of an 90° phase-matched ammonium dihydrogen arsenate (ASA) nonlinear crystal. They used a single mode cw ring dye laser to produce single-mode UV powers of more than 10 mW in a wavelength range from 292 to 302 nm, corresponding to temperatures from 20° to 100 °C. The 18-mm-long crystal had surfaces cut at Brewster's angle and was placed in the auxiliary beam waist. The crystal was enclosed in a protective plastic housing, and a continuous flow of dry nitrogen gas helped to protect the crystal surfaces. Although relatively high frequency doubled output powers could be achieved, some problems are created by using intracavity frequency doubling. The insertion of an additional optical element into the ring laser cavity leads to mode instabilities, especially when the laser is being tuned. Therefore, Bloomfield et al. (1982) have produced tunable cw UV radiation by frequency doubling the output of a cw ring dye laser inside a passive enhancement cavity (Fig. 3.9). The experimental conditions used with the ADA crystal were similar to the experimental conditions reported by Couillaud et al. (1980). The single-mode output power of the ring dye laser was around 1.4 W at 589 nm. The enhancement cavity had the same configuration and dimensions as the ring dye laser. The external cavity had a finesse of 20 and was locked on resonance by analyzing the reflected light from the input mirror (Couillaud and Hänsch 1980). A dispersion shaped error signal is generated from the difference current of two photodiodes and this voltage is integrated and amplified to become a unipolar high voltage which controls a PZT and, through this, the

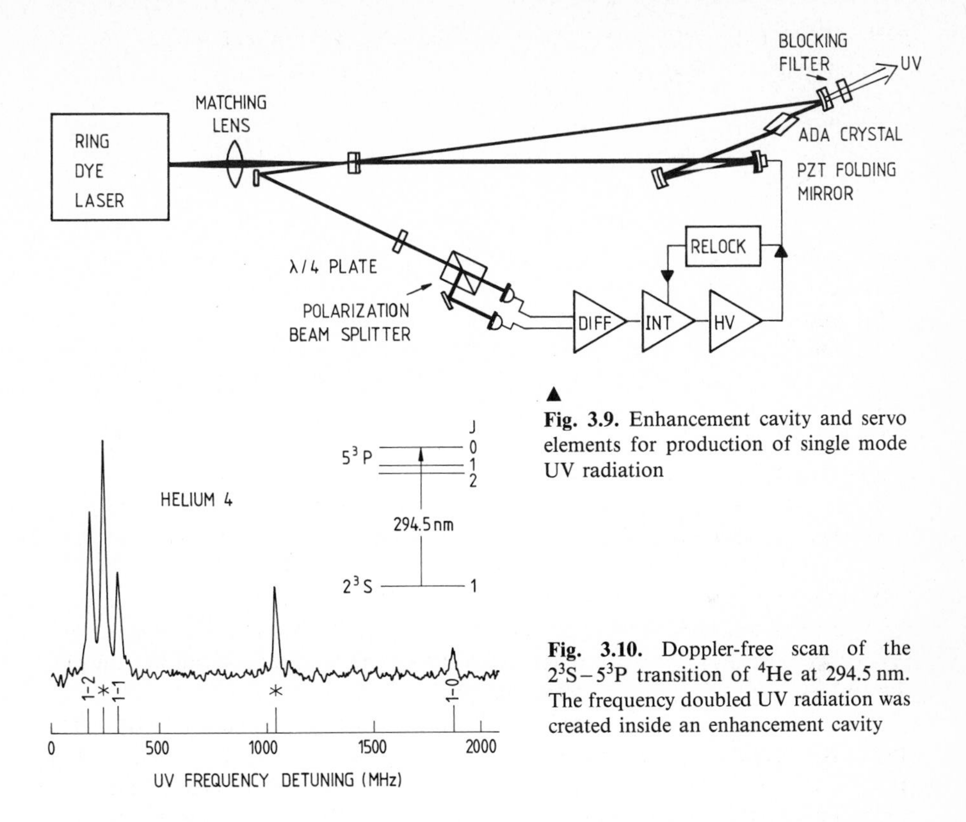


▲ **Fig. 3.9.** Enhancement cavity and servo elements for production of single mode UV radiation

**Fig. 3.10.** Doppler-free scan of the $2^3S-5^3P$ transition of $^4He$ at 294.5 nm. The frequency doubled UV radiation was created inside an enhancement cavity

laser frequency. The peak UV power obtained with this arrangement is over 50 mW. Figure 3.10 shows a Doppler-free scan of the $2^3S-5^3P$ transition of $^4He$ at 294.5 mm obtained with the UV radiation created inside the enhancement cavity. The observed linewidth (FWHM) of about 25 MHz is due to pressure broadening. The laser linewidth used in this experiment was around 1 MHz due to active stabilization techniques.

# 4. Ultrashort Pulse Dye Lasers

Charles V. Shank and Erich P. Ippen

With 12 Figures

Mode-locked dye lasers have become important sources of ultrashort optical pulses. The variety of commerical equipment available today has made it possible for researchers in almost any field to study dynamical processes with resolution well under a picosecond. In this chapter we will review the progress made in pulse generation with dye lasers and the important considerations for understanding the pulse generation process.

A good starting point for this discussion is to appreciate why organic dyes have proved to be one of the most useful laser media for generating ultrashort optical pulses. The same broad spectral features of dye systems that give rise to tunability also provide the bandwidth for generating an optical pulse. The homogeneously broadened gain spectrum that permits efficient tunability is a consequence of the rapid thermalization of the vibrational and rotational manifolds of the ground and excited electronic states of dye molecules (Shank 1975). In fact, thermalization of the upper and lower laser levels is sufficiently rapid to permit the generation of optical pulses in the femtosecond time regime. Also, organic dyes span the frequency spectrum from the ultraviolet to the infrared and are readily optically pumped. With these virtues it is not surprising that organic dyes have been so widely used to generate and amplify optical pulses.

## 4.1 Progress in Ultrashort Pulse Generation with Dye Lasers

During the decade and a half since the first generation of ultrashort laser pulses, optical pulse generation techniques have continued to improve. In Fig. 4.1, the shortest reported optical pulsewidth is plotted against year. The subject really began in 1966 with the report of the first generation of optical pulses in the picosecond range (DeMaria et al. 1966). About two years later the importance of dye lasers as tools for generating short pulses was demonstrated with the first report of mode-locking the flashlamp-pumped dye laser (W. Schmidt and Schäfer 1968). In the early 1970s, continuously pumped dye lasers provided a new approach to generating even shorter pulses (Dienes et al. 1971a; Ippen et al. 1972). In 1974 the first optical pulses shorter than 1 ps (Shank and Ippen 1974) were generated with a passively mode-locked dye laser. With the introduction of the colliding pulse concept in 1981 (Fork et al. 1981), optical pulses in the femtosecond time regime became a reality. An optical

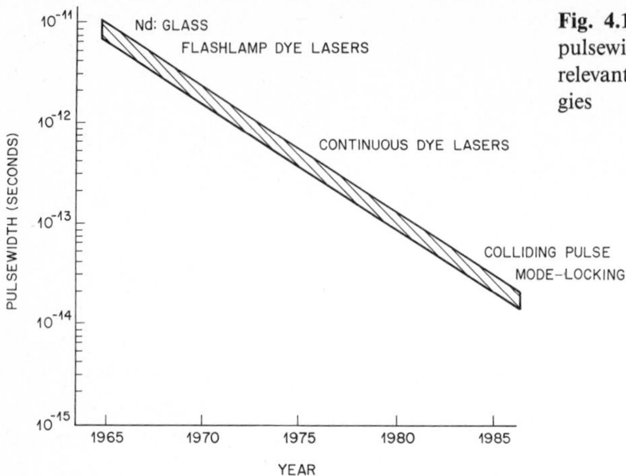


**Fig. 4.1.** Plot of minimum achieved pulsewidth as a function of year and relevant pulse generation technologies

pulse with a duration of 90 fs was initially reported and subsequently several laboratories around the world have been able to generate optical pulses in the range of 50 fs from this same laser by selecting the proper mirror coatings. In a parallel development, synchronous pumping (Soffer and Linn 1968) of the continuous dye laser first reported in 1972 (Shank and Ippen 1972) has continued to evolve as a useful technique for producing ultrashort pulses with commercially available equipment. These lasers are tunable and can be made to generate an optical pulse of duration less than 1 ps.

## 4.2 Mode-Locking

The process of mode-locking is central to the understanding of how ultrashort optical pulses are generated in a laser. In the most general sense a laser consists of an optical resonator formed by mirrors and a gain medium contained within the resonator. Although the gain medium and frequency-dependent loss in the optical resonator determine the center of the laser oscillating spectrum, the spacing between the laser's mirrors determines the precise spectral content of modes of oscillation. Unless some mode discriminating element is placed in the optical cavity, the laser output consists of a comb of resonator modes spaced in frequency by an amount $c/2L$ (Fig. 4.2). The electric field can be expressed as a sum of modes

$$E(\omega) = \sum_n \alpha_n \exp\left[(\omega_0 + n\delta\omega) + \phi_n\right] .$$

In general the mode phase is randomly fluctuating. The process of mode-locking fixes the phase relationship between the modes, resulting in a periodic train of pulses. For the condition when there is a single pulse in the optical cavity,

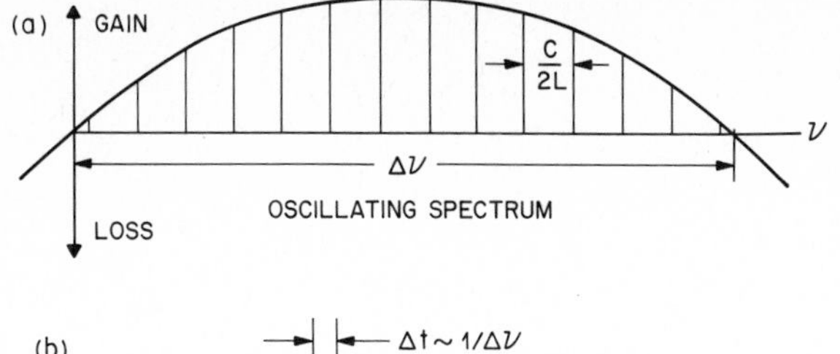


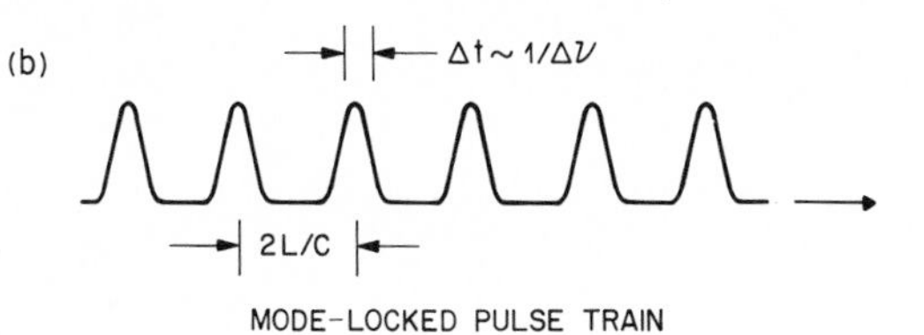


**Fig. 4.2.** (**a**) The spectral content of a laser operating in the fundamental transverse mode but in several longitudinal modes. (**b**) Temporal output of a laser when mode-locked

the period of the pulse train is $T_\mathrm{p} = 2L/c$ and the width of the pulse is inversely proportional to the width of the oscillating spectrum. For lasers based on sharp, precisely defined transitions, such as those in gas lasers and lasers based on rare earth transitions, the gain bandwidth limits the number of modes that oscillate. In a typical dye the gain bandwidth extends over nearly 100 nm, in which case if all the modes were locked in phase, the optical pulse would be approximately 10 fs. In practice it is difficult to achieve such a pulsewidth owing to cavity dispersion and frequency-dependent losses in the optical cavity.

## 4.3 Pulsewidth Determination

In general, the precise relationship between pulsewidth and frequency spectrum depends on the pulse shape. In Table 4.1 we have listed the product of FWHM pulsewidth, $\Delta\tau$, and the FWHM spectrum, $\Delta\nu$, for several pulse shapes.

Since optical pulsewidths are typically much shorter than the time resolution of electronic measuring systems, an optical technique is usually required to determine the optical pulsewidth. The most universal approach is to use a nonlinear optical technique to determine the pulse autocorrelation function. One method is to use second harmonic generation in a nonlinear crystal (Armstrong 1967) as diagrammed in Fig. 4.3. An optical pulse is split into two parts with one part delayed with respect to the other with a variable path delay. With the off-axis geometry shown in Fig. 4.3, the second harmonic is generated at an angle bisecting the angle between the two input beams. The generated second harmonic is proportional to $G^2(\tau)$, the second order autocorrelation function (Klauder et al. 1968; H. P. Weber 1968) given by

$$G^2(\tau) = \frac{\langle I(t)I(t+\tau)\rangle}{\langle I^2(t)\rangle} .$$

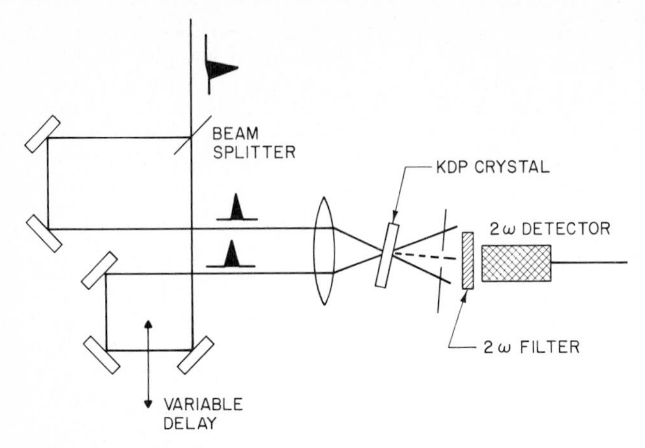


**Fig. 4.3.** Experimental arrangement for measuring the intensity autocorrelation function of a short optical pulse. The input pulse stream is split into two parts with a variable delay. Both pulses are focused into a KDP crystal and the second harmonic radiation generated as a function of variable delay is measured with a phototube

The time delay $\tau$ is related to the distance $z$ along the direction of propagation by $\tau = 2z/c$, where $c$ is the velocity of light. The half-width of the autocorrelation function can be related to the pulsewidth by making two important assumptions. First, the optical pulse must be assumed to be coherent, that is, the phases of the modes must be well defined. If not, a narrow spike can occur near $\tau = 0$ that corresponds to the width of the pulse spectrum and not the width of the intensity envelope. Such a pulse would have excess bandwidth and indicate that the pulse envelope is noisy and not smooth. Second, an optical pulse shape must be assumed in order to obtain the pulsewidth from the width of the autocorrelation function. For a Gaussian pulse shape the FWHM $\Delta\tau$ is related to the pulsewidth $\Delta t$ by $\Delta t = \Delta\tau/\sqrt{2}$. The appropriate factors for other pulse shapes are given in Table 4.1.

**Table 4.1.** Relationships between FWHM pulsewidth $\Delta\tau$, pulsewidth $\Delta t$, and frequency spectrum $\Delta\nu$ for various pulse shapes $I(t)$

| $I(t)$ | $\frac{\Delta\tau}{\Delta t}$ | $\Delta t \Delta\nu$ |
|---|---|---|
| $1\,(0 \leqslant t \leqslant \Delta t)$ | 1 | 0.886 |
| $\exp\left(-\frac{(4\ln 2)t^2}{\Delta t^2}\right)$ | $\sqrt{2}$ | 0.441 |
| $\mathrm{sech}^2\left(\frac{1.76t}{\Delta t}\right)$ | 1.55 | 0.315 |
| $\exp\left(-\frac{(\ln 2)t}{\delta t}\right)\ (t \geqslant 0)$ | 2 | 0.11 |

Optical pulse autocorrelation functions can also be measured using other nonlinear optical processes. For example, two-photon fluorescence in dye molecules (Giordmaine et al. 1967) has been successfully used to determine pulsewidths. Three-photon fluorescence (Rentzepis et al. 1970) can be used to determine the third-order autocorrelation functions and determine pulse asymmetry.

The ultimate limits of the above pulsewidth-determining techniques merit discussion as pulsewidths push down into the femtosecond time regime. The second harmonic generation technique has been successfully applied to measure optical pulses in the range of tens of femtoseconds. A number of possible sources of error should be considered when making measurements on this time scale. The first requirement is that the entire optical spectrum of the pulse must be up-converted by the nonlinear crystal. This condition can be assured by broadening the phase matching angle for the second harmonic at the focus of a lens. The second important point is to consider the influence of group velocity dispersion. Interestingly, the group velocity mismatch between the fundamental and second harmonic does not provide a limitation for the autocorrelation measurement. In this case the second harmonic is broadened in time, but the autocorrelation measurement is still accurate. More important, however, is group velocity dispersion within the fundamental pulse spectrum. An error can occur when two pulses that are separated in free space become overlapped inside the second-harmonic crystal due to group velocity dispersion. Since pulses that overlap in the crystal generate the second harmonic, this error tends to broaden the measured autocorrelation function. We can estimate the error due to dispersion for a crystal of length $l$ with the expression

$$\tau_d \approx l \left( \left.\frac{\partial k}{\partial \omega}\right|_{\omega^+} - \left.\frac{\partial k}{\partial \omega}\right|_{\omega^-} \right),$$

where $\omega^{\pm}$ are the highest and lowest frequency components of the pulse and $\tau_{\mathrm{d}}$ is the difference in propagation time through the crystal for frequency components $\omega^+$ and $\omega^-$. For KDP with $l = 0.2$ mm and a 30 fs pulse, $\tau_{\mathrm{d}} \sim 5$ fs. This error and errors due to lens aberrations cause the measured pulsewidth to be larger than the actual pulsewidth.

The off-axis autocorrelation method can also introduce errors simply due to geometric considerations. At the focus of a lens the two beams make an angle determined by the spacing between the beams and the focal length of the lens. This angle leads to an effective relative time smearing between the two beams because each component of the beam arrives at a time determined by its geometric path. This error is relatively unimportant in practice unless a resolution of better than a few femoseconds is required. In the case of such high resolution it is necessary to perform the autocorrelation with both beams collinear and the phase fronts interferometrically matched.

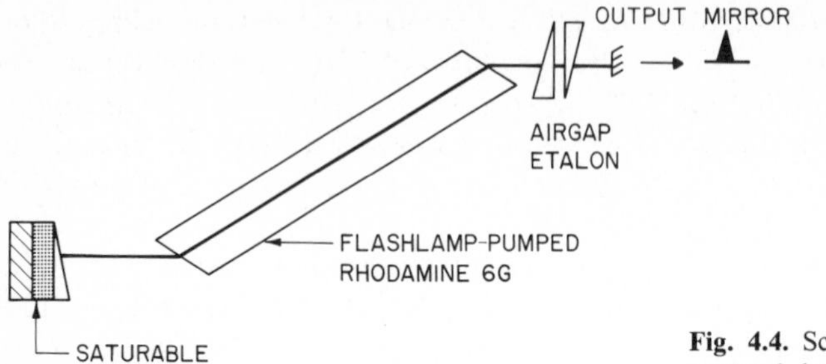


**Fig. 4.4.** Schematic of a flashlamp-pumped dye laser

## 4.4 Flashlamp-Pumped Dye Lasers

Historically, the first short pulse dye lasers were produced with flashlamp pumping. Passive mode-locking of the flashlamp-pumped organic dye laser was first reported by W. Schmidt and Schafer (1968). They observed mode-locking of a flashlamp-pumped rhodamine 6G dye laser using an organic dye as a saturable absorber. The emission was seen to consist of a train of equally spaced pulses with a $c/2L$ repetition frequency of 1 GHz but the pulsewidth determination in this early experiment was limited by the bandwidth of the photodiode and the oscilloscope (0.4 ns). The saturable absorber (DODCI, which is 3,3′-diethyloxadicarbocyanine iodide) is still used and remains one of the most effective mode-locking dyes for the rhodamine 6G dye laser. Bradley and O'Neill (1969) essentially reproduced these results with both rhodamine B and rhodamine 6G.

A diagram of a passively mode-locked dye laser that is pumped by a flashlamp is shown in Fig. 4.4. This is quite similar to the configurations reported by Bradley (Bradley and O'Neill 1969; Arthurs et al. 1972b). The gain dye cell is pumped with a flashlamp. The internal surfaces in the laser cavity are either set at Brewster's angle or wedged to prevent etalon resonances. The mode-locking saturable absorber dye is placed in contact with the mirror at the end of the cavity. This was found to be an optimum position for obtaining the minimum pulsewidth (Bradley et al. 1969). Pulsewidths measured for this system were about 6 ps. Tuning was achieved with an air gap etalon.

The wavelength range of passively mode-locked lasers is determined primarily by the availability of laser dyes and saturable absorbers. Arthurs et al. (1972b) have demonstrated tunable picosecond pulse generation over the range 584 nm to 704 nm. Various combinations of three laser dyes and four polymethine saturable absorbers were used to cover this range.

## 4.5 Continuously Mode-Locked Dye Lasers

The development of the cw dye laser (O. G. Peterson et al. 1970) has provided the impetus for a wide range of innovations in the development of ultrashort optical pulses. Two basic approaches have proved most successful. In the active approach some external source of modulation is applied to the laser in synchronism with the cavity round-trip time, while in the passive approach the nonlinear properties of a saturable gain and a saturable absorbing medium provide the pulse-forming mechanism.

## 4.6 Active Mode-Locking

The first attempts at mode-locking the cw dye laser utilized the insertion of either a loss modulator (Dienes et al. 1971 a) or a phase modulator (Kuizenga 1971) into the optical cavity. The frequency of the modulator was adjusted to equal the period of a round-trip time in the optical cavity. Optical pulses obtained with this approach were in the range of several tens of picoseconds. A more useful technique has been to insert the active modulator into the pump laser cavity and to use the resulting mode-locked pump laser to excite the dye laser cavity (Shank and Ippen 1972; Chan and Sari 1974). In this approach, termed synchronous pumping (Soffer and Linn 1968), the cavity is adjusted to match the pumping cavity length very accurately. Most of the commercially available mode-locked lasers employ synchronous pumping.

In Fig. 4.5, we show a diagram of a synchronously mode-locked dye laser. The dye laser is pumped with a mode-locked argon laser. Jain and Heritage (1968) have shown that the optical cavities of the dye laser and the mode-locked pump laser must be matched to within a few micrometers to obtain the minimum pulsewidth. Long-term stability of this system depends on careful mechanical and thermal design of the optical cavity. Lasers of this design have the advantage of being tunable over hundreds of angstroms while producing optical pulses of a few picoseconds or less.

The continuously mode-locked and frequency-doubled Nd: YAG laser has also proved to be a useful source for synchronously pumping dye lasers

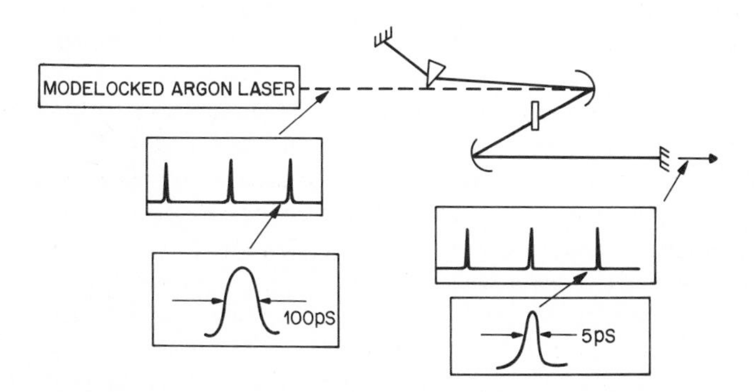


**Fig. 4.5.** Diagram of a synchronously pumped mode-locked dye laser. The pumping pulse is 100 ps and dye laser pulse is somewhat narrowed to 5 ps

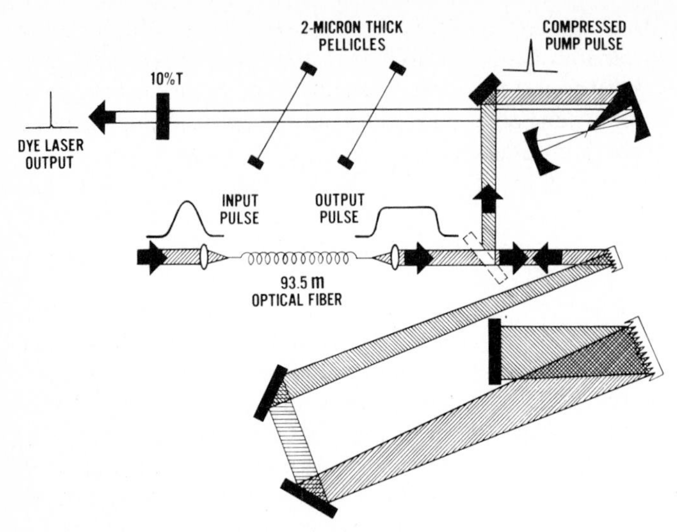


**Fig. 4.6.** Diagram of a short pulse dye laser synchronously pumped with compressed optical pulses

(Mourou and Sizer 1982). The frequency-doubled Nd: YAG produces pulses on the order of 40 ps with a power of several hundred watts at 532 nm. Using harmonic mode-locking, A. M. Johnson and Simpson (1985) have reduced the pulsewidth to 33 ps. This source has been used to synchronously pump a rhodamine 6G dye laser to produce optical pulses of less than one picosecond. Recently A. M. Johnson and Simpson (1985) have compressed the output of a passively mode-locked and frequency-doubled Nd: YAG laser and used these pulses to pump a dye laser. The experimental arrangement is shown in Fig. 4.6. Optical pulses as short as 210 fs have been obtained that are tunable over several hundred angstroms.

## 4.7 Passive Mode-Locking

Passive mode-locking is achieved with the insertion of a nonlinear absorbing medium inside the optical cavity of the laser. Typically, the nonlinear absorbing medium is an organic dye that has an absorption resonant with the dye laser emission wavelength. The first mode-locking of a dye laser was reported by W. Schmidt and Schäfer (1968), who observed mode-locking of a flash-lamp-pumped rhodamine 6G dye laser with the dye DODCI. This system was later optimized by Arthurs et al. (1972b) to produce optical pulses in the picosecond range.

The first report of passive mode-locking the continuous dye laser was made by Ippen et al. (1972). The optical cavity was a five-mirror cavity containing two points of sharp focus. At one point was a flowing cell containing rhodamine 6G, which was pumped with a cw argon laser, and at the second focal point was the saturable absorber DODCI. The pulsewidth was measured to be 1.5 ps. Shortly after this report, similar results were reported by O'Neill (1972), who measured a 4 ps pulse with a streak camera. Further improvements led to the report of the first optical pulses less than a picosecond by Shank and Ippen (1974). Letouzey and Sari (1973) replaced the dye cells with free flowing streams, adding to the simplicity of the optical design. Frequency sweep in the laser pulse was observed by Ippen and Shank (1975), who used a grating pair to compress the optical pulse to 0.3 ps.

In 1981 the report of the colliding pulse dye laser (Fork et al. 1981) opened the way to generating optical pulses of less than 100 fs. Following the initial report of 90 fs optical pulses, several laboratories reported optical pulses in the range 50–70 fs. Recently Valdmanis et al. (1985) have reported improvements in this laser design that have led to the production of pulses as short as 27 fs.

To understand the concept of the colliding pulse mode-locked dye laser, it is useful to discuss the mechanism for pulse formation in the passively mode-locked dye laser. The key concepts for understanding pulse generation in this laser were described by New and Rea (1976), who established the stability criteria for pulse generation. They defined a stability parameter $S$ given by $S = A_g \sigma_a / A_a \sigma_g$, where $\sigma_g$ and $\sigma_a$ are the gain and dye absorber cross sections. The quantities $A_g$ and $A_a$ are the mode cross-sectional areas in the gain and absorber media. Physically, the stability parameter measures the relative effectiveness of a photon in the laser cavity to saturate the absorber medium compared to saturating the gain medium. A stability parameter greater than 1 implies that loss occurs on both the leading and the trailing edges of the optical pulse with gain at the peak of the pulse. This condition is necessary for the formation of a short pulse in a passively mode-locked dye laser. To assure pulse-forming stability, in designing a passively mode-locked dye laser cavity the radii of curvature of mirrors focusing into the absorber medium are chosen to be smaller than the radii of the mirrors focusing into the gain medium. This assures that the spot size for the laser mode is smaller in the absorber medium than in the gain medium.

The colliding pulse passively mode-locked laser scheme provides an improvement in short pulse formation over previous configurations. The central idea is to utilize the interaction, or "collision" of two pulses in the optical cavity to enhance the effectiveness of the saturable absorber. Figure 4.7 illustrates this process in a general way for two, three, and four optical pulses in the laser cavity. With two optical pulses in a simple linear optical cavity, the saturable absorber must be placed precisely in the center of the cavity so that the two oppositely directed pulses will be able to interact in the saturable absorber at the same time. Since both pulses are coherent, they interfere with each other, creating a standing wave. At the nodes of the wave the intensity is greatest, more completely saturating the absorber and minimizing loss. At the antinodes of the field the absorber is unsaturated, but then of course the electric field is a minimum, again minimizing the loss. The net effect of using the standing wave field to saturate the absorber rather than the fields of the two

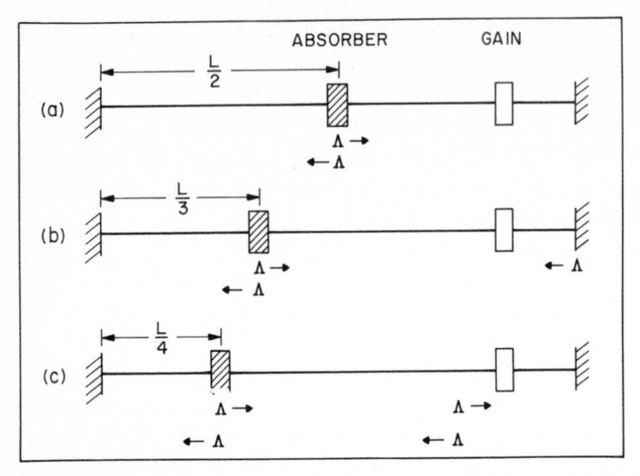


**Fig. 4.7.** The colliding pulse mode-locking configuration for **(a)** two, **(b)** three and **(c)** four optical pulses in the cavity

pulses separately is a reduction of the energy required to saturate the absorber by a factor of approximately 1.5. Since the gain medium is being pumped continuously, each pulse reaches the gain medium at a point in time when the gain medium is fully recovered. The net effect is that when the two pulses meet in the saturable absorber there is two times more energy to saturate the absorber than when there is only one pulse in the optical cavity. Thus, the effective saturation parameter is increased by approximately a factor of three over that for a conventional passively mode-locked dye laser (Stix and Ippen 1983).

The linear configuration shown in Fig. 4.7 has been demonstrated experimentally (Fork et al. 1981) with excellent results, but is difficult to align. The saturable absorber stream must be placed at precisely an integral submultiple of the cavity length for a pulse collision to take place. The precision required is on the order of 10 μm. Figure 4.8 shows an improved cavity configuration that assures proper timing for a pulse collision in the optical cavity

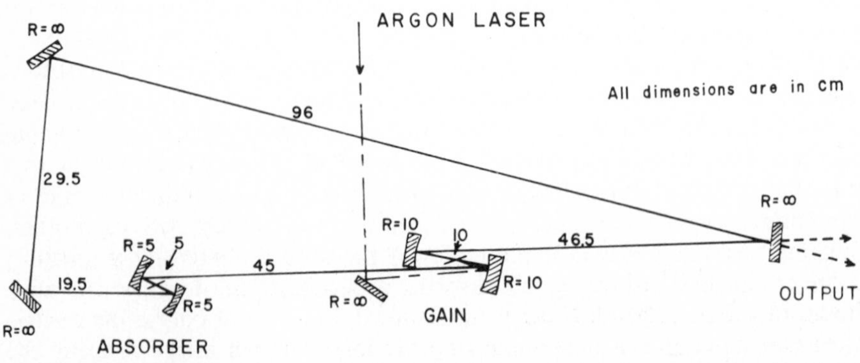


**Fig. 4.8.** Cavity configuration for a colliding pulse mode-locked laser pumped by an argon laser

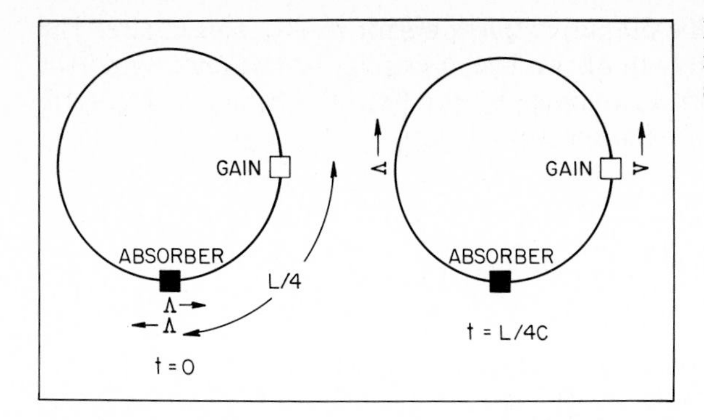


**Fig. 4.9.** The proper $L/4$ spacing between the gain and absorber media

without precise alignment. In this ring configuration the pulse collision occurs between oppositely directed pulses traveling around the cavity. To minimize the energy loss to the absorber, the pulses meet in the saturable absorber. In effect, the ring cavity provides automatic pulse synchronization and removes the requirement for the precise positioning of the absorber stream.

The stability of the laser operation can be enhanced by properly positioning the absorber and gain jets in the ring cavity. Figure 4.9 illustrates the advantage achieved by placing the absorber and gain medium at approximately one-fourth the round trip in the ring cavity. Since the two oppositely directed pulses meet in the absorber, they will draw power from the gain medium with the same time delay, corresponding to a one-half cavity round-trip time. In this manner both pulses see the same gain, since the continuously pumped dye has the same time to recover following each pulse. This mechanism stabilizes the laser train by reducing the formation of extra pulses in the cavity. The cavity configuration (Fork et al. 1981) shown in Fig. 4.8 for a colliding pulse mode-locked dye laser contains two pairs of mirrors, which create two points of focus. At the first point of focus a continuous argon laser pumps the gain medium, rhodamine 6G, in an ethylene glycol flow, and at the other point, the cavity radiation is focused into the absorber DODCI, also flowing in an

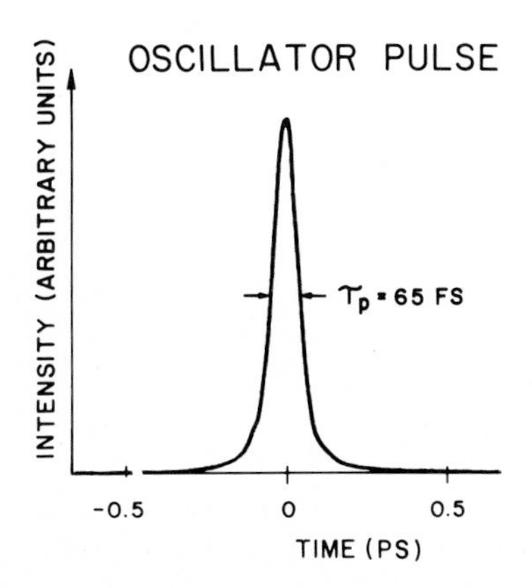


**Fig. 4.10.** Measured autocorrelation of a 65 fs optical pulse from a colliding pulse mode-locked dye laser

ethylene glycol stream. The output coupling mirror is 4% transmissive. The output power is 50 mW in each of two beams exiting the transmitting mirror with 5 W pump power from an argon laser operating at 514.5 nm. In Fig. 4.10, the measured pulse autocorrelation is plotted.

## 4.8 Amplification of Ultrashort Pulses

The high optical gain achievable with optically excited organic dyes immediately suggests their use as amplifiers. The large emission cross section of organic dyes allows the achievement of gains in excess of $10-100\,\mathrm{cm}^{-1}$ with modest pumping powers. This very high gain leads to a number of problems fundamental to dye laser amplifier design. Amplified noise or amplified spontaneous emission can rapidly deplete the gain of a dye amplifier. Further, the energy storage time is limited by the short spontaneous lifetime in organic dyes, which is typically in the nanosecond range.

The rate equation describing a pulse, $I(x,t)$, traveling through an amplifying medium with a population inversion $n(x,t)$ is described by

$$\frac{dn(x,t)}{dt}=\frac{[n_0-n(x,t)]}{T_1}-\sigma n(x,t)\mathrm{I}(x,t) \quad \text{and}$$

$$\frac{dI}{dx}=\sigma n(x,t)I(x,t) \ .$$

For an ultrashort pulse with pulsewidth $t_p \ll T$, the population inversion is given by

$$n(x,t)=n_1 \exp\left(-\int_{-\infty}^{t}\sigma I(x,t')dt'\right) .$$

From the above we see that the saturation energy density is given by

$$E_\mathrm{s}=1/\sigma \ .$$

If the input pulse energy is greater than $E_\mathrm{s}$, the stored energy will be swept out by the pulse and the amplification becomes nonlinear. Nonlinear amplification can give rise to distortion of the pulse by preferentially amplifying the leading edge of the pulse (Migus et al. 1982). This effect can be controlled by passing the pulse through a saturable absorber.

The storage time for a dye is determined by $T_1$ and is typically about 5 ns. This means that to efficiently pump the dye amplifier a pulse of this width or less is required. A laser source having a large energy storage capability is advantageous, although dye amplifiers have been flashlamp pumped (Bradley 1974). The high emission cross sections of amplifiers can lead to serious losses

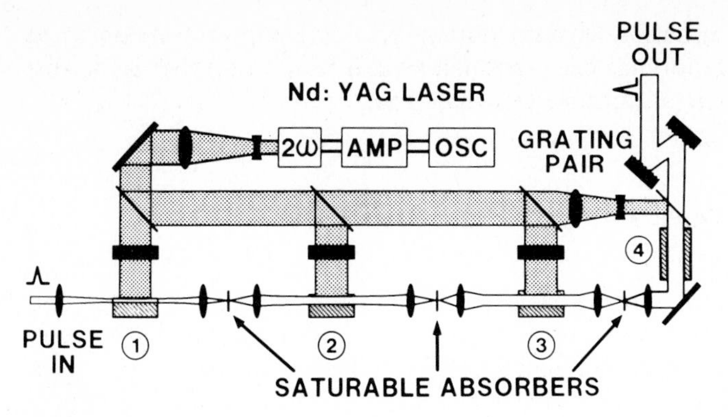


**Fig. 4.11.** Four-state pulsed dye amplifier pumped by a frequency-doubled Nd: YAG laser

through amplified spontaneous emission. For high gain amplifiers saturable absorbers are often required to isolate the stages.

A diagram of a Nd: YAG-pumped dye laser amplifier is shown in Fig. 4.11. The amplifier consists of four stages pumped by a frequency-doubled Nd: YAG laser and operates at a 10 Hz repetition rate. Each stage is isolated with a saturable absorber. At the output, a grating pair is used to compensate group velocity dispersion that arises when the pulse passes through the amplifying optics. This amplifier is capable of producing gigawatt optical pulses with a pulse duration of 70 fs (Fork et al. 1982).

## 4.9 Ultrashort Pulse Distributed Feedback Dye Lasers

Distributed feedback dye lasers (Kogelnik and Shank 1971) are distinguished from conventional lasers in that they do not use an optical cavity formed by mirrors. Instead, feedback is supplied by Bragg scattering from periodic modulation of either the refractive index or gain (Shank et al. 1971). The condition for oscillation is

$$\lambda_0 = 2\Lambda n \ ,$$

where $n$ is the average refractive index of the medium and $\Lambda$ is the period of the modulation. The laser is tunable by varying either the period of the modulation or refractive index.

Bor (1980) has shown that an optically pumped distributed feedback laser can be a useful source of tunable short optical pulses. The primary mechanism for pulse shortening is a type of relaxation oscillation or self-Q-switching. For a conventional laser the cavity lifetime is a constant determined by the cavity mirror reflectivities. For a distributed feedback laser pumped by two interfering pulses the cavity lifetime is a function of pumping intensity and time, which favors the formation of very short pulses. An improved version using traveling wave excitation has been reported (Szabo et al. 1984) that is capable of generating an optical pulse of less than 1 ps.

## 4.10 Traveling Wave Pumped Pulse Dye Laser

An ingenious pumping scheme has been developed for generating picosecond optical pulses using traveling wave excitation in a transverse pumping geometry (Bor et al. 1983; Szatmári and Schäfer 1984; Polland et al. 1983). This technique has a number of interesting advantages over conventional pumping methods when the pump source is a short optical pulse.

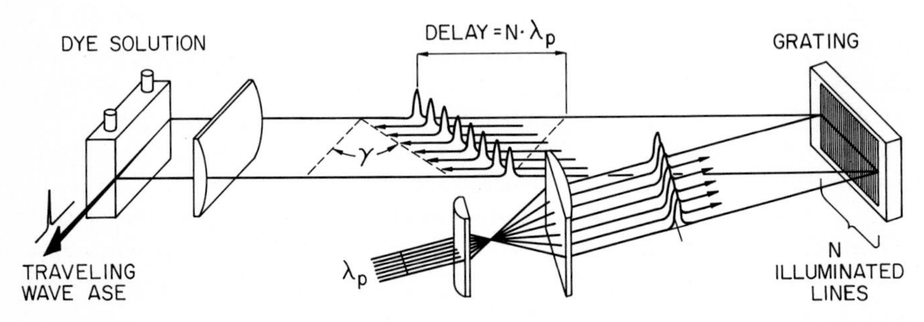


**Fig. 4.12.** Arrangement for generating picosecond optical pulses using traveling wave excitation

The experimental setup is shown in Fig. 4.12. A short pulse optical pumping beam is imaged onto a grating. For each pitch of the grating, an optical delay corresponding to the pump wavelength is introduced. In this way, a continuous spatial delay is created across the diffracted beam (G. Szabo et al. 1983; Wyatt and Marinero 1981). With proper adjustment of the grating angle, the transverse excitation pumping beam can be synchronized with the amplified spontaneous emission in the dye cell. With picosecond pulse excitation, the duration of the amplified spontaneous emission pulse was found to be typically 3 – 10 times shorter in duration than the pumping pulse (Bor et al. 1983).

Using a slight modification of this approach, a tunable highly monochromatic picosecond light source has been demonstrated (Szatmári and Schäfer 1984). In this case, two dye cells were used. The first cell was pumped using the traveling wave excitation described above. The broadband amplified spontaneous emission short pulse was then passed through a tunable grating filter to select a narrow frequency band. The filtered pulse was then amplified with a second traveling wave pumped cell. Picosecond optical pulses were obtained that were tunable across the entire gain bandwidth of the dye as the grating was rotated.

The traveling wave pumping scheme has also proved to be ideal for efficiently pumping dyes with low quantum efficiencies (Polland et al. 1983). This has proved to be especially important for dyes in the infrared, which typically have quantum efficiencies on the order of $10^{-4}$ and lifetimes between 5 and 12 ps. Using traveling wave picosecond excitation single picosecond optical pulses covering the range 1.4–1.8 μm were obtained.

## 4.11 Conclusion

Short pulse dye lasers have rapidly matured to highly developed devices that are commercially available and are useful for investigations in a broad range of disciplines. We expect the future to bring even further refinements and short pulse dye laser techniques to establish an important place for themselves in the pursuit of understanding ultrafast events.

# 5. Structure and Properties of Laser Dyes

Karl H. Drexhage

With 6 Figures

An essential constituent of any laser is the amplifying medium which, in the context of this book, is a solution of an organic dye. Since the beginning, the development of the dye laser has been closely tied in with the discovery of new and better laser dyes. The phthalocyanine solution employed for the original dye laser (Sorokin and Lankard 1966) is hardly used today, but the compound rhodamine 6G, found soon afterwards (Sorokin et al. 1967), is probably the most widely employed laser dye at the present time. In the years following the discovery of the dye laser various other compounds were reported for this purpose. Almost all were found by screening commercially available chemicals, but this source of new laser dyes soon became exhausted. It was reported in 1969 that a survey of approximately one thousand commercial dyes showed only four to be useful (Gregg and Thomas 1969), and three of these belonged to classes of laser dyes that were already well known. Considering the large number of available chemicals, it is perhaps surprising that so few good laser dyes have been found so far. The reason for this is that some very special requirements must be met by such dyes and this excludes the majority of organic compounds. This chapter is intended to give the reader some insight into the relations between molecular structure and the lasing properties of organic dyes, relations which have been applied in the planned synthesis of new laser dyes. In addition, the physicochemical properties of the most important laser dyes are discussed here in some detail.

## 5.1 Physical Properties of Laser Dyes

Organic dyes are characterized by a strong absorption band in the visible region of the electromagnetic spectrum. Such a property is found only in organic compounds which contain an extended system of conjugated bonds, i.e. alternating single and double bonds. Whereas the light absorption of dyes cannot be derived *rigorously* from their molecular structure owing to the complexity of the quantum-mechanical problem, more or less simple models have been found that are capable of explaining many experimental observations, at least within a given class of dyes. By adjusting the model parameters on the basis of empirical data, it is possible to predict the absorption properties of yet unknown dyes. For a more detailed discussion, the reader is referred to Chap. 1. The nonradiative processes in dye molecules (which are of the utmost

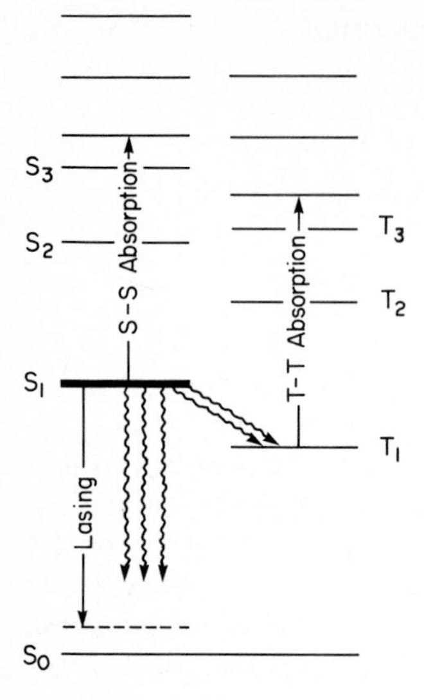


**Fig. 5.1.** Energy level diagram of an organic dye molecule

importance in dye lasers, as discussed below) are even less amenable than light absorption to an accurate quantum-mechanical treatment.

The long-wavelength absorption band of dyes is attributed to the transition from the electronic ground state $S_0$ to the first excited singlet state $S_1$. The transition moment for this process is typically very large, thus giving rise to an absorption band with an oscillator strength on the order of unity. The reverse process $S_1 \rightarrow S_0$ is responsible for the spontaneous emission known as fluorescence and for the stimulated emission in dye lasers. Because of the large transition moment the rate of spontaneous emission is rather high (radiative lifetime on the order of nanoseconds) and the gain of a dye laser may exceed that of solid-state lasers by several orders of magnitude.

When the dye laser is pumped with an intense light source (flashlamp or laser), the dye molecules are excited typically to some higher level in the singlet manifold, from which they relax within picoseconds to the lowest vibronic level of $S_1$, i.e., the upper lasing level (Fig. 5.1). For optimal lasing efficiency it would be desirable for the dye molecules to remain in this level until they are called on for stimulated emission. However, there are many nonradiative processes that can compete effectively with the light emission and thus reduce the fluorescence efficiency to a degree that depends in a complicated fashion on the molecular structure of the dye. These nonradiative processes can be grouped into those that cause a direct relaxation to the ground state $S_0$ (internal conversion) and those that are responsible for intersystem crossing to the triplet manifold. Because of the relatively long lifetime of the triplet molecules (in the order of microseconds) the dye accumulates during the pumping process in the triplet state $T_1$, which often has considerable absorption for the laser light. Thus not only are some of the dye molecules taken away from the lasing process but, owing to the triplet-triplet absorption, they cause an additional loss in the laser. The nonradiative decay to the ground state is comparatively less harmful to laser action. Nevertheless, in the ideal laser dye both processes should be negligible, so that the quantum yield of fluorescence has the highest possible value, 100%. In addition to these general requirements, an efficient laser dye in its first excited singlet state should have negligible absorption at the wavelength of the pump light and of the laser emission as well. Otherwise losses would occur, as in triplet-triplet absorption, because the decay to the first excited singlet or triplet level is nonradiative.

In addition to the aforementioned optical properties, the laser dye should have an absorption spectrum which matches the spectral distribution of the pump source. Since a substantial part of the light energy emitted by flashlamps is in the ultraviolet region, only dyes with moderate to strong absorption throughout this region can take full advantage of the pump light. If, on the other hand, the pump source is a laser with monochromatic emission, the dye should have a strong absorption at this wavelength. Although, in principle, a small absorption coefficient can be compensated for with a higher concentration, this is often not desirable because it also increases the absorption at the lasing wavelength, thus enhancing the cavity losses. In order to achieve a broad tuning range, dyes are required that have an unusually wide fluorescence band, or mixtures of dyes that absorb at the same wavelength but fluoresce with different Stokes shifts. Fluorescent dyes that react in the excited state to form a fluorescent product that is unstable in the ground state can be considered ideal for this purpose (see Sect. 5.5.5). Here the concentration may be adjusted so that the gain in the regions of fluorescence of the original dye and its reaction product are approximately equal. Following the emission of light, the reaction product dissociates immediately and no additional absorption interfering with the lasing process is encountered, such as is likely to be the case in a mixed solution of several dyes.

Inevitably during the lasing process a certain amount of thermal energy is released, giving rise to temperature gradients in the solution that may cause optical inhomogeneities. In this respect, the best media appear to be water and its deuterated relative, heavy water. Therefore the ideal laser dye should be soluble in water and still maintain its lasing properties, which generally means that it must not form dimers at the lasing concentrations or that it responds readily to disaggregating agents (see Sect. 5.5.4).

Finally, we should like to mention photochemical stability as a property relevant to laser dyes. Whereas this is of lesser importance in dye lasers employing liquid solutions, the poor stability of many laser dyes is a serious problem in lasers where the dye is incorporated in a solid matrix and therefore cannot be circulated.

## 5.2 Internal Conversion $S_1 \rightarrow S_0$

The nonradiative decay of the lowest excited singlet state $S_1$ directly to the ground state $S_0$ is mostly responsible for the loss of fluorescence efficiency in organic dyes. Depending on the molecular structure of the dye and the properties of the solvent, the rate of the relaxation can vary by many orders of magnitude. Since there are several different structural features that contribute to the nonradiative decay $S_1 \to S_0$, the relation between molecular structure and fluorescence efficiency can be puzzling. For a general survey the reader is referred to [Förster 1951 (p. 94); Wehry 1967; Parker 1968 (p. 428); Birks 1970 (p. 142)]. In this section we dwell on some processes that are especially relevant to laser dyes.

### 5.2.1 Structural Mobility

It has been known for a long time that a rigid, planar molecular structure favors high fluorescence efficiency, as may be illustrated by the dyes phenolphthalein and fluorescein [Förster 1951 (p. 109)]. Whereas the former is practically nonfluorescent[1] in alkaline solution, the introduction of the oxygen bridge causes the fluorescence quantum yield to assume the value of 90% in fluorescein. A similarly strong effect has been observed in a number of other cases. In the laser dye rhodamine B the fluorescence efficiency in ethanol is dependent on the temperature (Huth et al. 1969). About 40% at 25 °C, the quantum yield increases to nearly 100% if the temperature is lowered, but it drops to only a few percent in boiling ethanol. These effects can be attributed to the mobility of the diethylamino groups, which is enhanced with increasing temperature. If these groups are rigized, as in rhodamine 101, the quantum yield of fluorescence is found to be virtually 100%, independent of the temperature (Drexhage 1972a, 1973a; Drexhage and Reynolds 1972). Still another structural mobility that increases the rate of internal conversion is illustrated by the dyes rosamine 4 and rhodamine 110. In the latter, as in other rhodamine dyes, the carboxyphenyl substituent is held in a position nearly perpendicular to the xanthene chromophore by the bulky carboxyl group, thus allowing little, if any, mobility. In the case of rosamine 4, however, the phenyl substituent can pivot to a certain degree, thus reducing the fluorescence efficiency in ethanol at 25 °C from 85% in rhodamine 110 to 60% (Drexhage 1972a, 1973b). The fluorescence efficiency of rosamine 4 is also temperature-dependent, but to a lesser degree than in rhodamine B.

One must not conclude from the foregoing discussion that any dye with a completely rigid structure will have a very high fluorescence efficiency. The dye may, in fact, be nonfluorescent if one of the quenching processes to be discussed below is operative. On the other hand, it is also possible for a dye with a nonrigid structure to be highly fluorescent. This is the case, for instance, with rhodamine 110 and rhodamine 6G which in alcoholic solution have a fluores-

[1] Since there is always a finite probability of light emission by a molecule in an excited state, a compound can never be truly nonfluorescent. Even with very weakly fluorescing compounds the fluorescence can be observed when extremely intense light sources are used for excitation. Likewise, a quantum efficiency of 100% cannot be realized because the rate of nonradiative processes cannot be zero.

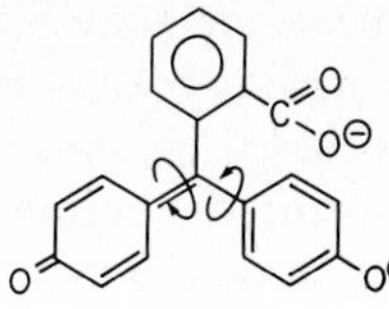

phenolphthalein

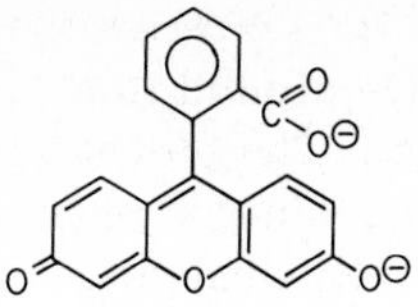

fluorescein

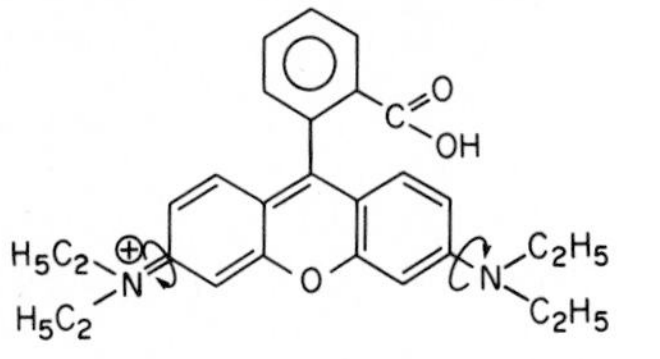

rhodamine B

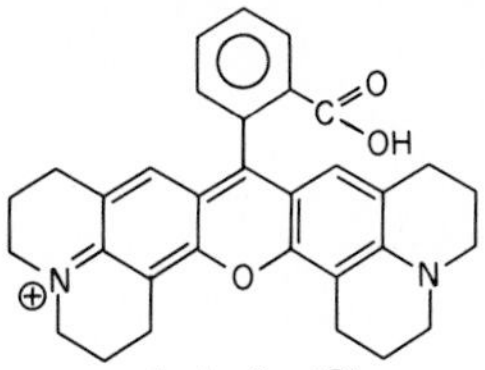

rhodamine 101

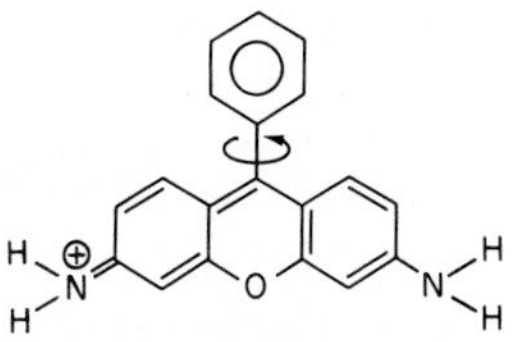

rosamine 4

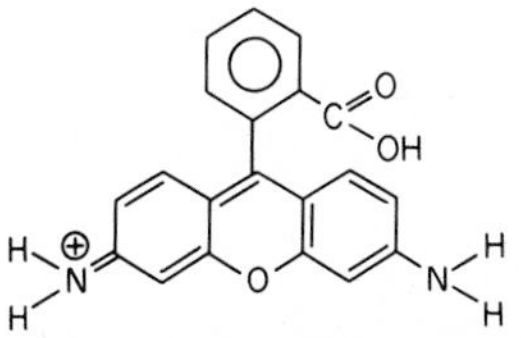

rhodamine 110

cence quantum yield of 85% and 95%, respectively, independent of temperature, although their amino groups are potentially mobile as in the case of rhodamine B. The reason for this different behavior lies in the excited-state $\pi$-electron density within the C=N bond, which should be dependent on the number of electron-donating alkyl substituents (Drexhage 1972a, 1973a). If the $\pi$-electron density is high, the thermal energy of the solvent molecules is not sufficient to twist the amino groups out of planarity (rhodamine 110, rhodamine 6G). It is interesting that rhodamine B has a quantum efficiency of nearly 100% in viscous solvents like glycerol. This suggests that the chromophore is fully rigid in the ground state and loosens up only after excitation, provided the solvent is of low viscosity. In glycerol the viscosity is sufficiently high to prevent thermal equilibrium being reached during the radiative lifetime of a few nanoseconds. Hence the planarity of the ground state is not lost before light emission takes place. The fluorescence efficiency of rhodamine B and related dyes with dialkylamino end groups depends also in a peculiar way on solvent properties other than viscosity, and is moreover sensitive to other influences, as will be discussed in Sect. 5.7.

### 5.2.2 Hydrogen Vibrations

A pathway of internal conversion, which may be distinguished from the processes discussed above, can occur in certain dyes, even if their chromophore is fully rigid and planar. In contrast to other relaxation processes, this one is, in first approximation, independent of temperature and solvent viscosity. It involves the conversion of the lowest vibronic level of the excited state $S_1$ to a higher vibronic level of the ground state $S_0$, which then rapidly relaxes to the lowest vibronic level of $S_0$. The probability for this process is inversely proportional to the change in vibronic quantum number during the conversion. Because of the comparatively small mass of the hydrogen atom, the quanta of hydrogen stretching vibrations have the highest energies in organic dyes, and thus hydrogen vibrations are very likely to contribute to the mechanism considered here.

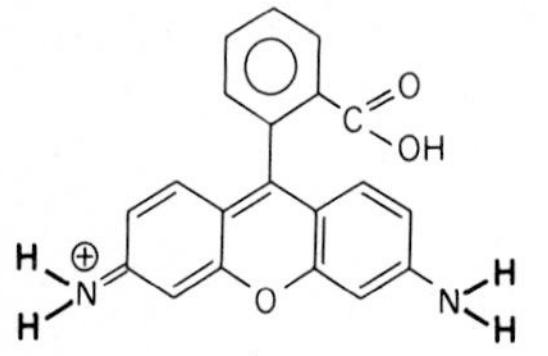

rhodamine 110 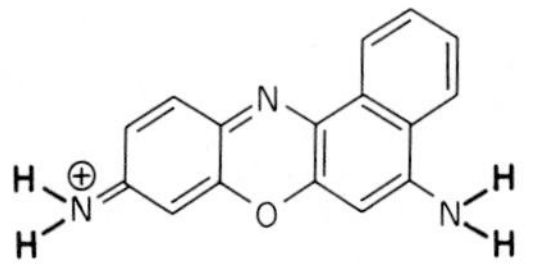 cresyl violet

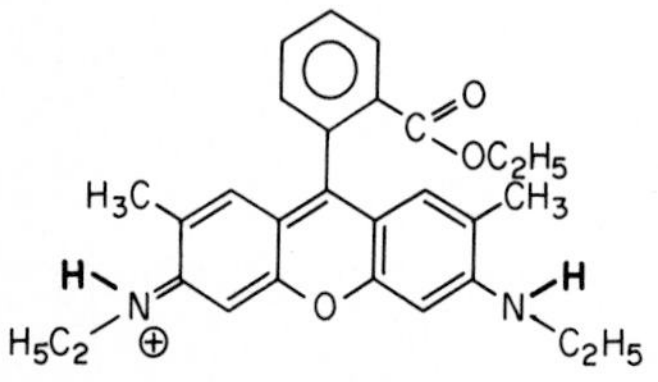

rhodamine 6G 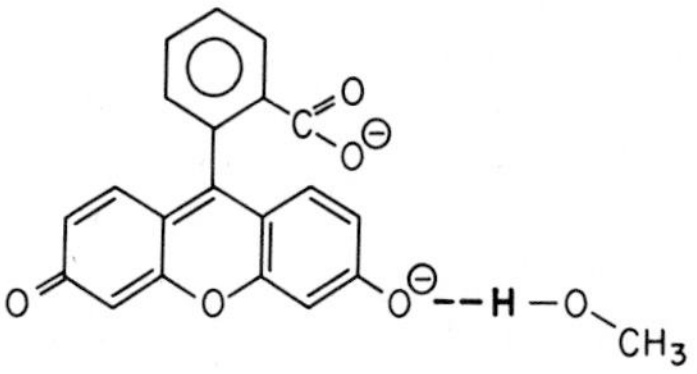 fluorescein

It can be expected that only those hydrogen atoms that are directly attached to the chromophore of the dye will influence the nonradiative process $S_1 \rightarrow S_0$. Furthermore, this mechanism should become increasingly effective with decreasing energy difference between $S_1$ and $S_0$. On the other hand, a replacement of hydrogen by deuterium should reduce the rate of nonradiative decay by this mechanism, and thus increase the fluorescence efficiency. This was indeed observed, for instance, in the dyes rhodamine 110 and Cresyl Violet (Drexhage 1972a, 1973a). When these dyes were dissolved in monodeuterated methanol ($CH_3OD$), their quantum yields of fluorescence increased from 85% to 92% and from 70% to 90%, respectively, compared with the yields when normal methanol was the solvent. Since in $CH_3OD$ solution the protons of the amino groups are readily exchanged with deuterons from the solvent, which is in large excess, the dye molecules exist here with deuterated amino groups. Among other xanthene dyes, the fluorescence quantum yield of fluorescein in alkaline alcoholic solution is 10% below and that of rhodamine 6G in alcohol 5% below the maximum value of 100%. As with rhodamine 110, these effects can be ascribed for the most part to hydrogen vibrations in the end groups of the chromophores, involving solvent molecules in the case of fluorescein. This mechanism, though of little importance in dyes that fluoresce in the visible range, can be expected to seriously reduce the fluorescence efficiency of infrared dyes.

### 5.2.3 Other Intramolecular Quenching Processes

In addition to the mechanisms of internal conversion discussed already, there are several other intramolecular processes that may cause quenching of fluorescence. For instance, if a part of the dye molecule is strongly electron-donating or withdrawing, a reversible charge transfer may occur between this group and the excited chromophore, resulting in the loss of electronic excitation. Thus when a nitro group is introduced into the carboxyphenyl substituent of a rhodamine dye, a nonfluorescent derivative is obtained. Likewise, a substituent with a low-lying singlet or triplet state may quench the fluorescence via energy transfer. Furthermore, it is possible that under certain circumstances the intersystem crossing process $S_1 \rightarrow T_1$ effectively drains the state $S_1$ before the emission of fluorescence. Apart from the processes discussed in Sect. 5.3, this is observed with molecules where a $n-\pi^*$ transition occurs at lower energy than the $\pi-\pi^*$ transition of the main absorption band. Since these processes are at present of little importance with regard to laser dyes, they are not treated here in detail.

### 5.2.4 Determination of Fluorescence Yields

The fact that a great variety of nonradiative relaxation processes may compete with the emission of fluorescence can make it very difficult to elucidate the relations between molecular structure and the fluorescence quantum yield of organic dyes. The problem is aggravated by the many pitfalls encountered in the measurement of quantum yields. It has proven to be immensely difficult to determine an absolute quantum yield with an accuracy of only $\pm 5\%$. Most authors have overestimated the accuracy of their methods, even after careful evaluation of potential errors. The review by Demas and Crosby (1971) of the measurement of fluorescence quantum yields gives an excellent account of the problems.

Easier than an absolute determination is the measurement of fluorescence efficiency relative to that of a standard solution. For dyes that fluoresce in the center of the visible region, one may use an alkaline fluorescein solution whose fluorescence quantum yield has been determined carefully with several independent methods and is generally assumed now to have the value $90 \pm 5\%$. The quantum yield values given by me are based on this figure. In an extensive investigation of the fluorescence of xanthene dyes I did not observe any quantum yields apparently higher than 100% when the value for fluorescein was taken as 90%, although the ceiling of 100% was reached by some derivatives, e.g. rhodamine 101. I believe that this lends further support to the accuracy of the value 90% for fluorescein. Whereas the quantum yield of this dye seems to be most reliable, its use as a standard solution is hampered by poor chemical stability. Other compounds proposed for this purpose also have more or less serious drawbacks. Rhodamine 6G perchlorate seems to be superior to all the compounds proposed so far for the wavelength region between 500 and 600 nm.

These sketchy remarks on a seemingly simple, yet very difficult problem are intended to warn the reader to use caution when he encounters fluorescence quantum yields. I hasten to point out that, with a little experience, relative quantum yields can be estimated quite well when the solutions are compared visually under an intense light source.

## 5.3 Intersystem Crossing $S_1 \rightarrow T_1$

Besides the nonradiative decay directly to the ground state, a molecule excited to the state $S_1$ may enter the system of triplet states and relax to the lowest level $T_1$. As was pointed out earlier, this is undesirable in a laser dye for various reasons. In this section some structural features are discussed that influence the rate of intersystem crossing in organic dyes. For additional information the reader may turn to [Förster 1951 (p. 261); Wehry 1967; Parker 1968 (p. 303); Birks 1970 (p. 193)].

### 5.3.1 Dependence on $\pi$-Electron Distribution

The intersystem crossing from the singlet to the triplet manifold, being a forbidden process, is comparatively slow considering the small energy gap between $S_1$ and $T_1$. Its rate varies, however, in a peculiar fashion with the molecular structure of the dye. Whereas quantum-mechanical concepts have not been applied yet to the classes of dyes which concern us in connection with dye lasers, a simple rule has been found that explains many experimental observations (Drexhage 1972a). This rule states that in a dye where the $\pi$-electrons of the chromophore can make a loop when oscillating between the end groups, the triplet yield will be higher than in a related compound where this loop is blocked. It may be said that the circulating electrons create an orbital magnetic moment which couples with the spin of the electron. This increased spin-orbit coupling then enhances the rate of intersystem crossing, thus giving rise to a higher triplet yield.

The above rule may be illustrated by the class of rhodamine and acridine dyes, whose $\pi$-electron distribution can be described in terms of mesomeric resonance structures in the following way[2]:

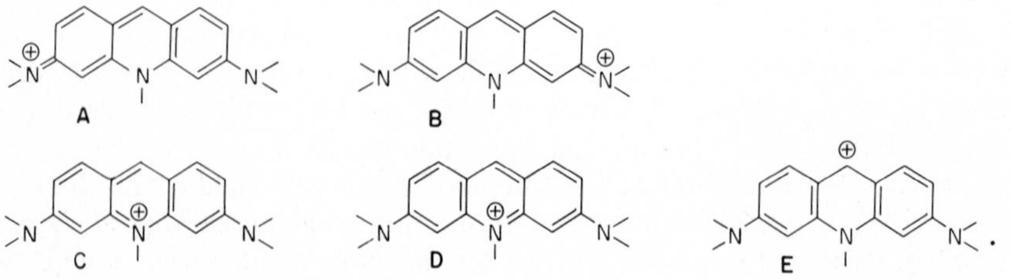

A B C D E

[2] A number of additional mesomeric forms that are of lesser importance are left out here for the sake of clarity.

While none of these mesomeric structures alone depicts the real $\pi$-electron distribution in the chromophore, they all contribute according to their relative energies. Particular attention must be paid to structures C and D because they symbolize the path the electrons follow along the nitrogen bridge. In the acridine dyes depicted here, structures C and D contain an ammonium configuration, as do structures A and B, so that all of them have approximately the same importance. In the case of rhodamine dyes, however, the central nitrogen atom is replaced by oxygen. Therefore structures C and D now involve an oxonium configuration, which renders them energetically less favored than structures A and B, and they do not contribute as much to the real $\pi$-electron distribution as they do in the case of the acridine dyes. Hence, according to the above rule, a lower triplet yield is expected in rhodamine than in acridine dyes. This is confirmed experimentally, as triplet yields of 1% and 10% have been found for rhodamine 6G (Webb et al. 1970) and acridine orange (Soep et al. 1972), respectively.

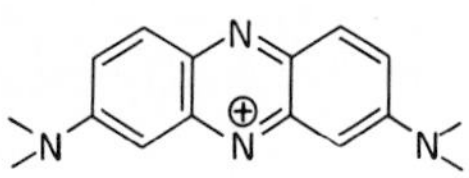

safranin 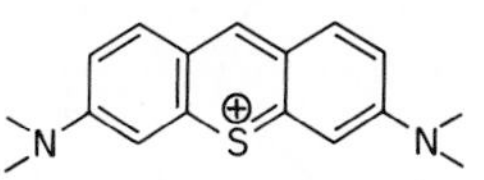 thiopyronin 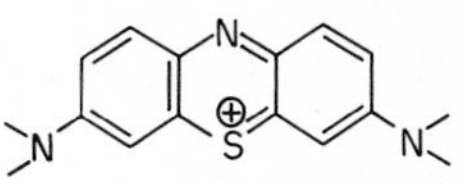 thiazine

The loop rule has proven very useful in the design of new laser dyes (Drexhage 1972a, 1973a). It predicts, for instance, a high triplet yield in safranin, thiopyronin, and thiazine dyes, in full agreement with experimental observations (Drexhage 1971). For fluorescein, it predicts a low triplet yield in alkaline solution, yet a high triplet yield in acidic solution, where structures A, B, C, and D all contain the oxonium configuration. These predictions are also borne out by experiments (Drexhage 1971; Soep et al. 1972). Furthermore, the rule predicts relatively high triplet yields in dyes with a ringlike chromophore, as is the case, e.g. with the chlorophyll and phthalocyanines. On the other hand, a very low triplet yield is expected in oxazine dyes and in compounds that contain a tetrahedral carbon atom in place of the oxygen bridge, and thus do not allow any $\pi$-electrons to pass over the bridge (see Sect. 5.8).

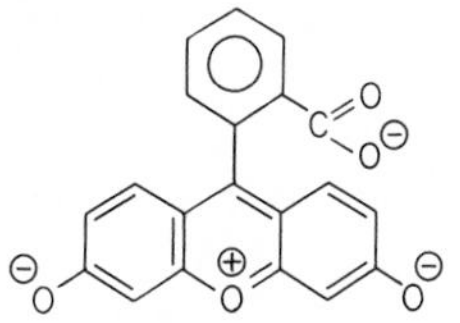

fluorescein, basic 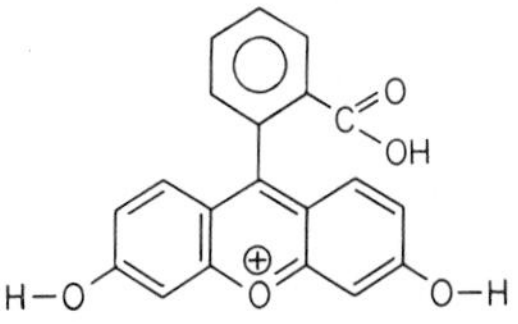 fluorescein, acidic

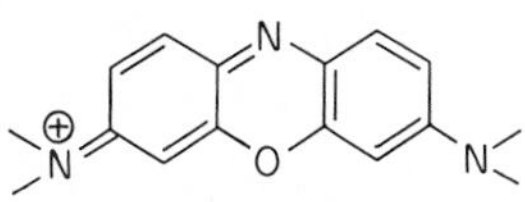

oxazine 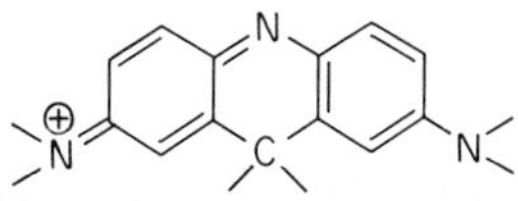 carbazine

### 5.3.2 Heavy-Atom Substituents

Above we have discussed an intersystem crossing mechanism whose magnitude depends essentially on the $\pi$-electron distribution in the chromophore. In the examples given, the molecules were composed exclusively of light atoms, i.e. hydrogen and elements that appear in the first row of the periodic table. However, the intersystem crossing rate, intrinsic to the chromophore, can be greatly enhanced if the dye is substituted with heavier elements, which increase the spin-orbit coupling (Turro 1965 (p. 50); Wehry 1967; Birks 1970 (p. 208)]. This undesirable effect for a laser dye can be demonstrated on the fluorescein derivatives eosin and erythrosin. The triplet yield of eosin in an alkaline solution was found to be 76%, which is to be compared with the value 3% for fluorescein (Soep et al. 1972). While substitution with chlorine has very little effect on the intersystem crossing rate of fluorescein, the replacement of the oxygen atoms in 3- and 6-position by sulfur, a direct neighbor of chlorine in the periodic table, yields a dye (dithiofluorescein, absorption maximum in alkaline ethanol 635 nm) that is absolutely nonfluorescent (Drexhage 1972b). Obviously, in any planned synthesis of efficient laser dyes, heavy-atom substituents are to be avoided.

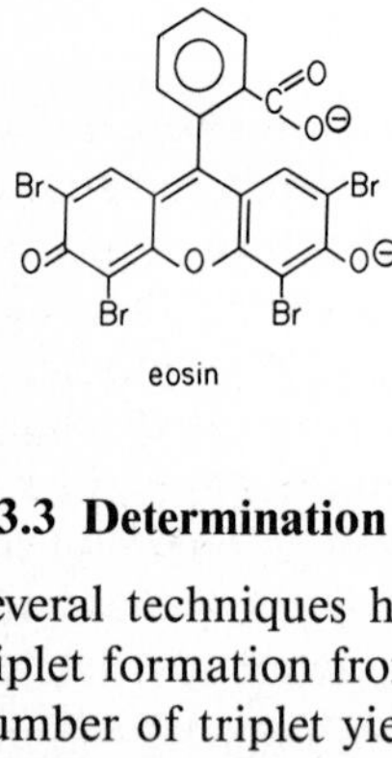

eosin 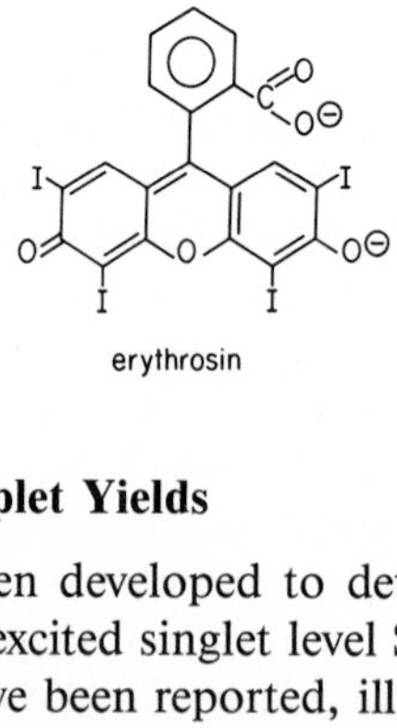 erythrosin 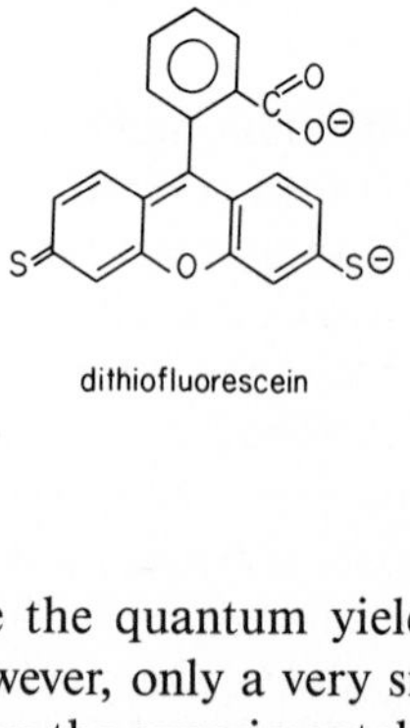 dithiofluorescein

### 5.3.3 Determination of Triplet Yields

Several techniques have been developed to determine the quantum yield of triplet formation from the excited singlet level $S_1$. However, only a very small number of triplet yields have been reported, illustrating the experimental difficulties encountered in such measurements. For a review the interested reader is referred to [Parker 1968 (p. 303); Birks 1970 (p. 193); Soep et al. 1972]. While all the methods for the measurement of triplet yields published so far require rather elaborate equipment and are very time consuming, a reasonably good estimate of this quantity can be obtained in a few minutes with a new technique which requires only test tubes, an ultraviolet lamp, and a few organic solvents (Drexhage 1971). As this method is particularly convenient in the study of laser dyes, it is described here briefly.

The technique is based on the well-known enhancement of the intersystem crossing rate $k_{ST}$ by solvents containing heavy atoms (see Sect. 5.5.3). It was found that solvents like iodomethane or iodobenzene increase the intrinsic value of $k_{ST}$ in first approximation by a constant factor independent of the magnitude of $k_{ST}$. Hence in the case of a dye in which the triplet yield is small ($k_{ST} \ll$ rate of fluorescence emission $k_{fl}$), the enhancement of $k_{ST}$ by a factor of, say, $10^3$ will not change the fluorescence efficiency appreciably. However, in the case of a dye where the values of $k_{ST}$ and $k_{fl}$ are comparable, the same enhancement of $k_{ST}$ will cause a reduction of the fluorescence efficiency by three orders of magnitude. Thus, for an approximate determination of the triplet yield, one simply adds an equal amount of iodomethane[3] to the ethanolic dye solution and observes the concomitant reduction in fluorescence intensity. The results obtained with this technique were in agreement with published triplet-yield data in all cases studied, including, among others, aromatic compounds, cyanine dyes, and typical laser dyes like rhodamines. For example, the fluorescence of acriflavine is completely quenched on addition of iodomethane, whereas that of rhodamine 6G remains almost unchanged, demonstrating the large difference in the intersystem crossing rates (see Sect. 5.3.1). Since it is independent of the fate of the triplet-state molecules after their production, this is a most direct technique which enables the determination of an intersystem crossing, even if no triplet molecules can be detected. It is applicable to such fugitive compounds as excimers and exciplexes (see Sect. 5.5.5).

## 5.4 Light Absorption in the States $S_1$ and $T_1$

As with molecules in the ground state $S_0$, there is a well-defined absorption spectrum associated with molecules that are in the excited states $S_1$ or $T_1$. Unfortunately, these spectra are very difficult to measure and accordingly very few data on laser dyes or related compounds have been reported. This is the case particularly with the $S_1$-absorption, which, owing to the short lifetime of this state, can usually be measured only by laser techniques. While some data are available on aromatic compounds (Novak and Windsor 1968; Nakato et al. 1968; Bonneau et al. 1968), cyanine (Müller 1968), and phthalocyanine dyes (Müller and Pflüger 1968), the $S_1$-absorption spectra of the widely used laser dyes are not known. That this absorption gives rise to appreciable losses when a coumarin laser is pumped at high power densities was suggested recently (Wieder 1972).

The absorption spectrum of molecules in the lowest triplet state $T_1$ is more easily accessible if the compound is embedded in a solid matrix, where the triplet lifetime is very long (in some cases up to several seconds). If the solid solution is excited with a highly intense light source, an appreciable population of the triplet level is achieved, and the spectrum can be determined without much difficulty, in many cases simply with a commercial spectrometer. With such a technique the $T_1$-absorption spectra of several acridine dyes and of fluorescein in acidic solution have been carefully determined (Zanker and Miethke 1957a, 1957b; Nouchi 1969). Similarly, the $T_1$-absorption of the laser dyes rhodamine B and rhodamine 6G is known for the red region of the

[3] Some compounds are alkylated by iodomethane. In such cases the less reactive iodobenzene is a suitable heavy-atom solvent.

spectrum (Buettner et al. 1969; Morrow and Quinn 1973). It was found that, in the region where these dyes lase, the molar decadic extinction coefficient of the triplet state has a value of $\approx 1.5 \times 10^4$ l mole$^{-1}$ cm$^{-1}$. The $T_1$-absorption of 7-diethylamino-4-methylcoumarin has been determined in the region from 400–650 nm (Morrow and Quinn 1973). In the lasing region of this compound (450–500 nm) the extinction coefficient was found to have a value of $\approx 0.3 \times 10^4$ l mole$^{-1}$ cm$^{-1}$. In addition, the triplet absorption of other compounds of interest for dye lasers is given in this paper.

## 5.5 Environmental Effects

In Sects. 5.2 and 5.3 the nonradiative processes in dye molecules have been related to their molecular structure. In the discussion given there, little attention was paid to the environment of the dye molecules, i.e. the solvent and other solute molecules. While such a simplification is often practical, there are also many instances where the surroundings of the dye molecules affect the rates of nonradiative processes to a degree that cannot be neglected.

### 5.5.1 Fluorescence Quenching by Charge Transfer Interactions

It has been known for a long time that the fluorescence of dyes is quenched by certain anions [Pringsheim 1949 (p. 322); Förster 1951 (p. 181)]. The quenching efficiency depends strongly on the chemical nature of the anion. The quenching ability, which is very strong in the case of iodide ($I^-$), decreases in the order: Iodide ($I^-$), thiocyanate ($SCN^-$), bromide ($Br^-$), chloride ($Cl^-$), perchlorate ($ClO_4^-$). While the mechanism of the phenomenon has not yet been elucidated in detail, this succession suggests that the excited state of the dye is quenched by a charge-transfer interaction. Since this effect is undesirable in lasing solutions, the use of perchlorate as the anion is often preferable to the common chloride, as was demonstrated in the acidic umbelliferone laser (Bergman et al. 1972). Because in many laser dyes the chromophore carries a positive charge, these dyes invariably have an anion, commonly chloride or iodide. Whether the fluorescence efficiency is affected by the anion accompanying the chromophore depends on the concentration and the polarity of the solvent. It was found that the fluorescence efficiencies of rhodamine 6G iodide and perchlorate in $10^{-4}$ molar solutions in ethanol were identical and very high (Drexhage 1972a, 1973a), indicating no quenching by the anions in either case. However, the fluorescence of rhodamine 6G iodide at the same concentration in the nonpolar solvent chloroform was almost completely quenched, whereas the perchlorate was as efficient as in ethanol. Apparently, the dye salts are fully dissociated in the polar solvent ethanol, but practically undissociated in chloroform. Hence, in ethanol, the quenching anions do not have sufficient time to reach the excited dye molecules during their lifetime, whereas, in chloroform, they are immediately available for a reaction. It is therefore advisable to use the dye perchlorate in lasing solutions of low polarity and when high concentrations are required. In addition to the anions mentioned here, a number of other compounds quench the fluorescence by a similar mechanism [Pringsheim 1949 (p. 322); Förster 1951 (p. 181); Leonhardt and Weller 1962].

### 5.5.2 Quenching by Energy Transfer

Another mechanism by which excited states, singlet as well as triplet, are quenched externally can operate if the quenching molecule has a level of energy equal to or lower than that of the state to be quenched. Under favorable conditions such energy transfer can occur over distances up to about 10 nm [Förster 1951 (p. 83), 1959; Kellogg 1970; Birks 1970 (p. 518)]. In liquid solution, where the reactants can approach each other very closely, energy-transfer processes are very efficient, provided the diffusion time is shorter than the lifetime of the excited state. The main interest in this kind of process in connection with dye lasers arises from its application to the quenching of undesired triplet states. Most widely utilized for this purpose is the ubiquitous molecular oxygen, which is available in sufficient concentration in laser solutions that are exposed to the atmosphere (Snavely and Schäfer 1969; Marling et al. 1970a). It is not clear yet whether the triplet quenching of oxygen is due to its low-lying excited singlet states or to its paramagnetic properties [Wehry 1967 (p. 91); Becker 1969 (p. 230); Birks 1970 (p. 492)]. However, the quenching of dye triplets by cyclooctatetraene and cycloheptatriene (Pappalardo et al. 1970a) must certainly be attributed to the low-lying triplet levels of these molecules. Several other compounds have been reported to quench triplets of laser dyes (Marling et al. 1970b, 1971). However, in some of these cases it is doubtful whether the quenching action was caused by the compound itself or by some impurity.

### 5.5.3 External Heavy-Atom Effect

In Sect. 5.3.2 we discussed how substitution with heavier elements enhances the rate of intersystem crossing in organic compounds and thus increases the triplet yield following optical excitation. It is remarkable that the same effect occurs if an organic compound is merely dissolved in a solvent that contains heavy-atom substituents, e.g. iodomethane or iodobenzene. For more information on such phenomena the reader is referred to the literature [Turro 1965 (p. 57); Wehry 1967 (p. 86); Birks 1970 (p. 208)]. Here it may suffice to mention that heavy-atom solvents are not well suited for lasing solutions owing to the increased triplet build-up. How such solvents can be used for the determination of triplet yields is described in Sect. 5.3.3.

### 5.5.4 Aggregation of Dye Molecules

Since water is a highly desirable solvent for laser dyes, their behavior in this medium has attracted the attention of many workers in this field. It has long been known that organic dyes in aqueous solution have a tendency to form dimers and higher aggregates which make themselves known through a distinctly different absorption spectrum. The dimers usually have a strong absorption band at shorter wavelengths than the monomers and often an additional weaker band at the long-wavelength side of the monomer band [Förster 1951 (p. 254); Förster and König 1957; Rohatgi and Mukhopadhyay 1971; Selwyn and Steinfeld 1972]. Furthermore, they generally are only weakly fluorescent or not at all. The equilibrium between monomers and dimers shifts to the side of the latter with increasing dye concentration and with decreasing temperature. At ambient temperature and $10^{-4}$ molar concentration, the dimerization of dyes like rhodamine B and rhodamine 6G is severe enough to prevent laser action. Not only is part of the pump light absorbed by the nonfluorescent dimers, but the dimers also increase the cavity losses owing to their long-wavelength absorption band, which is in the same region as the fluorescence of the monomers.

A number of factors have been suggested as responsible for the aggregation of organic dyes. It has been assumed that an attractive dispersion force between the highly polarizable dye chromophores plays an important role, while the high dielectric constant of water reduces the Coulombic repulsion between the identically charged molecules (Rabinowitch and Epstein 1941). However, it was found that the dimerization tendency was much less pronounced when the solvent was formamide, which has an even higher dielectric constant than water (Arvan and Zaitseva 1961). It has also been suggested that hydrogen bonding between the dye molecules (Levshin and Gorshkov 1961; Rohatgi and Singhal 1966), and an interaction with the accompanying anions (Lamm and Neville 1965; Larsson and Norden 1970) may be responsible for the dimerization in certain cases. A systematic study of the class of rhodamine dyes has shown that the dimerization tendency increases with the number and size of alkyl substituents (Drexhage and Reynolds 1972). In agreement with earlier suggestions, this indicates a tendency on the part of the hydrophobic dye molecules to shed water in much the same way that organic compounds like benzene avoid water, so that they become insoluble. While aggregation in water is common with most dyes, it usually does not occur in *organic* solvents, even at very high concentration. I do not agree with the claim that aggregation of xanthene dyes occurs in ethanol and other organic solvents (Selwyn and Steinfeld 1972). The dependence of the absorption spectra on concentration and temperature observed by these authors is more probably due to equilibria between different monomeric forms of the dyes (see Sect. 5.7) (Drexhage 1972a, 1973a).

It is possible to suppress the aggregation of dyes in aqueous solution by the addition of organic compounds. While any compound that is soluble in water will work to a certain degree (Crozet and Meyer 1970), there are some materials that are exceptionally efficient in this respect, e.g. N,N-dimethyldodecylamine-N-oxide[4] (Tuccio and Strome 1972), hexafluoroisopropanol (Drexhage

[4] An aqueous solution of this compound is on the market under the tradename Ammonyx-LO.

1972a, 1973a), and N,N-dipropylacetamide (Drexhage 1973a). These compounds can also be used to solubilize dyes that are insoluble in water (Tuccio et al. 1973). Presumably these additives form a cage around the hydrophobic dye molecules and thus shield them from each other and from the water. In agreement with such a cluster formation, it was observed that the rotational relaxation time of the dye molecules is lengthened in such solutions (Siegman et al. 1972).

### 5.5.5 Excited State Reactions

Although no interaction between ground-state molecules of organic dyes is evident in organic solvents, there is usually a strong interaction between excited molecules and those in the ground state, and this shows up at high concentrations. While the absorption spectrum of, e.g. rhodamine 6G in ethanol is unchanged even at concentrations as high as $10^{-2}$ molar, the fluorescence at such concentrations is strongly quenched owing to collisions of the excited dye molecules with those in the ground state [Förster 1951 (p. 230); Baranova 1965]. It is not known if in cases like this dimeric dye molecules exist for a short time; however, if they do, they are nonfluorescent. Although the fluorescence of the majority of organic compounds is quenched at high concentrations by this mechanism, it has been found that with pyrene and some other compounds a new fluorescence band appears when the concentration is increased [Parker 1968 (p. 344); Förster 1969; Birks 1970 (p. 301)]. This new band is due to dimers that exist only in the excited state (excimers). Following the emission of a photon, they immediately dissociate into ground-state monomers.

Similarly, an excited molecule may react with a molecule of a different compound (solvent or other solute) to form an excited complex (exciplex) which, on radiative de-excitation, decomposes immediately into the components [Knibbe et al. 1968; Birks 1970 (p. 403)]. Since the ground state of excimers and exciplexes is unstable, these species are ideal lasing compounds, provided the fluorescence efficiency is high and no disturbing triplet effects occur (Schäfer 1968, 1970). Because some compounds become more basic or acidic on optical excitation, they may pick up a proton from the solution or lose one to the solution (protolysis) [Weller 1958; Parker 1968 (p. 328); Becker 1969 (p. 239)]. If the new forms are fluorescent, they have the same advantage as exciplexes (Schäfer 1968, 1970; Srinivasan 1969; Shank et al. 1970a). The acid-base equilibria of some coumarin derivatives are discussed in Sect. 5.6.2.

## 5.6 Coumarin Derivatives

A group of widely used laser dyes emitting in the blue-green region of the spectrum are derived from coumarin by substitution with an amino or hydroxyl group in the 7-position. Since some members of this class rank among the most efficient laser dyes known today, we give here a summary of their pertinent optical properties. The marked change in basicity that occurs on optical excitation causes a shift of the fluorescence to longer wavelengths, a property which may be utilized to achieve a particularly wide tuning range in a dye laser.

### 5.6.1 Absorption and Fluorescence

The chromophore of such compounds can be described essentially by the mesomeric forms A and B depicted here. In the electronic ground state the $\pi$-electron distribution of the molecule closely resembles form A. The comparatively small weight of the polar form B is increased by factors that reduce its energy with respect to form A and a shift of the main absorption band to longer wavelengths occurs. If the weight of structures A and B becomes equal, the coumarin attains the character of a symmetrical cyanine dye and absorption occurs at the longest wavelength possible for this system. The positive charge at the N atom in form B is stabilized, for instance, by electron-donating

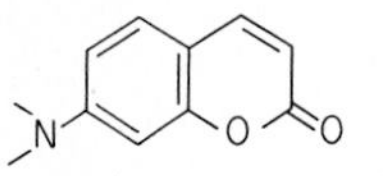

A 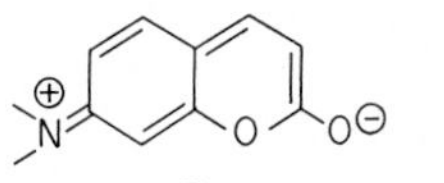 B

alkyl groups. Accordingly, a successive shift of the absorption band to longer wavelengths is found in the series, coumarin 120, coumarin 2, coumarin 1, coumarin 102 (Drexhage 1972a, 1973a). The absorption maximum of these dyes in methanol occurs at 351 nm, 364 nm, 373 nm, and 390 nm, respectively. The more polar mesomeric form B is also stabilized if the dye molecule is sur-

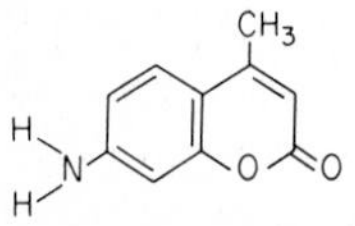

Coumarin 120 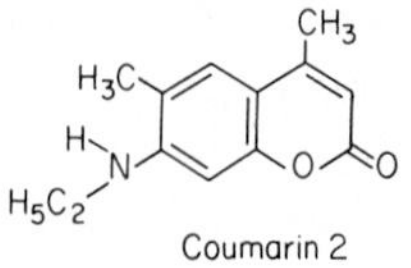 Coumarin 2

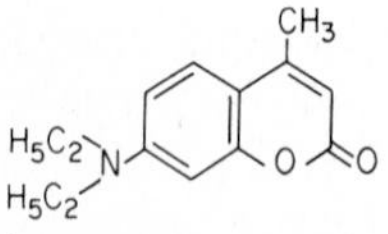

Coumarin 1 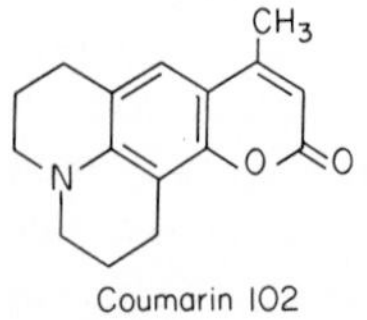 Coumarin 102

rounded by the molecules of a polar solvent. Therefore, the absorption maximum should shift to longer wavelengths with increasing polarity of the solvent. The effect can best be seen in the case of coumarin 102 (Table 5.1), where the absorption maximum shifts from 383 nm in N-methyl-pyrrolidinone (NMP) to 418 nm in hexafluorosiopropanol (HFIP). This phenomenon is obscured, however, by a counteracting effect in those coumarins that carry one or two hydrogen atoms at the amino group. Here, polar solvents like water or HFIP exert a specific influence on the amino group, reducing its ability to donate electrons and thereby reducing the weight of structure B. As a consequence,

**Table 5.1.** Coumarin dyes. $\lambda_{abs}$ maximum of main absorption band; $\lambda_{las}$ approximate lasing wavelength (flashlamp-pumped, untuned)

| Dye | Solvent[a] | $\lambda_{abs}$ [nm] | $\lambda_{las}$ [nm] |
|---|---|---|---|
| Coumarin 120 | HFIP | 326 | |
| | TFE | 338 | |
| | $H_2O$ | 343 | |
| | MeOH | 351 | 440 |
| Coumarin 2 | $CH_2Cl_2$ | 355 | |
| | NMP | 362 | |
| | MeOH | 364 | 450 |
| | HFIP | 378 | 460 |
| Coumarin 1 | MeOH | 373 | 460 |
| | TFE | 388 | 470 |
| | HFIP | 398 | 480 |
| Coumarin 102 | NMP | 383 | 470 |
| | $CH_2Cl_2$ | 386 | 470 |
| | MeOH | 390 | 480 |
| | TFE | 405 | 500 |
| | HFIP | 418 | 510 |
| Coumarin 30 | MeOH | 405 | 510 |
| Coumarin 6 | MeOH | 455 | 540 |

[a] HFIP hexafluoroisopropanol, TFE trifluoroethanol, NMP N-methyl-2-pyrrolidinone.

the absorption of coumarin 120 occurs at shorter wavelengths in HFIP than in methanol (Table 5.1). A similar phenomenon is found in the class of xanthene dyes (see Sect. 5.7).

Whereas in the electronic ground state $S_0$ of the coumarins the mesomeric structure A is predominant and structure B makes only a minor contribution to the actual $\pi$-electron distribution, the opposite is true for the first excited singlet state $S_1$, in which the polar form B is predominant. Therefore on optical excitation the (static) electric dipole moment increases, and a major rearrangement of the surrounding solvent molecules takes place immediately after excitation. Thus the energy of the excited state is markedly lowered before light emission occurs. This is the reason for the large energy difference between absorption and fluorescence (Stokes shift) in the coumarin derivatives shown in Table 5.1 as compared with, e.g. the xanthene dyes. Without going into detail, we note that the quantum yield of fluorescence has values above 70% in the compounds in Table 5.1.

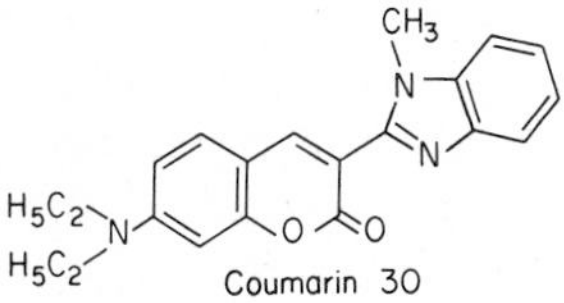

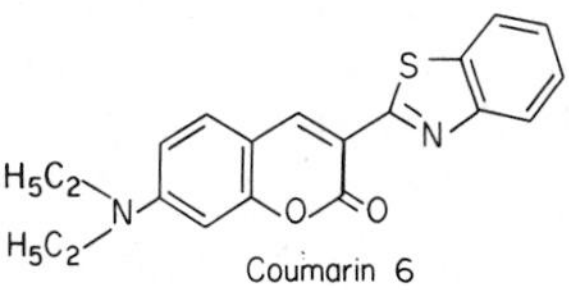

It was found that the lasing range covered by coumarin dyes can be extended appreciably if a heterocyclic substituent is introduced into the 3-position

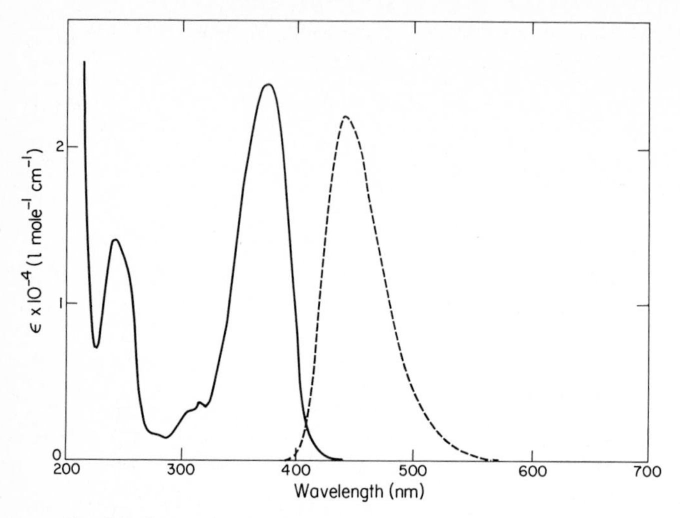


**Fig. 5.2.** Coumarin 1 in ethanol. (———) Absorption spectrum ($\varepsilon$ molar decadic extinction coefficient); (– – –) quantum spectrum of fluorescence (arbitrary units)

(Drexhage 1972a, 1973a). Coumarin 6 in alcoholic solution lases untuned at 540 nm; it is unmatched in efficiency and photochemical stability by other dyes which lase at this wavelength.

An important advantage in flashlamp pumping is the relatively strong absorption of coumarin dyes below 300 nm (Fig. 5.2). It was found that under flashlamp pumping conditions, where the coumarin dyes listed in Table 5.1 lased very efficiently, other dyes having approximately the same long-wavelength absorption and fluorescence efficiency, but very little UV absorption, did not lase at all (Drexhage et al. 1972). These same dyes did lase very efficiently, however, when excited with a monochromatic pump source matching their main absorption band. Recently, continuous laser operation with all coumarin derivatives in Table 5.1 was achieved by using aqueous media in which the dyes were rendered soluble by means of suitable additives (Sect. 5.5.4) (Tuccio et al. 1973).

### 5.6.2 Acid-Base Equilibria

Owing to the predominance of structure B in the excited state of 7-aminocoumarins, the carbonyl group in 2-position becomes markedly basic following optical excitation and has then an enhanced tendency to pick up a proton from the solution. In the case of coumarin 2, the protonated form created in its excited state exhibits a green fluorescence instead of the blue emission of the unprotonated dye (Srinivasan 1969). Since the carbonyl group is much less basic in the ground state, the protonated molecules dissociate immediately following the light emission. A dye like coumarin 2 does show some basicity in the ground state, but it is associated with the amino group rather than with the

EXCITED STATE EQUILIBRIUM:

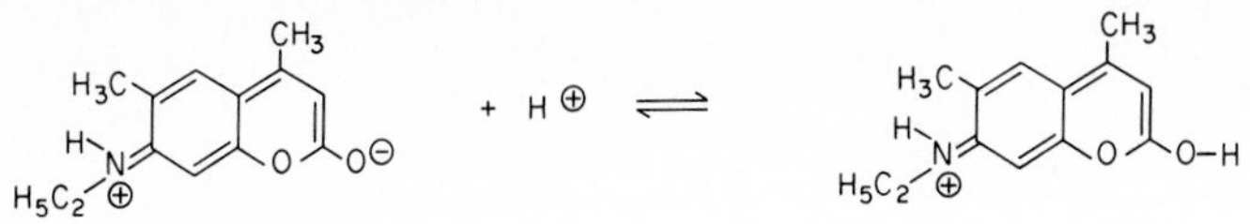

GROUND STATE EQUILIBRIUM:

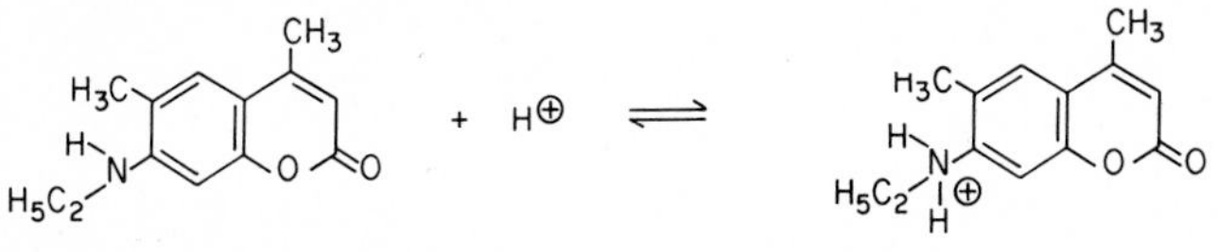

carbonyl group. Therefore, on addition of a few percent of a strong acid (e.g., HCl or $HClO_4$) to an alcoholic solution of coumarin 2, some protonation on the amino group occurs, which causes a reduction of the optical density at 364 nm, as the protonated form does not absorb at this wavelength (Drexhage 1971). Whereas the protonation in the ground state reaches an equilibrium depending on the basicity of the dye and on the $H^{\oplus}$-ion concentration, this is generally not the case in the excited state owing to its short lifetime, on the order of a few nanoseconds. Protonation in the excited state, therefore, is diffusion-controlled. It will increase with the acid concentration, but the undesirable protonation in the ground state increases too, because the equilibrium is dependent on the acid concentration.

While in coumarin 2 protonation in both ground and excited state takes place when its alcoholic solution is acidified with HCl or $HClO_4$, these reactions may be observed separately in other coumarin derivatives (Drexhage 1971). Since the basicity of the amino group is enhanced by electron-donating alkyl substituents, it is highest in a compound carrying two such substituents. Accordingly, it was found that, under the same conditions of acidification as before, coumarin 1 loses its absorption at 373 nm completely owing to protonation of the amino group. On the other hand, in the derivative coumarin 102 no protonation of the amino group is possible for geometric reasons. Hence, on acidification as above, there is no reduction of the optical density and a nearly complete protonation in the excited state is observed. In this case the fluorescence of the protonated form is yellow with a spectrum extending well beyond 600 nm (Fig. 5.3).

Basically, the same protonation-deprotonation reactions as with the 7-aminocoumarins are found for 7-hydroxycoumarins, e.g. 4-methylumbelliferone (coumarin 4) (Shank et al. 1970a; Dienes et al. 1970). The anion of this compound, which is formed in alkaline solution, corresponds closely in its electron distribution to the neutral forms of the 7-aminocoumarins. As with coumarin 2, the anion of coumarin 4 absorbs near 360 nm and fluoresces strongly at $\approx$ 450 nm. Since in the ground state the mesomeric structure A predominates (see Sect. 5.6.1) and in the excited state structure B, the location of the negative charge and thus of the basicity of the molecule changes as indicated. While this is of no consequence in alkaline solution, it becomes im-

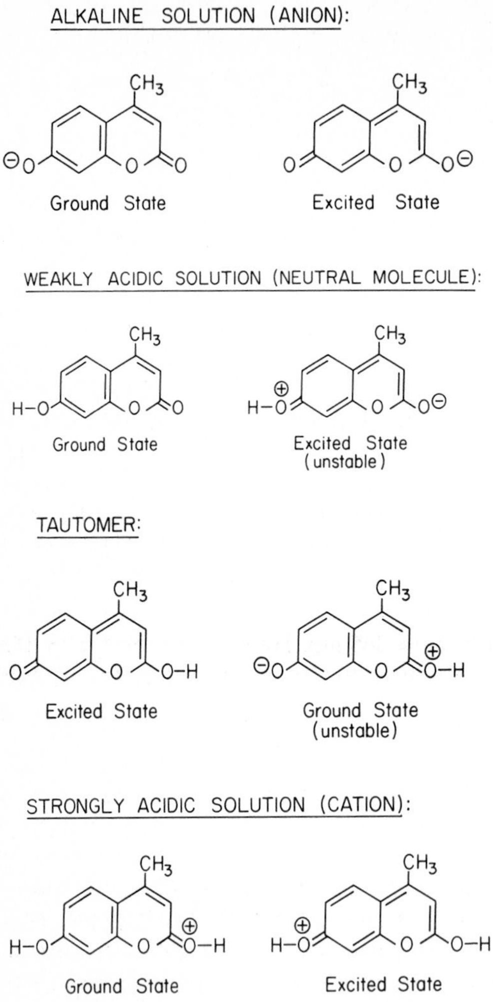

ALKALINE SOLUTION (ANION):

Ground State

Excited State

WEAKLY ACIDIC SOLUTION (NEUTRAL MOLECULE):

Ground State

Excited State
(unstable)

TAUTOMER:

Excited State

Ground State
(unstable)

STRONGLY ACIDIC SOLUTION (CATION):

Ground State

Excited State

portant in neutral or weakly acidic solutions where the compound exists in its neutral form which absorbs near 320 nm (Hammond and Hughes 1971; Nakashima et al. 1972). Following optical excitation, the proton of the hydroxyl group finds itself suddenly at an oxygen with a positive charge and dissociates away (most easily in solvent mixtures containing water), leaving behind the anion in its excited state. This anion can now either return to the ground state, emitting its characteristic blue fluorescence or, at sufficiently high acid concentration, can pick up a proton from the carbonyl group and form the tautomer of the dye. The latter returns to its ground state, emitting green fluorescence like coumarin 2 in acidic solution. Since the ground state of the tautomer is unstable, the proton will change position, and the stable neutral form of the dye is regenerated. In *strongly* acidic solution the molecule

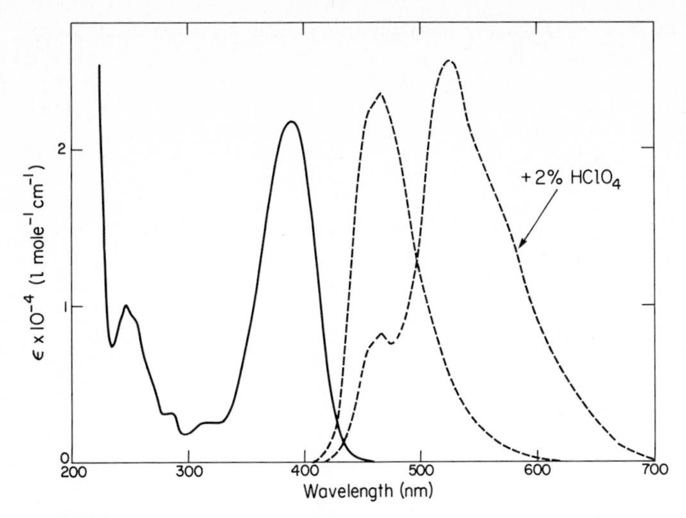


**Fig. 5.3.** Coumarin 102 in ethanol. (——) Absorption spectrum ($\varepsilon$ molar decadic extinction coefficient); (– – –) quantum spectrum of fluorescence (arbitrary units) in neutral and acidic solution. Quantum yield in acidic solution weaker by a factor of $\approx 2$

exists as the cation, which absorbs at $\approx$ 345 nm. Here, no reaction takes place in the excited state, and the cation returns to the ground state, emitting violet fluorescence.

## 5.7 Xanthene Dyes

Most dye lasers today operate with materials that belong to the class of xanthene dyes. They cover the wavelength region from 500–700 nm and are generally very efficient. Unlike most coumarin derivatives, the xanthene dyes are soluble in water, but tend to form aggregates in this solvent (see Sect. 5.5.4). Fortunately for the development of dye laser applications, derivatives like rhodamine 6G, rhodamine B, and fluorescein have been on the market in good quality (see Sect. 5.10), and give excellent lasing results even without further purification.

### 5.7.1 Absorption Spectra

The $\pi$-electron distribution in the chromophore of the xanthene dyes can be described approximately by the two identical mesomeric structures, A and B. Several other structures of lesser weight (see Sect. 5.3.1) may be neglected here. Unlike the coumarin dyes, forms A and B have the same weight, and thus in the xanthene dyes there is no static dipole moment parallel to the long axis of the molecule in either ground or excited state. For a simplified quantum-mechanical treatment of the main absorption features of these dyes the reader

**Table 5.2.** Rhodamine dyes. $\lambda_{abs}$ maximum of main absorption band; $\lambda_{las}$ approximate lasing wavelength (flashlamp-pumped, untuned)

| Dye | Solvent[a] | $\lambda_{abs}$ [nm] | $\lambda_{las}$ [nm] |
|---|---|---|---|
| Rhodamine 110 | HFIP | 487 | 540 |
| | TFE | 490 | 550 |
| | EtOH, basic | 501 | 560 |
| | EtOH, acidic | 510 | 570 |
| | DMSO | 518 | 575 |
| Rhodamine 19 | EtOH, basic | 518 | 575 |
| | EtOH, acidic | 528 | 585 |
| Rhodamine 6G | HFIP | 514 | 570 |
| | TFE | 516 | 575 |
| | EtOH | 530 | 590 |
| | DPA | 537 | 595 |
| | DMSO | 540 | 600 |
| Rhodamine B | EtOH, basic | 543 | 610 |
| | EtOH, acidic | 554 | 620 |
| Rhodamine 3B | HFIP | 550 | 610 |
| | TFE | 550 | 610 |
| | EtOH | 555 | 620 |
| | DMSO | 566 | 630 |
| Rhodamine 101 | HFIP | 572 | 625 |
| | TFE | 570 | 625 |
| | EtOH, basic | 564 | 630 |
| | EtOH, acidic | 577 | 640 |
| | DMSO | 586 | 650 |

[a] HFIP hexafluoroisopropanol, TFE trifluoroethanol, DMSO dimethyl sulfoxide, DPA N,N-dipropylacetamide.

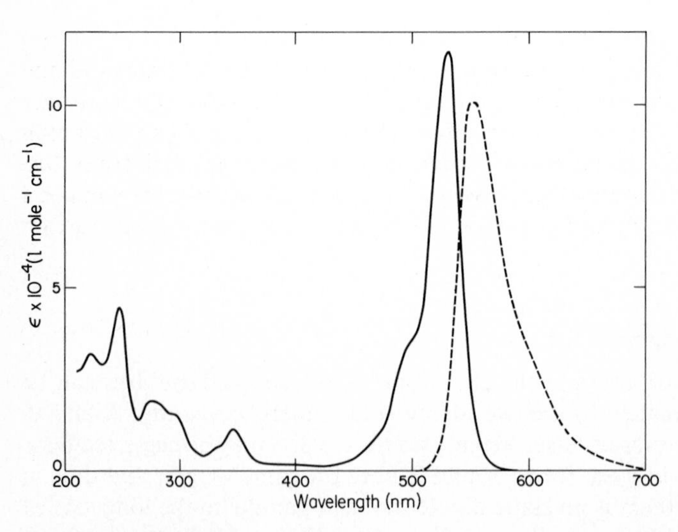


**Fig. 5.4.** Rhodamine 6G in ethanol. (———) Absorption spectrum (ε molar decadic extinction coefficient); (– – –) quantum spectrum of fluorescence (arbitrary units)

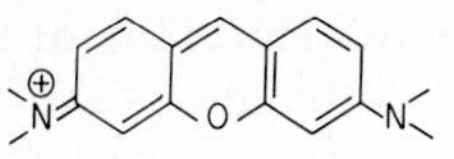

A 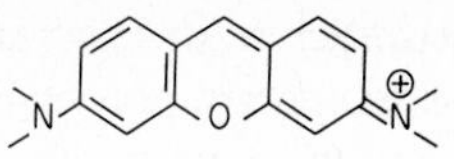 B

is referred to Kuhn [1959 (p. 411)]. The transition moment of the main absorption band, which occurs between 450 nm and 600 nm (Table 5.2), is oriented parallel to the long axis of the molecule. Some of the transitions at shorter wavelengths (Fig. 5.4), however, are oriented perpendicular to the long axis (Jakobi and Kuhn 1962).

The position of the long-wavelength band depends markedly on the substituents in 3- and 6-positions of the xanthene nucleus. The absorption maximum in ethanol shifts from 500 nm for the fluorescein dianion to 577 nm for the

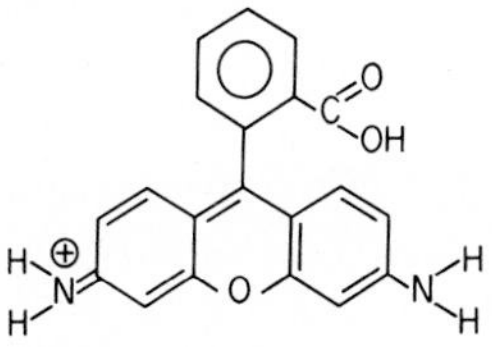

rhodamine 110 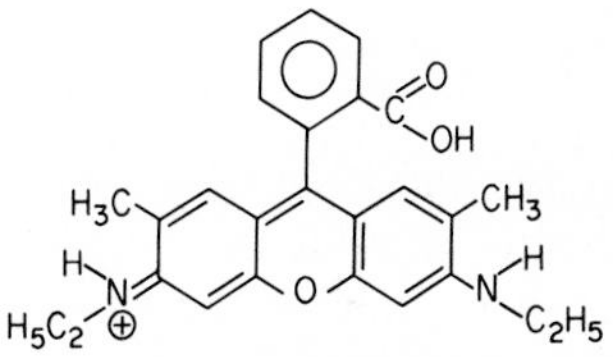 rhodamine 19

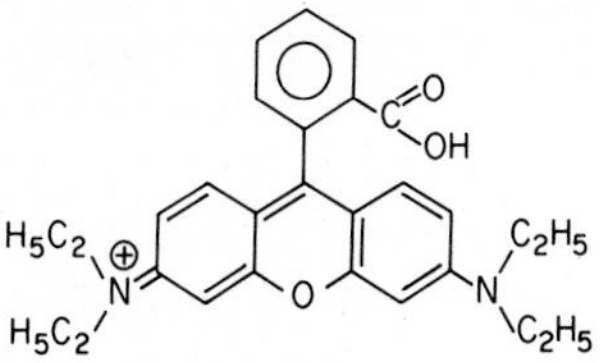

rhodamine B 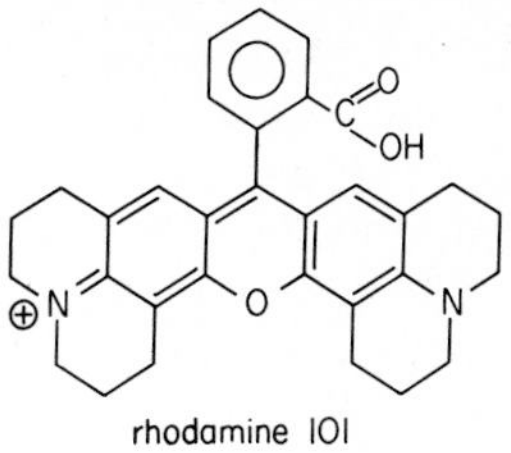 rhodamine 101

rhodamine 101 cation (Table 5.2). Since the carboxyphenyl substituent is not part of the chromophore of these dyes (see Sect. 5.2.1), it has only a minor influence on the absorption spectrum. Hence the dyes pyronin 20 and pyronin B absorb in ethanol at 527 nm and 552 nm, respectively, i.e. within 2 nm of the corresponding rhodamine cations (Table 5.2). The ultraviolet absorption of the

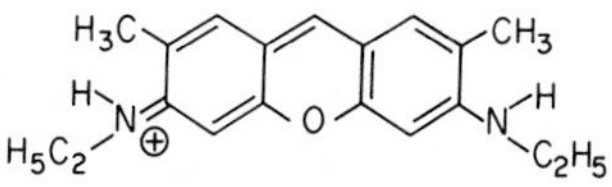

pyronin 20 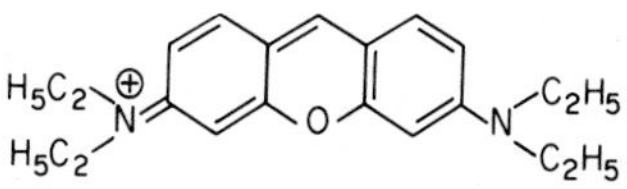 pyronin B

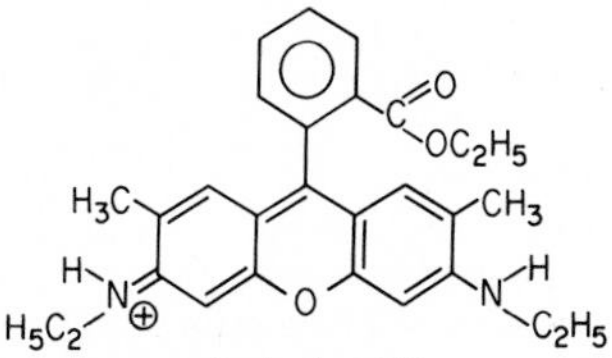

rhodamine 6G 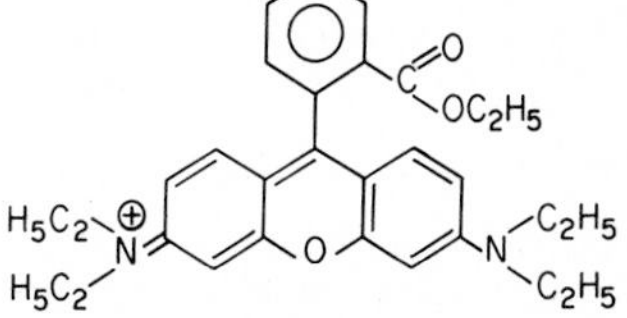 rhodamine 3B

rhodamines is slightly stronger than that of the pyronins. The esters of the rhodamines have absorption spectra practically identical with those of the free acids (Table 5.2). The methyl substituents commonly found in rhodamine 6G[5] have no influence on either absorption or fluorescence of this dye (Drexhage and Reynolds 1972). As can be seen from Table 5.2, the absorption maximum of the rhodamines is surprisingly dependent on the solvent, in particular with those dyes whose amino groups are not fully alkylated, e.g. rhodamine 110 and rhodamine 6G (Drexhage 1972a, 1973a).

POLAR SOLVENT:

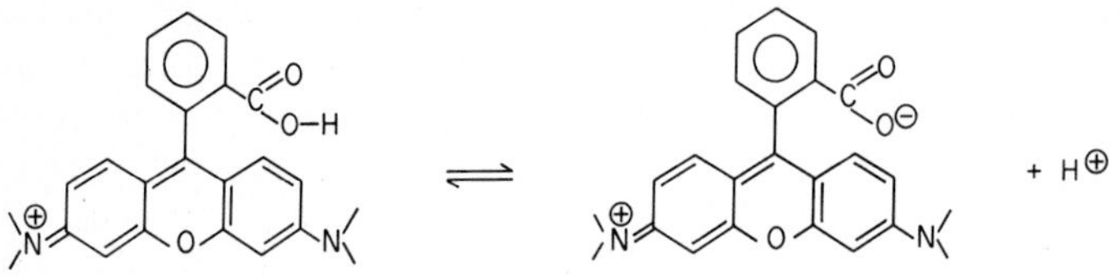

$\rightleftharpoons$ + $H^{\oplus}$

Rhodamine dyes that carry a free (nonesterified) COOH group can exist in several forms. In polar solvents like ethanol or methanol the carboxyl group participates in a typical acid-base equilibrium. The dissociation is enhanced by dilution or, most easily, by adding a small amount of a base. It can be followed spectroscopically, since absorption and fluorescence of the zwitterionic form of the dye are shifted to shorter wavelengths. While this shift amounts to only 3 nm in water (Ramette and Sandell 1956), it is about 10 nm in ethanol (Table 5.2) (Drexhage 1972a, 1973a). If a solution of such a dye is prepared in a neutral solvent, it will usually contain both forms of the dye. However, on addition of a small amount of either acid or base, a solution containing essentially only one of the two forms is obtained. In nonpolar solvents, e.g. acetone, the zwitterionic form is not stable. Instead, lactone is formed in a reversible

NONPOLAR SOLVENT:

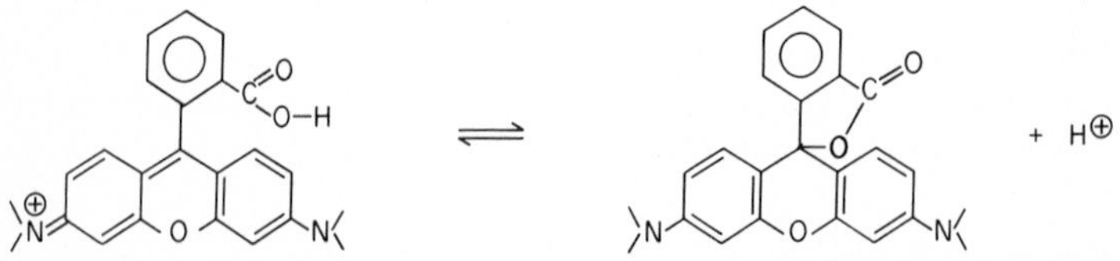

$\rightleftharpoons$ + $H^{\oplus}$

reaction; this is colorless because the $\pi$-electron system of the dye chromophore is interrupted. It may be mentioned that the pyronin dyes also become colorless on addition of a base. However, this reaction is generally not reversible, since in the presence of molecular oxygen a xanthone derivative is rapidly formed, indicated by its strong blue fluorescence.

[5] The name rhodamine 6G has also been associated in the literature with a compound that lacks the methyl groups. This ambiguity has been of no consequence, as the optical properties of both dyes are virtually identical.

### 5.7.2 Fluorescence Properties

The fluorescence spectra of xanthene dyes closely resemble the mirror image of the long-wavelength absorption band (Fig. 5.4). The fluorescence maximum is shifted by about 10 nm in the pyronins and by about 20 nm in the rhodamines with respect to the absorption maximum. It has been shown that the fluorescence quantum yield of rhodamine B in glycerol and in ethylene glycol is independent of the excitation wavelength down to 250 nm (Melhuish 1962; Yguerabide 1968). It is reasonable to assume that this is also true for other xanthene dyes. How the fluorescence efficiency depends on the character of the end groups of the chromophore has been discussed in Sect. 5.2. I do not agree with earlier interpretations (Viktorova and Gofman 1965) which were derived on the assumption that the fluorescence efficiency of the rhodamine dyes would increase with the energy difference between $S_1$ and $S_0$. Such an interpretation cannot be upheld, since the new compound rhodamine 101 has the smallest energy difference between $S_1$ and $S_0$ of all rhodamines discussed here and yet a fluorescence quantum yield of virtually 100% (Drexhage 1972a, 1973a).

In those xanthene dyes that contain mobile dialkylamino substituents the fluorescence efficiency is influenced by a variety of factors. As already mentioned in Sect. 5.2, the temperature and viscosity of the solvent affect this important property. It is particularly interesting also that certain low-viscosity solvents appreciably increase the fluorescence efficiency of these dyes above the value found in ethanol (Table 5.3) (Drexhage 1972a, 1973a). The fluorinated alcohols, e.g. trifluoroethanol and hexafluoroisopropanol, provide the additional advantage of disaggregating these dyes in aqueous solutions. Furthermore, it was found that the fluorescence quantum yield of rhodamine B has a value of 40% in acidic ethanol, but of 60% in basic ethanol, whereas

**Table. 5.3.** Fluorescence quantum yield of rhodamine 3B perchlorate in various solvents at room temperature

| Solvent | Qu. Yield [%] |
|---|---|
| Acetonitrile | 27 |
| Acetone | 30 |
| Ethanol | 42 |
| Chloroform | 50 |
| Trifluoroethanol | 72 |
| Benzonitrile | 75 |
| Dibromomethane | 82 |
| Benzaldehyde | 82 |
| Benzyl alcohol | 87 |
| Hexafluoroacetone hydrate | 87 |
| Hexafluoroisopropanol | 87 |
| Dichloromethane | 91 |
| Glycerol | 99 |

the fluorescence efficiencies of rhodamine 110, rhodamine 19, and rhodamine 101 are independent of the acidity (Drexhage 1972b). Apparently a negative charge at the carboxyl group of rhodamine B increases the $\pi$-electron density near the remote amino groups sufficiently to cause a reduction in their mobility. This is a particularly intriguing example of the subtle interplay of the various factors determining the fluorescence efficiency of dyes. It demonstrates how difficult it may be in some cases to unravel the relation between structure and nonradiative processes in organic molecules.

## 5.8 Related Classes of Efficient Laser Dyes

In the earlier sections we have outlined a number of structural features that are to be met by efficient laser dyes. On the basis of such principles, novel classes of useful laser dyes can be predicted and prepared by planned synthesis. Here we discuss several classes of dyes that have been found very useful in lasers emitting in the red and near infrared.

### 5.8.1 Oxazine Dyes

If the central $=CH-$ group of a pyronin dye is formally replaced by $=N-$, a compound is obtained whose absorption is shifted by about 100 nm to longer wavelengths [Kuhn 1959 (p. 411)]. Such an oxazine or phenoxazine dye is planar and rigid like its xanthene relative. Furthermore, the loop rule (see Sect. 5.3.1) predicts a very low triplet yield, as was confirmed experimentally (Drexhage 1971, 1972a, 1973a). The position of the absorption maximum depends on the end groups of the chromophore in the same fashion as in the xanthene dyes (Table 5.4). The fluorescence maximum shows a Stokes shift of $\approx$ 30 nm in ethanol (Fig. 5.5), and it was found that these dyes lase quite efficiently at the wavelengths given in Table 5.4. All oxazine dyes are photochemically much more stable than the pyronins and rhodamines.

**Table 5.4.** Oxazine dyes. $\lambda_{abs}$ maximum of main absorption band; $\lambda_{las}$ approximate lasing wavelength (laser-pumped, untuned); solvent ethanol

| Dye | $\lambda_{abs}$ [nm] | $\lambda_{las}$ [nm] |
|---|---|---|
| Resorufin | 578 | 610 |
| Oxazine 118 | 588 | 630 |
| Oxazine 4 | 611 | 650 |
| Oxazine 1 | 645 | 715 |
| Oxazine 9 | 601 | 645 |
| Nile blue | 635 | 690 |

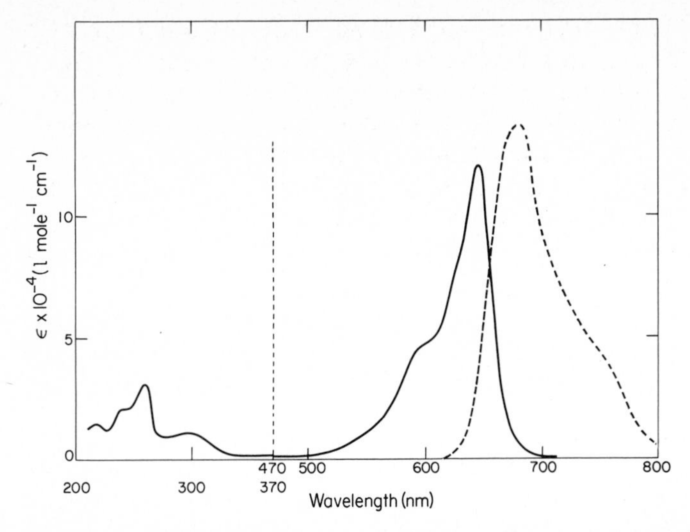


**Fig. 5.5.** Oxazine 1 in ethanol. (———) Absorption spectrum (ε molar decadic extinction coefficient); (– – –) quantum spectrum of fluorescence (arbitrary units)

resorufin

oxazine 118

oxazine 4

oxazine 1

Whereas no triplet problems are encountered in these dyes, it is comparatively difficult to suppress the processes of internal conversion. Owing to the smaller energy difference between $S_1$ and $S_0$, the influence of hydrogen vibrations at the end groups is more pronounced than in xanthene dyes (see Sect. 5.2.2). Hence the use of deuterated alcohol as a solvent markedly increases the fluorescence efficiency in these dyes (Drexhage 1972a, 1973a). As in the case of xanthene derivatives, dyes such as oxazine 1 have a reduced fluorescence efficiency in ethanol owing to the mobility of the dialkylamino substituents. However, the fluorescence efficiency of oxazine 1 perchlorate is high in the solvents dichloromethane, 1,2-dichlorobenzene, or $\alpha,\alpha,\alpha$-trifluorotoluene. In such a solvent the dye lases, flashlamp pumped, near 740 nm with threshold values comparable to that of rhodamine 6G (Drexhage 1972a, 1973a). Fluorinated alcohols are not useful solvents, since they interact with the central nitrogen atom and so reduce the fluorescence efficiency of oxazine dyes.

Some dyes of this class have a structure modified by an additional benzo group (oxazine 9, nile blue). These have been reported to be efficient laser dyes

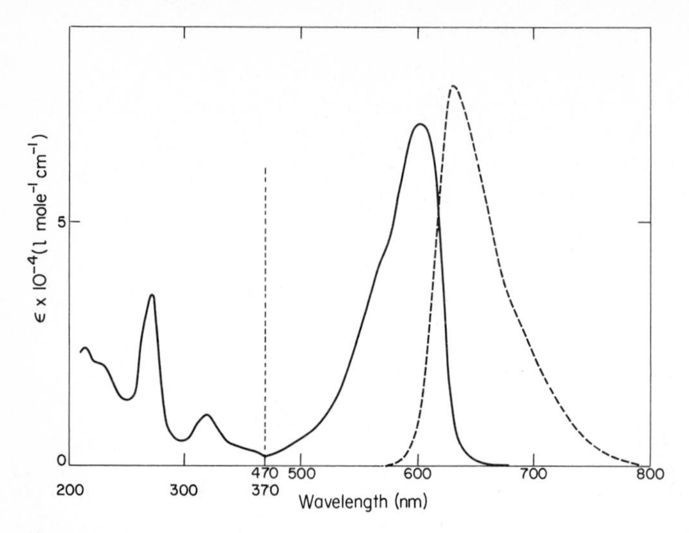


**Fig. 5.6.** Cresyl violet in ethanol. (———) Absorption spectrum (ε molar decadic extinction coefficient); (– – –) quantum spectrum of fluorescence (arbitrary units)

oxazine 9
(Cresyl Violet)

Nile Blue

too (Marling et al. 1970a; Runge 1971)[6]. The additional benzo group causes only a slight shift of absorption and fluorescence toward longer wavelengths but has a distinct influence on the shape of the absorption spectrum (Fig. 5.6). The extinction at the absorption maximum is reduced, and the spectrum is broadened compared with the spectra of other oxazine dyes (Fig. 5.5). This distortion of the absorption is caused by a steric interference between the amino group and the hydrogen atom indicated in the structural formulas of these dyes (Drexhage 1971). The amount of twist exerted on the amino group depends on the temperature. Hence the familiar shape of the absorption spectrum is restored at low temperatures (Gacoin and Flamant 1972).

### 5.8.2 Carbon-Bridged Dyes

The xanthene and oxazine dyes discussed above can be considered as diphenylmethane derivatives which are rigidized by an oxygen bridge. Although such a bridge is a good insulator for π-electrons, it still permits some

[6] It is not certain whether all reports on cresyl violet lasers were actually based on the dye with the structure given here (see Sect. 5.10).

passage of these electrons (see Sect. 5.3.1). To prevent conjugation along the bridge entirely, a tetrahedral carbon atom may be chosen. Whereas such dyes are in general tedious to synthesize, they are very stable once prepared (Drexhage and Reynolds 1972). As examples, the dyes carbopyronin 149 and carbazine 122 can be mentioned here. Their absorption maxima in ethanol are at 605 nm and 610 nm, respectively; i.e., they are shifted toward longer wavelengths as compared with the corresponding oxygen-bridged dyes pyronin B and resorufin.

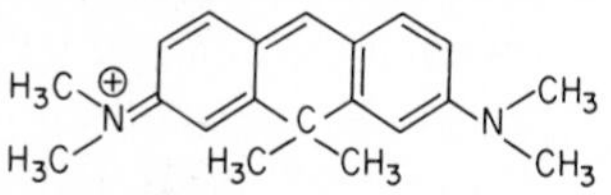

carbopyronin 149

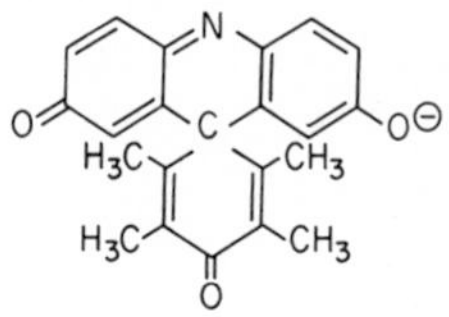

carbazine 122

The fluorescence of these compounds is very strong but the quantum efficiency is slightly reduced by the factors operating in xanthene and oxazine dyes. In the case of carbopyronin 149 the only nonradiative process encountered is associated with the mobility of the dimethylamino end groups. Its influence is reduced if trifluoroethanol is employed as the solvent. The dye in this solvent lases flashlamp-pumped at 650 nm with a threshold and efficiency comparable to those of rhodamine 6G. Carbazine 122 has a similarly high fluorescence efficiency. It lases flashlamp-pumped at 720 nm in basic ethanol and at 740 nm in basic dimethyl sulfoxide. It is also soluble in water containing a small amount of base and lases at 700 nm in this solvent.

### 5.8.3 Naphthofluorescein and Related Dyes

A new class of laser dyes is closely related to the well-known xanthene dyes in that it is derived from $\alpha$-naphthols instead of phenols (Drexhage and Reynolds

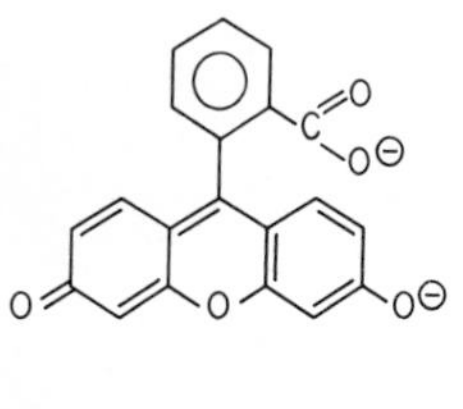

fluorescein

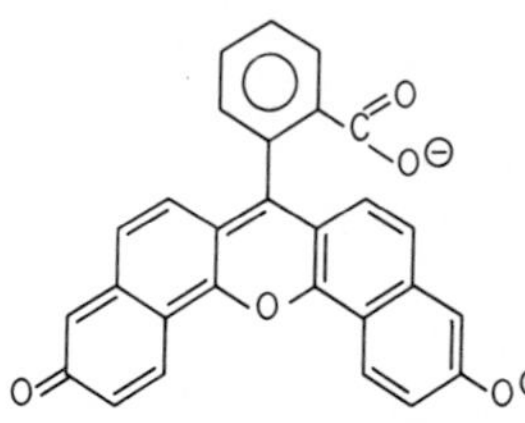

naphthofluorescein 126

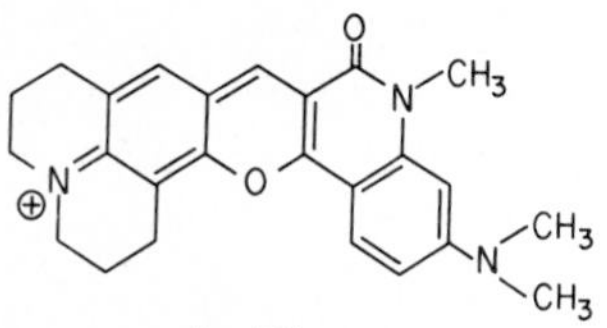

dye 141

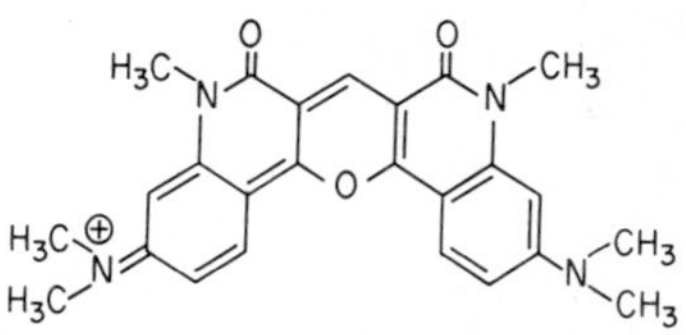

dye 140

1972). These dyes have a longer chromophore than the xanthene derivatives. Therefore they absorb and fluoresce at appreciably longer wavelengths. For instance, the compound naphthofluorescein 126 absorbs in basic ethanol at 615 nm and lases with good efficiency at 700 nm.

Another way of constructing a long, rigid $\pi$-electron system is illustrated by the amidopyrylium dyes 141 and 140 (Reynolds and Drexhage 1972). These compounds can be considered as pyrylium dyes made rigid by amide groups. The fluorescence efficiency of these dyes is reduced in most solvents (e.g. ethanol) because they carry at least one mobile dimethylamino end group. As in other cases discussed earlier, this mobility can be virtually eliminated by the use of hexafluoroisopropanol as a solvent. In this solvent, dye 141 and dye 140 have an absorption maximum at 590 nm and 660 nm, respectively, and lase very efficiently at 650 nm and 720 nm, respectively.

## 5.9 Other Efficient Laser Dyes

It is important to note that the structural principles outlined in this article do not constitute *necessary* requirements for efficient laser dyes. In particular, it is quite possible – and not contradictory to the foregoing discussions – that a dye *lacking* structural rigidity has a high fluorescence efficiency, and thus may be a good laser dye. For instance, it has been found that the compounds *p*-terphenyl, *p*-quaterphenyl, and some scintillator dyes, none of which is rigid, are efficient laser dyes in the near-ultraviolet region of the spectrum (Furumoto and Ceccon 1970). The finding that the threshold for laser action is higher than in rhodamine 6G can be ascribed in part to the fact that these dyes cannot utilize as much of the flash energy as rhodamine 6G, whose absorption spectrum covers a wider range of frequencies. Maeda and Miyazoe (1972) found that the nonrigid cyanine dyes are valuable laser materials, in particular, when dimethyl sulfoxide is used as a solvent.

Although it is not our intention to provide a comprehensive review of all efficient laser dyes, a few compounds other than those already discussed can be mentioned here. Besides the coumarins, the compounds acridone (Furumoto and Ceccon 1969b) and 9-aminoacridine hydrochloride (Gregg and Thomas 1969) have been reported to lase efficiently in the blue region of the spectrum. Some quinoline derivatives, which are closely related to the coumarins, are the most efficient compounds known to lase between 400 nm and 430 nm (Srinivasan 1969). The compounds brilliant sulphaflavine (Marling et al. 1970a) and 6-acetylaminopyrene-1,3,8-trisulfonic acid in alkaline solution (Schäfer 1968, 1970) are excellent lasing materials in the green and yellow. A number of other valuable laser dyes belonging to various classes have been discovered and can be found in the list at the end of this article.

## 5.10 Purity and Chemical Identity of Dyes

Though they may appear mystical to some, organic dyes are, in principle, well-defined compounds. Unfortunately, the purification of dyes is sometimes rather difficult. However, if one succeeds, one may be rewarded with a beautiful fluorescence of the crystalline material, which before was quenched by tiny amounts of impurities. For instance, highly pure rhodamine 6G chloride exhibits a strong red fluorescence, whereas the commercial material is almost nonfluorescent in the crystalline state (Drexhage 1971). Even so, compared with other commercially available organic dyes, this compound, like rhodamine B, is unusually pure. The quality of many other commercial dyes, however, is not sufficient for laser applications, and purification then yields a material of improved efficiency; see e.g. Gacoin and Flamant (1972).

It is little known, even among scientists working with these materials, that organic dyes are often sold under false names. A recent study has shown that, out of four commercial samples of pyronin B, three were actually rhodamine B and one rhodamine 6G (Horobin and Murgatroyd 1969). Likewise, all four samples of acridine red investigated by the same authors were incorrectly labeled. A dye of this structure should have an absorption maximum of $\approx 530$ nm in ethanol and would have lasing properties almost identical with those of rhodamine 6G. Instead, all commercial samples studied by this author absorb at $\approx 550$ nm and have optical properties fairly similar to those of pyronin B (Drexhage 1973a). Another example of an ill-defined dye frequently used in dye-lasers is cresyl violet. It is likely that most lasing samples sold under this name have the structure of oxazine 9, which is given in Sect. 5.8.1. But a different structure has also been suggested [Colour Index 1971 (p. 5406)].

## 5.11 Concluding Remarks

It was proposed in this chapter to give a critical review of the knowledge on laser dyes. Emphasis was placed on the general concepts relevant to an understanding of their physicochemical properties, and many details had to be omitted. Likewise, in several instances references to the original work were not given; they can be found, however, in the review articles quoted. Although some progress has been made recently in the understanding of nonradiative deactivation processes, this important topic is still far from understood. Many unexplained observations had to be omitted here for the sake of brevity. Furthermore, it must be stressed that present theory is unable to predict the rates of nonradiative processes in dye molecules with any useful degree of accuracy.

The more or less heuristic principles outlined in this article have proven to be very efficient tools in the planned synthesis of new laser dyes and can be expected to lead to further improvements in the future. One goal will be the extension of the lasing range to shorter as well as to longer wavelengths.

In the ultraviolet region, the lasing efficiency decreases with *increasing* energy difference between the levels $S_1$ and $S_0$ owing to increased absorption of the laser light by molecules in $S_1$. In the infrared, on the other hand, the lasing efficiency decreases with *decreasing* energy difference between $S_1$ and $S_0$ owing to increased internal conversion between $S_1$ and $S_0$. Hence, it is no coincidence that the most efficient laser dyes known today emit in the visible region of the spectrum. However, efforts are being made to construct molecules with high lasing efficiency down to $\approx 300$ nm and up to $\approx 1000$ nm. Furthermore, it can be expected that in the future dyes will become available that lase very efficiently in aqueous solution without the addition of any additives. It also should be possible to synthesize good laser dyes that exhibit much greater photochemical stability than, e.g., rhodamines and most coumarins known today.

In addition to the accompanying list of laser dyes, which was intended to be complete up to 1972, there are many other organic compounds that have been found to show stimulated emission. An extensive list of laser dyes known until 1980 was published by Maeda (1984). The reader should bear in mind that most compounds which fluoresce also will lase, when exposed to a pump source of sufficient intensity. Only relatively few dyes lase very efficiently. Detailed information on their lasing properties is available from manufactures (e.g. Brackmann 1986, Exciton 1989).

## List of Laser Dyes

| Dye | Solvent[a] | Excit.[b] | Lasing Wavelength [nm] | Ref. |
|---|---|---|---|---|
| p-terphenyl | $C_6H_{12}$ | L | 341 | Abakumov et al. (1969a) |
| | EtOH | L | 340 | Abamukov et al. (1969b) |
| | DMF | F | 341 | Furumoto and Ceccon (1970) |
| | Dioxane | F | 343–356 | Furumoto and Ceccon (1970) |
| p-quaterphenyl | DMF | F | 374 | Furumoto and Ceccon (1970) |
| | $C_6H_5CH_3$ | L | 362–390 | Myer et al. (1970) |
| 1,4-diphenyl-1,3-butadiene | $C_6H_5CH_3$ | L | 383 | Abakumov et al. (1969a) |
| p-distyrylbenzene | $C_6H_5CH_3$ | L | 415 | Kotzubanov et al. (1968a) |
| p-bis(o-methylstyryl)benzene, bis-MSB | $C_6H_5CH_3$ | F | 424 | Furumoto and Ceccon (1969b) |
| | EtOH | L | 419 | Deutsch and Bass (1969) |
| | $C_6H_6$ | L | 425 | Deutsch and Bass (1969) |
| 1-(o-methoxystyryl)-4-styrylbenzene | $C_6H_5CH_3$ | L | 425 | Kotzubanov et al. (1968b) |
| p-bis(o-methoxystyryl)benzene | $C_6H_5CH_3$ | L | 430 | Kotzubanov et al. (1968a) |
| 1-(m-methoxystyryl)-4-styrylbenzene | $C_6H_5CH_3$ | L | 415 | Kotzubanov et al. (1968b) |
| 1-(p-methoxystyryl)-4-styrylbenzene | $C_6H_5CH_3$ | L | 425 | Kotzubanov et al. (1968b) |
| 1-(o-chlorostyryl)-4-styrylbenzene | $C_6H_5CH_3$ | L | 420 | Kotzubanov et al. (1968b) |
| 1-(p-chlorostyryl)-4-styrylbenzene | $C_6H_5CH_3$ | L | 420 | Kotzubanov et al. (1968b) |
| p-bis(p-chlorostyryl)benzene | $C_6H_5CH_3$ | L | 420 | Kotzubanov et al. (1968a) |
| 4,4′-diphenylstilbene, 1,2-bis(4-biphenylyl)ethylene | $C_6H_5CH_3$ | L | 408 | Kotzubanov et al. (1968a) |
| | $C_6H_6$ | L | 409 | Deutsch and Bass (1969) |
| | $C_6H_6$ | F | 409 | Furumoto and Ceccon (1970) |
| | DMF | F | 409 | Furumoto and Ceccon (1970) |
| 1-(p-phenylstyryl)-4-styrylbenzene, 1-styryl-4-[2-(4-biphenylyl)vinyl]-benzene | $C_6H_5CH_3$ | L | 432 | Kotzubanov et al. (1968a) |

[a] $CH_2Cl_2$ dichloromethane, $CH_3CN$ acetonitrile, $C_5H_5N$ pyridine, $C_5H_{11}OH$ isoamyl alcohol, $C_6H_4Cl_2$ o-dichlorobenzene, $C_6H_5CH_3$ toluene, $C_6H_5CF_3$ $\alpha,\alpha,\alpha$-trifluorotoluene, $C_6H_5NO_2$ nitrobenzene, $C_6H_6$ benzene, $C_6H_{10}O$ cyclohexanone, $C_6H_{12}$ cyclohexane, $C_7H_{16}$ n-heptane, DCE dichloroethane, DMF N,N-dimethylformamide, DMSO methyl sulfoxide, DPA N,N-dipropylacetamide, EG ethylene glycol, EtOH ethanol, HFIP hexafluoroisopropanol, $H_2O$ water, $H_2SO_4$ sulfuric acid, IP isopentane, MCH methylcyclohexane, MeCs methyl cellosolve, MeOD monodeuterated methanol, MeOH methanol, NMP 1-methyl-2-pyrrolidinone, PMMA polymethylmethacrylate, PSt polystyrene, TFE trifluoroethanol, THF tetrahydrofuran.

[b] L laser-pumped, F flashlamp-pumped, C continuous (laser-pumped).

List of Laser Dyes (cont.)

| Dye | Solvent[a] | Excit.[b] | Lasing Wavelength [nm] | Ref. |
|---|---|---|---|---|
| 4,4′-(p-phenylenedivinylene)bis[N,N-dimethylaniline] | PMMA | L | 478 | Naboikin et al. (1970) |
| 1,2-di-1-naphthylethylene | $C_6H_5CH_3$ | L | 426 | Kotzubanov et al. (1968a) |
| 1-[2-(2-naphthyl)vinyl]-4-styrylbenzene, 1-styryl-4-[2-(2-naphthyl)vinyl]-benzene | $C_6H_5CH_3$ | L | 425 | Kotzubanov et al. (1968b) |
| 2,5-diphenylfuran, | $C_6H_6$, EtOH | L | 369 – 380 | Broida and Haydon (1970) |
| PPF | EtOH | L | 365 – 371 | Myer et al. (1970) |
| | DMF | F | 371 | Furumoto and Ceccon (1970) |
| | Dioxane | F | 371 | Furumoto and Ceccon (1970) |
| 1,3-diphenylisobenzofuran | EtOH | F | 484 – 518 | Marling et al. (1970a) |
| 2,5-diphenyloxazole, | $C_6H_{12}$ | L | 357 | Lidholt and Wladimiroff (1970) |
| PPO | $C_6H_5CH_3$, $C_6H_6$, dioxane | L | 359 – 391 | Broida and Haydon (1970) |
| | Dioxane | F | 381 | Furumoto and Ceccon (1970) |
| 2,5-bis(4-biphenylyl)oxazole, | $C_6H_5CH_3$ | L | 409 | Abakumov et al. (1969a) |
| BBO | $C_6H_6$ | L | 409 | Deutsch and Bass (1969) |
| | $C_6H_6$ | F | 410 | Furumoto and Ceccon (1970) |
| | DMF | L | 413 | Lidholt and Wladimiroff (1970) |
| 2-(1-naphthyl)-5-phenyloxazole, | $C_6H_5CH_3$ | L | 400 | Abakumov et al. (1969a) |
| $\alpha$-NPO | $C_6H_{12}$ | L | 393 | Deutsch and Bass (1969) |
| | EtOH | L | 398 | Deutsch and Bass (1969) |
| | EtOH | F | 400 | Furumoto and Ceccon (1970) |
| | PMMA | L | 396 | Naboikin et al. (1970) |
| | PSt | L | 411 | Naboikin et al. (1970) |
| 2,2′-p-phenylenebis(5-phenyloxazole), | $C_6H_5CH_3$ | L | 417 | Kotzubanov et al. (1968a) |
| 1,4-bis[2-(5-phenyloxazolyl)]benzene, | $C_6H_{12}$ | L | 411 | Deutsch and Bass (1969) |
| POPOP | EtOH | L | 421 | Deutsch and Bass (1969) |
| | $C_6H_5CH_3$ | F | 419 | Furumoto and Ceccon (1969b) |
| | THF | L | 415 – 430 | Myer et al. (1970) |
| | PMMA | L | 415 | Naboikin et al. (1970) |
| | PSt | L | 420 | Naboikin et al. (1970) |

| | | | | |
|---|---|---|---|---|
| 2,2′-p-phenylenebis(4-methyl-5-phenyl-oxazole), | $C_6H_{12}$ | L | 423 | Deutsch and Bass (1969) |
| 1,4-bis[2-(4-methyl-5-phenyloxa-zolyl)]benzene, | EtOH | L | 431 | Deutsch and Bass (1969) |
| dimethyl-POPOP | $C_6H_6$, $C_6H_5CH_3$, $C_6H_{12}$ | L | 424–441 | Broida and Haydon (1970) |
| 2,2′-p-phenylenebis[5-(4-biphenylyl)oxazole], BOPOB | THF | L | 428–450 | Myer et al. (1970) |
| 2,2′-p-phenylenebis[5-(1-naphthyl)oxazole], $\alpha$-NOPON | $C_6H_6$ | L | 430–455 | Myer et al. (1970) |
| 5-phenyl-2-(p-styrylphenyl)oxazole | $C_6H_5CH_3$ | L | 417 | Naboikin et al. (1970) |
| | PMMA | L | 414 | Naboikin et al. (1970) |
| 5-phenyl-2-[p-(p-phenylstyryl)phenyl]oxazole | $C_6H_5CH_3$ | L | 432 | Naboikin et al. (1970) |
| | PMMA | L | 428 | Naboikin et al. (1970) |
| 2-(4-biphenylyl)-5-(1-naphthyl)oxazole, | $C_6H_5CH_3$ | L | 413 | Naboikin et al. (1970) |
| B$\alpha$NO | PMMA | L | 410 | Naboikin et al. (1970) |
| 5-(4-biphenylyl)-2-[p-(4-phenyl- | $C_6H_5CH_3$ | L | 448 | Naboikin et al. (1970) |
| 1,3-butadienyl)phenyl]oxazole | PMMA | L | 443 | Naboikin et al. (1970) |
| 2,2′-[p-phenylenebis(vinylene- | $C_6H_5CH_3$ | L | 460 | Naboikin et al. (1970) |
| p-phenylene)]bis[5-(4-biphenylyl)oxazole] | PMMA | L | 455 | Naboikin et al. (1970) |
| 2-[p-[2-(2-naphthyl)vinyl]phenyl]- | $C_6H_5CH_3$ | L | 428 | Naboikin et al. (1970) |
| 5-phenyloxazole | PMMA | L | 425 | Naboikin et al. (1970) |
| 2-[p-[2-(9-anthryl)vinyl]phenyl]-5-phenyloxazole | PMMA | L | 480 | Naboikin et al. (1970) |
| 2,2′-vinylenebis[5-phenyloxazole] | $C_6H_5CH_3$ | L | 447 | Naboikin et al. (1970) |
| | PMMA | L | 444 | Naboikin et al. (1970) |
| 2,2′-(2,5-thiophenediyl)bis(5-*tert*-butyl-benzoxazole), 2,5-bis[5-*tert*-butylbenzoxazolyl(2)] thiophene, BBOT | $C_6H_6$ | F | 437 | Furumoto and Ceccon (1970) |
| 2,5-diphenyl-1,3,4-oxadiazole, | EtOH | L | 347 | Abakumov et al. (1969b) |
| PPD | Dioxane | F | 348 | Furumoto and Ceccon (1970) |
| 2-(p-methoxyphenyl)-5-phenyl-1,3,4-oxadiazole | EtOH | L | 365 | Abakumov et al. (1969b) |

List of Laser Dyes (cont.)

| Dye | Solvent[a] | Excit.[b] | Lasing Wavelength [nm] | Ref. |
|---|---|---|---|---|
| 2,5-bis(p-methoxyphenyl)-1,3,4-oxadiazole | EtOH | L | 359, 372 | Abakumov et al. (1969b) |
| 2-[p-(dimethylamino)phenyl]-5-phenyl-1,3,4-oxadiazole | $C_6H_5CH_3$ | L | 408 | Naboikin et al. (1970) |
| 2-(p-chlorophenyl)-5-[p-(dimethyl-amino)phenyl]-1,3,4-oxadiazole | $C_6H_5CH_3$ | L | 420 | Naboikin et al. (1970) |
| 2-(4-biphenylyl)-5-phenyl-1,3,4-oxadiazole, | EtOH | F | 363 | Furumoto and Ceccon (1970) |
| 2-phenyl-5-(4-biphenylyl)- | EtOH | L | 355–382 | Myer et al. (1970) |
| 1,3,4-oxadiazole, PBD | $C_6H_5CH_3$ | L | 377–415 | Broida and Haydon (1970) |
| 2-(4-biphenylyl)-5-p-cumenyl-1,3,4-oxadiazole, | $C_6H_{12}$ | F | 361 | Furumoto and Ceccon (1970) |
| isopropyl-PBD | EtOH | F | 370 | Furumoto and Ceccon (1970) |
| | $C_6H_5CH_3$ | L | 363–385 | Turek and Yardley (1971) |
| 2,5-bis(4-biphenylyl)-1,3,4-oxadiazole, BBD | $C_6H_5CH_3$ | L | 374–398 | Turek and Yardley (1971) |
| 2-(4-biphenylyl)-5-styryl-1,3,4-oxadiazole | $C_6H_5CH_3$ | L | 391 | Abakumov et al. (1969a) |
| 2,5-di-1-naphthyl-1,3,4-oxadiazole, $\alpha$-NND | $C_6H_5CH_3$ | L | 391 | Abakumov et al. (1969a) |
| 2-(1-naphthyl)-5-styryl-1,3,4-oxadiazole | $C_6H_5CH_3$ | L | 399 | Naboikin et al. (1970) |
| | PMMA | L | 416 | Naboikin et al. (1970) |
| 2-(4-biphenylyl)-5-(p-styrylphenyl)- | $C_6H_5CH_3$ | L | 403 | Naboikin et al. (1970) |
| 1,3,4-oxadiazole | PMMA | L | 400 | Naboikin et al. (1970) |
| 2-phenyl-5-[p-(4-phenyl-1,3-butadienyl) | $C_6H_5CH_3$ | L | 424 | Naboikin et al. (1970) |
| phenyl]-1,3,4-oxadiazole | PMMA | L | 418 | Naboikin et al. (1970) |
| 1,5-diphenyl-3-styryl-2-pyrazoline | PMMA | L | 450 | Naboikin et al. (1970) |
| 3-(p-chlorostyryl)-1,5-diphenyl-2-pyrazoline | PMMA | L | 457 | Naboikin et al. (1970) |
| Salicylic acid (sodium salt) | EtOH | L | 395–418 | Myer et al. (1970) |
| Aminobenzoic acid | $C_6H_5CH_3$ | L | 398–406 | Myer et al. (1970) |
| Amino G acid | $H_2O$ | F | 459 | Maeda and Miyazoe (1972) |
| Anthracene | Fluorene (crystal) | L | 408 | Karl (1972) |
| 9-methylanthracene | EtOH-MeOH(glass) | L | 414 | Ferguson and Mau (1972a) |

| | | | | |
|---|---|---|---|---|
| 9,10-dimethylanthracene | MCH-$C_6H_5CH_3$(glass) | L | 432 | Ferguson and Mau (1972a) |
| 9-chloroanthracene | EtOH-MeOH(glass) | L | 416 | Ferguson and Mau (1972a) |
| 9-phenylanthracene | EtOH-MeOH(glass) | L | 417 | Ferguson and Mau (1972a) |
| 9,10-diphenylanthracene | $C_6H_{12}$ | L | 433 | Huth and Farmer (1968) |
| | EtOH | L | 435–450 | Myer et al. (1970) |
| | EtOH-MeOH(glass) | L | 430 | Ferguson and Mau (1972a) |
| 2-amino-7-nitrofluorene | $C_6H_4Cl_2$ | L | 540–585 | Gronau et al. (1972) |
| 3-aminofluoranthene | EtOH | F | 548–580 | Marling et al. (1970a) |
| 8-hydroxypyrene-1,3,6-trisulfonic acid (sodium salt) | $H_2O$(basic) | F | 550 | Drexhage (1971) |
| 8-ethylaminopyrene-1,3,6-trisulfonic acid (sodium salt) | $H_2O$ | L | 441 | Schäfer et al. (1967) |
| 8-acetamidopyrene-1,3,6-trisulfonic acid | $H_2O$(neutral) | L, F | 441–453 | Schäfer (1970) |
| (sodium salt) | $H_2O$(basic) | L, F | 566–574 | Schäfer (1970) |
| Perylene | DMF | L | 472 | Lidholt and Wladimiroff (1970) |
| | DCE | L | 472–476 | Hammond and Hughes (1971) |
| Coronene | MCH-IP(glass) | F | 444 | Kohlmannsperger (1969) |
| 7-amino-2-hydroxy-4-methylquinoline, carbostyril 124 | EtOH | F | 413 | Srinivasan (1969) |
| 7-dimethylamino-2-hydroxy-4-methyl-quinoline, carbostyril 165 | EtOH | F | 425 | Srinivasan (1969) |
| 7-hydroxycoumarin, | $H_2O$(basic) | F | 457 | Snavely and Peterson (1968) |
| umbelliferone | EtOH(acidic) | L | 405–568 | Dienes et al. (1970) |
| 7-hydroxy-4-methylcoumarin, | $H_2O$(basic) | F | 454 | Snavely et al. (1967) |
| 4-methylumbelliferone, | EtOH(acidic) | L | 391–567 | Shank et al. (1970a) |
| coumarin 4 | $H_2O$(basic) | C | 460 | Tuccio et al. (1973) |
| Esculin | $H_2O$(basic) | F | blue | Sorokin et al. (1968) |
| 4,8-dimethyl-7-hydroxycoumarin, | EtOH(basic) | F | 455–505 | Gregg et al. (1970) |
| 4,8-dimethylumbelliferone | EtOH(acidic) | L | 447–569 | Hammond and Hughes (1971) |
| Benzyl-$\beta$-methylumbelliferone | EtOH-$H_2O$(basic) | F | 464–468 | Gregg and Thomas (1969) |
| 4-methylumbelliferone-methylene-iminodiacetic acid | EtOH(basic) | F | 459–464 | Marling et al. (1970a) |
| Calcein blue | EtOH(basic) | F | 449–490 | Marling et al. (1970a) |

List of Laser Dyes (cont.)

| Dye | Solvent[a] | Excit.[b] | Lasing Wavelength [nm] | Ref. |
|---|---|---|---|---|
| 7-hydroxy-6-methoxycoumarin | $H_2O$(basic) | F | blue | Crozet et al. (1971) |
| 3,4-cyclopentano-7-hydroxycoumarin | $H_2O$(basic) | F | blue | Crozet et al. (1971) |
| 7-acetoxy-4-methylcoumarin | EtOH(basic) | F | 441–486 | Gregg et al. (1970) |
| 7-butyroxy-4-methylcoumarin | $H_2O$-$CH_3CN$(basic) | F | blue | Crozet et al. (1971) |
| 7-acetoxy-5-allyl-4,8-dimethylcoumarin | EtOH(basic) | F | 458–515 | Gregg et al. (1970) |
| 7-amino-4-methylcoumarin, | MeOH | F | 440 | Drexhage (1972a, 1973a) |
| coumarin 120 | $H_2O$-DPA | C | 450 | Tuccio et al. (1973) |
| 4,6-dimethyl-7-methylaminocoumarin, | EtOH | F | 443 | Srinivasan (1969) |
| coumarin 9 | EtOH(acidic) | F | 484 | Srinivasan (1969) |
| | EtOH(acidic) | L | 430–530 | Gutfeld et al. (1970) |
| 4,6-dimethyl-7-ethylaminocoumarin, | EtOH | F | 446 | Srinivasan (1969) |
| coumarin 2 | EtOH(acidic) | F | 487 | Srinivasan (1969) |
| | HFIP | F | 460 | Drexhage (1973a) |
| | $H_2O$-DPA | C | 460 | Tuccio et al. (1973) |
| 7-diethylamino-4-methylcoumarin, | EtOH | F | 460 | Kagan et al. (1968) |
| coumarin 1 | EtOH | F | 460 | Ferrar (1969b) |
| | HFIP | F | 480 | Drexhage (1972a, 1973a) |
| | TFE | F | 470 | Drexhage (1973a) |
| Coumarin 102 | MeOH | F | 480 | Drexhage (1972a, 1973a) |
| | HFIP | F | 510 | Drexhage (1972a, 1973a) |
| | NMP | F | 470 | Drexhage (1973a) |
| | TFE | F | 500 | Drexhage (1973a) |
| | $H_2O$-DPA | C | 480 | Tuccio et al. (1973) |
| 7-amino-3-phenylcoumarin, coumarin 10 | $H_2O$-DPA | C | 480 | Tuccio et al. (1973) |
| Coumarin 24 | MeOH | F | 510 | Drexhage (1972a) |
| Coumarin 30 | MeOH | F | 510 | Drexhage (1973a) |
| | $H_2O$-DPA | C | 505 | Tuccio et al. (1973) |
| Coumarin 7 | $H_2O$-DPA | C | 525 | Tuccio et al. (1973) |
| Coumarin 6 | MeOH | F | 540 | Drexhage (1972a, 1973a) |
| | $H_2O$-DPA | C | 540 | Tuccio et al. (1973) |

| | | | | |
|---|---|---|---|---|
| 2,4,6-triphenylpyrylium fluoborate | MeOH | L | 485 | Schäfer et al. (1967) |
| 2,4,6-tri-p-tolylpyrylium perchlorate | MeOH | L | 492 | Schäfer (1970) |
| 2,6-bis(p-methoxyphenyl)-4-phenyl-pyrylium fluoborate | MeOH | L, F | 566–573 | Schäfer (1970) |
| 9(10H)-acridone | EtOH | L | 439 | Sorokin et al. (1967) |
| | EtOH | F | 435 | Furumoto and Ceccon (1969b) |
| 9-aminoacridine hydrochloride | MeOH | F | 449–453 | Schäfer (1970) |
| | EtOH, $H_2O$(acidic) | F | 457–460 | Gregg and Thomas (1969) |
| N-methylacridinium perchlorate | MeCs | L | 535–553 | Hammond and Hughes (1971) |
| Lucigenin, bis-N-methylacridinium nitrate | $H_2O$(acidic) | L | 600 | Stepanov and Rubinov (1968) |
| Acriflavine, | EtOH | L | 510 | McFarland (1967) |
| trypaflavine | EtOH | F | 517 | Furumoto and Ceccon (1970) |
| Acridine yellow | EtOH | F | 514 | Maeda and Miyazoe (1972) |
| 3-aminophthalimide | $C_5H_{11}OH$ | L | 500 | Stepanov and Rubinov (1968) |
| 3-amino-N-methylphthalimide | EtOH | L | 519 | Neporent and Shilov (1971) |
| 3-amino-6-dimethylamino-N-methyl-phthalimide | $C_5H_5N$ | L | 560–588 | Sevchenko et al. (1968a) |
| 3,6-bis(methylamino)-N-methylphthalimide | $C_5H_5N$ | L | 578–605 | Sevchenko et al. (1968a) |
| 3-dimethylamino-6-methylamino- | $C_5H_5N$ | L | 580–593 | Sevchenko et al. (1968a) |
| N-methylphthalimide | $C_6H_5CH_3$ | L | 578 | Aristov et al. (1970) |
| | EtOH | L | 595 | Aristov et al. (1970) |
| | Glycerol | L | 610 | Aristov et al. (1970) |
| | $C_7H_{16}$-EtOH | L | 560–590 | Aristov and Kuzin (1971) |
| 3,6-bis(dimethylamino)-N-methylphthalimide | $C_5H_5N$ | L | 575–590 | Sevchenko et al. (1968a) |
| | $C_6H_6$ | L | 577 | Aristov et al. (1970) |
| | $C_6H_{10}O$ | L | 578 | Aristov et al. (1970) |
| | EtOH | L | 584 | Aristov et al. (1970) |
| 3-acetamido-6-amino-N-methylphthalimide | $C_5H_5N$ | L | 570–600 | Sevchenko et al. (1968a) |
| Brilliant sulphaflavine | EtOH | F | 508–573 | Marling et al. (1970a) |
| Fluorescein (sodium salt), | $H_2O$(basic) | L | 539 | Schäfer et al. (1967) |
| uranine | EtOH | L | green | Sorokin et al. (1967) |
| | $H_2O$, EtOH | F | 550 | Sorokin and Lankard (1967) |
| | EtOH | C | 555 | Strome and Tuccio (1971) |
| | $H_2O$-(Ammonyx LO) | C | 522–570 | Hercher and Pike (1971c) |

List of Laser Dyes (cont.)

| Dye | Solvent[a] | Excit.[b] | Lasing Wavelength [nm] | Ref. |
|---|---|---|---|---|
| 6-carboxyfluorescein | EtOH(basic) | F | 539 – 548 | Marling et al. (1970a) |
| Fluorescein isothiocyanate | EtOH(basic) | F | 546 | Maeda and Miyazoe (1972) |
| Fluorescein diacetate, | EtOH(basic) | F | 541 – 571 | Gregg et al. (1970) |
| diacetylfluorescein | | | | |
| 2′,7′-dichlorofluorescein | EtOH(basic) | F | green | Sorokin et al. (1968) |
| | EtOH(basic) | C | 575 | Strome and Tuccio (1971) |
| 2′,4′,5′,7′-tetrachlorofluorescein | EtOH(basic) | C | 582 | Strome and Tuccio (1971) |
| Monobromofluorescein | Glycerol | L | 560 | Stepanov and Rubinov (1968) |
| Dibromofluorescein | Glycerol | L | 568 | Stepanov and Rubinov (1968) |
| Eosin (sodium salt), | EtOH | L | yellow | Sorokin et al. (1967) |
| 2′,4′,5′,7′-tetrabromofluorescein | MeOH(basic) | L | 553 – 557 | Schäfer (1970) |
| (sodium salt) | DMF | L | 568 | Lidholt and Wladimiroff (1970) |
| Rhodamine 110, | | C | 525 – 585 | Tuccio and Strome (1972) |
| unsubstituted rhodamine | HFIP | F | 540 | Drexhage (1972a, 1973a) |
| | TFE | F | 550 | Drexhage (1973a) |
| | EtOH(basic) | F | 560 | Drexhage (1973a) |
| | EtOH(acidic) | F | 570 | Drexhage (1973a) |
| | DMSO | F | 575 | Drexhage (1973a) |
| Rhodamine 19 | EtOH(basic) | F | 575 | Drexhage (1973a) |
| | EtOH(acidic) | F | 585 | Drexhage (1973a) |
| Rhodamine 6G | EtOH | L | green-orange | Sorokin et al. (1967) |
| | EtOH | F | 585 | Sorokin and Lankard (1967) |
| | MeOH | F | 598 | W. Schmidt and Schäfer (1968) |
| | Glycerol | L | 557 | Kotzubanov et al. (1968a) |
| | PMMA | F | 601 | O. G. Peterson and Snavely (1968) |
| | $H_2O$-(Triton X100) | C | 597 | O. G. Peterson et al. (1970) |
| | $H_2O$-HFIP | C | 540 – 605 | Tuccio and Strome (1972) |
| | $H_2O$-(Ammonyx LO) | C | 565 – 640 | Tuccio and Strome (1972) |
| | $H_2O$-(Ammonyx LO) | C | 590 – 610 | Ippen et al. (1972) |
| | HFIP | F | 570 | Drexhage (1972a, 1973a) |
| | DMSO | F | 600 | Drexhage (1972a, 1973a) |

| | | | | |
|---|---|---|---|---|
| | TFE | F | 575 | Drexhage (1973a) |
| | DPA | F | 595 | Drexhage (1973a) |
| N,N′-bis(β-phenylethyl)rhodamine | EtOH | L | 569 | Aristov et al. (1970) |
| Rhodamine B, | EtOH | L | 608 | Schäfer et al. (1967) |
| N,N,N′N′-tetraethylrhodamine | MeOH | F | rot | W. Schmidt and Schäfer (1967) |
| | PMMA | F | 625 | O. G. Peterson and Snavely (1968) |
| | $H_2O$-HFIP | C | 580–655 | Tuccio and Strome (1972) |
| | EtOH | F | 605–645 | Arthurs et al. (1972b) |
| | EtOH(basic) | F | 610 | Drexhage (1972a, 1973a) |
| | EtOH(acidic) | F | 620 | Drexhage (1972a, 1973a) |
| Rhodamine 3B | $C_5H_{11}OH$ | L, F | 615 | Rubinov and Mostovnikov (1968) |
| | HFIP | F | 610 | Drexhage (1972a, 1973a) |
| | EtOH | F | 620 | Drexhage (1972a, 1973a) |
| | TFE | F | 610 | Drexhage (1973a) |
| | DMSO | F | 630 | Drexhage (1973a) |
| N,N′-bispentamethylenerhodamine | EtOH | L | 599 | Aristov et al. (1970) |
| Rhodamine 101 | EtOH(basic) | F | 630 | Drexhage (1972a, 1973a) |
| | EtOH(acidic) | F | 640 | Drexhage (1972a, 1973a) |
| | DMSO | F | 650 | Drexhage (1972a, 1973a) |
| | HFIP | F | 625 | Drexhage (1973a) |
| | TFE | F | 625 | Drexhage (1973a) |
| Rhodamine S, | $C_5H_{11}OH$ | L, F | 620 | Rubinov and Mostovnikov (1968) |
| rhodamine C | EtOH | L | 570 | Kotzubanov et al. (1968a) |
| | EtOH | F | 578–595 | Gregg and Thomas (1969) |
| Lissamine rhodamine B-200 | EtOH | F | 575–645 | Marling et al. (1970a) |
| Xylene red B | EtOH(basic) | F | 584–645 | Gregg et al. (1970) |
| Kiton red S | EtOH | F | 589–642 | Gregg et al. (1970) |
| Rhodamine 6Y | EtOH | L | 572 | Aristov et al. (1970) |
| Acridine red | EtOH | L | orange | Sorokin et al. (1967) |
| | EtOH | F | 602 | Sorokin and Lankard (1967) |
| Pyronin G, | $C_5H_{11}OH$ | L, F | 600 | Rubinov and Mostovnikov (1968) |
| pyronin Y | $C_5H_{11}OH$ | F | 585 | Stepanov and Rubinov (1968) |
| | EtOH(acidic) | F | 590–635 | Marling et al. (1970a) |
| Pyronin B | MeOH | F | gelb | W. Schmidt and Schäfer (1967) |
| | PMMA | L | 576 | Kotzubanov et al. (1968a) |
| | EtOH | F | 615–632 | Gregg and Thomas (1969) |

List of Laser Dyes (cont.)

| Dye | Solvent[a] | Excit.[b] | Lasing Wavelength [nm] | Ref. |
|---|---|---|---|---|
| Safranin T | EtOH | L | 610 | Kotzubanov et al. (1968a) |
| | MeOH | L | 621 – 625 | Schäfer (1970) |
| Resorufin | EtOH(basic) | L | 610 | Drexhage (1972a, 1973a) |
| Resazurin | MeOH, EG | C | | Runge (1972) |
| Oxazine 118 | EtOH | L | 630 | Drexhage (1972a, 1973a) |
| Oxazine 4 | EtOH | L | 650 | Drexhage (1972a, 1973a) |
| Oxazine 1 | EtOH | L | 715 | Drexhage (1972a, 1973a) |
| | $C_6H_5CF_3$ | F | 730 | Drexhage (1972a) |
| | $C_6H_4Cl_2$ | F | 740 | Drexhage (1972a, 1973a) |
| | $CH_2Cl_2$ | F | 740 | Drexhage (1972a, 1973a) |
| Cresyl violet, | EtOH | F | 646 – 709 | Marling et al. (1970a) |
| oxazine 9 | MeOH | C | 650 | Runge (1971) |
| | EtOH | F | 644 – 704 | Arthurs et al. (1972b) |
| | MeOD | F | 680 | Drexhage (1972a, 1973a) |
| Nile blue | EtOH | L | 690 | Drexhage (1972a, 1973a) |
| Nile blue A | MeOH | C | | Runge (1971) |
| Carbopyronin 149 | TFE | F | 650 | Drexhage and Reynolds (1972) |
| Carbazine 122 | EtOH(basic) | F | 720 | Drexhage and Reynolds (1972) |
| | DMSO(basic) | F | 740 | Drexhage and Reynolds (1972) |
| Thionin | $H_2SO_4$ | L | 850 | Rubinov and Mostovnikov (1967) |
| Methylene blue | $H_2SO_4$ | L | 835 | Stepanov et al. (1967b) |
| Toluidine blue | $H_2SO_4$ | L | 848 | Rubinov and Mostovnikov (1967) |
| Naphthofluorescein 126 | EtOH(basic) | F | 700 | Drexhage and Reynolds (1972) |
| Amidopyrylium dye 141 | HFIP | F | 650 | Reynolds and Drexhage (1972) |
| Amidopyrylium dye 140 | HFIP | F | 720 | Reynolds and Drexhage (1972) |
| Lachs | Glycerol | L | 540 | Stepanov and Rubinov (1968) |
| Pinaorthol | EtOH | L | 565 | Stepanov and Rubinov (1968) |
| Violettrot | $C_5H_{11}OH$ | L | 620 | Stepanov and Rubinov (1968) |
| | $C_5H_{11}OH$ | F | 610 | Stepanov and Rubinov (1968) |
| Isoquinoline red | $H_2O$ | L | 620 | Stepanov and Rubinov (1968) |

| | | | | |
|---|---|---|---|---|
| Rapid-filter gelb | $C_5H_{11}OH$ | L | 620 | Stepanov and Rubinov (1968) |
| | $C_5H_{11}OH$ | F | 610 | Stepanov and Rubinov (1968) |
| Echtblau B, pure blue | Glycerol | L | 753 | Rubinov and Mostovnikov (1967) |
| Naphthalene green | Glycerol | L | 756 | Rubinov and Mostovnikov (1967) |
| Rhoduline blue 6G | Glycerol | L | 758 | Rubinov and Mostovnikov (1967) |
| Brilliant green | Glycerol | L | 759 | Rubinov and Mostovnikov (1967) |
| Victoria blue | Glycerol | L | 809 | Rubinov and Mostovnikov (1967) |
| Victoria blue R | Glycerol | L | 814 | Rubinov and Mostovnikov (1967) |
| Methylene green | $H_2SO_4$ | L | 829 | Rubinov and Mostovnikov (1967) |
| Phthalocyanine | $H_2SO_4$ | L | 863 | Stepanov et al. (1967b) |
| Chloroaluminum phthalocyanine, | EtOH | L | 755 | Sorokin and Lankard (1966) |
| aluminum phthalocyanine chloride | DMSO | L | 762 | Sorokin et al. (1967) |
| Magnesium phthalocyanine | Quinoline | L | 759 | Stepanov et al. (1967b) |
| Chlorophyll | $H_2SO_4$ | L | 800 | Rubinov and Mostovnikov (1967) |
| 3,3′-diethyloxacarbocyanine iodide, DOC iodide | Glycerol | F | 541 | Maeda and Miyazoe (1972) |
| 3,3′-diethylthiacarbocyanine iodide, DTC iodide | Glycerol | F | 625 | Maeda and Miyazoe (1972) |
| 1,3,3,1′,3′,3′-hexamethylindocarbocyanine iodide | Glycerol | F | 614 | Maeda and Miyazoe (1972) |
| Dicyanine | Glycerol | L | 756 | Rubinov and Mostovnikov (1967) |
| | Quinoline | L | 723, 752 | Rubinov and Mostovnikov (1967) |
| 1,1′-diethyl-4,4′-carbocyanine bromide | Glycerol | L | 754 | Miyazoe and Maeda (1968) |
| 1,1′-diethyl-4,4′-carbocyanine iodide, cryptocyanine | Glycerol | L | 745 | Spaeth and Bortfeld (1966) |
| 1,1′-dimethyl-4,4′-carbocyanine iodide | Glycerol | L | 749 | Miyazoe and Maeda (1968) |
| 3,3′diethyloxadicarbocyanine iodide, | MeOH | L | 658 | Schäfer et al. (1967) |
| DODC iodide | DMSO | F | 662 | Maeda and Miyazoe (1972) |
| | MeOH, EG | C | | Runge (1972) |
| 3,3′-diethylthiadicarbocyanine iodide, | MeOH | L | 731 | Schäfer et al. (1966) |
| DTDC iodide | Acetone | L | 711 | Miyazoe and Maeda (1968) |
| | EG | L | 728 | Miyazoe and Maeda (1970) |
| | MeOH | C | | Runge (1971) |
| | DMSO | F | 759 | Maeda and Miyazoe (1972) |

List of Laser Dyes (cont.)

| Dye | Solvent [a] | Excit. [b] | Lasing Wavelength [nm] | Ref. |
|---|---|---|---|---|
| 3,3′-diethyl-5,5′-dimethoxy-6,6′-bis (methylmercapto)-10-methyl-thiadicarbocyanine bromide | EtOH | L | 727–739 | Derkacheva et al. (1968a) |
| 3,3′-diethyl-10-chloro-(4,5,4′,5′-dibenzo)-thiadicarbocyanine iodide | Acetone | L | 774 | Miyazoe and Maeda (1968) |
| 3,3′-diethyl-10-chloro-(5,6,5′,6′-dibenzo)-thiadicarbocyanine iodide | Acetone | L | 714 | Miyazoe and Maeda (1968) |
| 1,3,3,1′,3′,3′-hexamethylindodicarbocyanine iodide | DMSO | F | 740 | Maeda and Miyazoe (1972) |
| 1,1′-diethyl-2,2′-dicarbocyanine iodide | Glycerol | L | 760–790 | Spaeth and Bortfeld (1966) |
| | EG | L | 812 | Miyazoe and Maeda (1970) |
| 1,1′-diethyl-11-bromo-2,2′-dicarbocyanine iodide | Glycerol | L | 815 | Miyazoe and Maeda (1968) |
| 1,1′-dimethyl-11-bromo-2,2′-dicarbocyanine iodide | Glycerol | L | 745 | Miyazoe and Maeda (1968) |
| 1,1′-diethyl-11-cyano-2,2′-dicarbocyanine fluoborate | MeOH | L | 740 | Schäfer et al. (1966) |
| | $C_5H_5N$ | L | 768 | Bradley et al. (1968b) |
| 1,1′-diethyl-11-acetoxy-2,2′-dicarbocyanine fluoborate | MeOH | L | 797 | Schäfer et al. (1966) |
| 1,1′-diethyl-4,4′-dicarbocyanine iodide | EG | L | 930 | Miyazoe and Maeda (1970) |
| 1,1′-diethyl-11-bromo-4,4′-dicarbocyanine iodide | MeOH | L | 830 | Miyazoe and Maeda (1968) |
| 1,1′-diethyl-11-nitro-4,4′-dicarbocyanine fluoborate | MeOH | L | 796 | Schäfer et al. (1966) |
| | Acetone | L | 814 | Schäfer et al. (1966) |
| | DMF | L | 815 | Schäfer et al. (1966) |
| | $C_5H_5N$ | L | 821 | Schäfer et al. (1966) |
| 3,3′-diethyloxatricarbocyanine iodide, DOTC iodide | Acetone | L | 742 | Miyazoe and Maeda (1968) |
| | EtOH | L | 718–739 | Derkacheva et al. (1968a) |
| | EG | L | 771 | Miyazoe and Maeda (1970) |
| 3,3′-dimethyloxatricarbocyanine iodide | Acetone | L | 744 | Miyazoe and Maeda (1968) |
| | DMSO | F | 809 | Maeda and Miyazoe (1972) |

| | | | | |
|---|---|---|---|---|
| 3,3′-diethylthiatricarbocyanine bromide, DTTC bromide | MeOH | L | 813, 835 | Schäfer et al. (1966) |
| | Acetone | L | 808 | Schäfer et al. (1966) |
| 3,3′-diethylthiatricarbocyanine iodide, DTTC iodide | MeOH | L | 816 | Sorokin et al. (1966a) |
| | DMF | L | 808 | Sorokin et al. (1967) |
| | Glycerol | L | 810 | Sorokin et al. (1967) |
| | DMSO | L | 816 | Sorokin et al. (1967) |
| | Acetone | L | 829 | Miyazoe and Maeda (1968) |
| | DMSO | F | 863, 889 | Maeda and Miyazoe (1972) |
| 3,3′-diethyl-11-methoxythiatricarbocyanine iodide | EtOH | L | 773–798 | Derkacheva et al. (1968a) |
| 3,3′-diethyl-5,6,5′,6′-tetramethoxythiatricarbocyanine iodide | Acetone | L | 853 | Miyazoe and Maeda (1968) |
| 3,3′-diethyl-(4,5,4′,5′-dibenzo)-thiatricarbocyanine iodide | Acetone | L | 860 | Miyazoe and Maeda (1968) |
| 3,3′-diethyl-6,7,6′,7′-dibenzothiatricarbocyanine iodide | EtOH | L | 824–853 | Derkacheva et al. (1968a) |
| 3,3′-diethyl-6,7,6′,7′-dibenzo-11-methylthiatricarbocyanine iodide | EtOH | L | 843–869 | Derkacheva et al. (1968a) |
| 3,3′-diethylselenatricarbocyanine iodide | Acetone | L | 826 | Miyazoe and Maeda (1968) |
| | EG | L | 850 | Miyazoe and Maeda (1970) |
| 1,3,3,1′,3′,3′-hexamethylindotricarbocyanine iodide | EtOH | L | 779–808 | Derkacheva et al. (1968a) |
| | Acetone | L | 819 | Miyazoe and Maeda (1968) |
| | | F | 800 | Furumoto and Ceccon (1970) |
| | EG | L | 836 | Miyazoe and Maeda (1970) |
| 1,3,3,1′,3′,3′-hexamethyl-4,5,4′,5′-dibenzoindotricarbocyanine perchlorate | EtOH | L | 816–833 | Derkacheva et al. (1968a) |
| 1,1′-diethyl-2,2′-tricarbocyanine iodide | Acetone | L | 898 | Miyazoe and Maeda (1968) |
| | EtOH | L | 886–898 | Derkacheva et al. (1968a) |
| 1,1′-diethyl-4,4′-tricarbocyanine iodide, xenocyanine | Acetone | L | 1000 | Miyazoe and Maeda (1968) |
| 3,3′-diethyl-2,2′-(5,5′-dimethyl)-thiazolinotricarbocyanine iodide | Glycerol | L | 717 | Miyazoe and Maeda (1968) |
| 3-ethyl-3′-methylthiathiazolinotricarbocyanine iodide | EtOH | L | 738–801 | Derkacheva et al. (1968a) |

List of Laser Dyes (cont.)

| Dye | Solvent [a] | Excit. [b] | Lasing Wavelength [nm] | Ref. |
|---|---|---|---|---|
| 3,3′-diethyl-12-ethylthiatetracarbo-cyanine iodide | EtOH | L | 916–924 | Derkacheva et al. (1968a) |
| 1,1′-diethyl-13-acetoxy-2,2′-tetra-carbocyanine iodide | DMSO | L | 1100 | Miyazoe and Maeda (1970) |
| 3,3′-diethyl-9,11 : 15,17-dineopenty-lenethiapentacarbocyanine iodide | $C_6H_5NO_2$ | L | 1093 | Derkacheva et al. (1968b) |
| | MeOH | L | 1100 | Schäfer (1970) |
| Pentacarbocyanine dye | MeOH | L | 1120 | Varga et al. (1968) |
| | $C_6H_5NO_2$ | L | 1175 | Varga et al. (1968) |

[a] $CH_2Cl_2$ dichloromethane, $CH_3CN$ acetonitrile, $C_5H_5N$ pyridine, $C_5H_{11}OH$ isoamyl alcohol, $C_6H_4Cl_2$ o-dichlorobenzene, $C_6H_5CH_3$ toluene, $C_6H_5CF_3$ $\alpha,\alpha,\alpha$-trifluorotoluene, $C_6H_5NO_2$ nitrobenzene, $C_6H_6$ benzene, $C_6H_{10}O$ cyclohexanone, $C_6H_{12}$ cyclohexane, $C_7H_{16}$ n-heptane, DCE dichloroethane, DMF N,N-dimethylformamide, DMSO methyl sulfoxide, DPA N,N-dipropylacetamide, EG ethylene glycol, EtOH ethanol, HFIP hexafluoroisopropanol, $H_2O$ water, $H_2SO_4$ sulfuric acid, IP isopentane, MCH methylcyclohexane, MeCs methyl cellosolve, MeOD monodeuterated methanol, MeOH methanol, NMP 1-methyl-2-pyrrolidinone, PMMA polymethylmethacrylate, PSt polystyrene, TFE trifluoroethanol, THF tetrahydrofuran.

[b] L laser-pumped, F flashlamp-pumped, C continuous (laser-pumped).

# 6. Laser Wavemeters

James J. Snyder and Theodor W. Hänsch

With 11 Figures

The advent of dye lasers that are not only widely tunable but also highly monochromatic (Hänsch 1972) has had a profound impact on many areas of optical spectroscopy. However, the power and potential of tunable laser sources cannot be fully exploited unless the laser wavelength is accurately known as it is tuned. Unfortunately, conventional methods for measuring optical wavelengths tend to be either inadequate in their resolution and accuracy or else cumbersome and slow. Laser spectroscopists and other scientists soon recognized the need for new measuring instruments comparable in utility to the frequency counters commonly used in the microwave region. Research efforts directed at the development of such instruments, which became known generically as "laser wavemeters", began in the mid 1970s and rapidly expanded as dye lasers and, more recently, other tunable laser sources became available commercially. In this chapter we attempt to review the more successful laser wavemeters developed to date. It should be noted, however, that many problems related to laser wavelength metrology remain to be solved, and that wavemeter technology continues to develop and improve.

An ideal wavemeter should be able to determine the exact laser wavelength quickly, conveniently, and accurately at any time. By "quickly" we mean that the measured wavelength should be available in a time short enough to permit the user to interact with the experiment. Optimally, the measurement and processing time for the wavemeter should be shorter than the human reaction time of a few hundredths of a second in order to facilitate operations such as tuning the laser to a particular wavelength. By "conveniently" we mean that the wavemeter should be reasonably compact and easy to acquire (by either building or purchasing) and to use. "Accurate" means different things depending on the laser and on the application, but in practice generally implies a measurement uncertainty below $10^{-6}$ of the optical frequency (about 500 MHz for visible light).

Most approaches to wavelength metrology are based either on comparison with a reference spectrum, on a spectrograph, on an interferometer, or on direct frequency measurement. We shall consider each of these approaches separately, but with emphasis on the interferometric techniques. These seem to provide the most likely candidates for practical wavemeters for the near-term future, although the long-term prospects for wavemeters based on direct frequency measurements are promising.

## 6.1 Wavelength Measurements Based on Reference Spectra

It is a time-honored technique in classical spectroscopy to use reference spectra to calibrate grating spectrograph measurements in situ. The reason is that spectrographs are generally better suited to the accurate measurement of small wavelength differences rather than absolute wavelengths. Similarly, a reference spectrum can be used in laser spectroscopy for accurate wavelength measurements, as seen for example in (Berg et al. 1976). Generally, this approach requires a continuous, smooth scan of the laser frequency that encompasses both the spectral feature in question and a known reference line, followed by a measurement of their separation in order to determine the wavelength. The only fundamental requirements are that a catalogued reference spectrum exist for the wavelength region of interest, including specific lines that lie within the scanning range of the laser, and that the reference wavelengths be known with sufficient accuracy. The extrapolation or interpolation from the reference line(s) to the spectral feature is accomplished by assuming a linear frequency or wavelength scan of the laser and, possibly, by simultaneously recording transmission fringes of a reference Fabry-Perot interferometer of known spacing. The major difficulty with this approach, other than its slowness, lies in the initial determination of the approximate laser wavelength. If there is no other source of wavelength information, the spectroscopist is often reduced to scanning the laser over several reference lines and then comparing the recorded spectral patterns with the catalogued spectra.

In addition to being useful in direct wavelength measurements by interpolation or extrapolation, reference spectra have a vital role in the calibration of interferometer-based wavemeters. As discussed below, a crucial test for any interferometric wavemeter is the accuracy with which it can measure the wavelength of a known reference line. Spectral catalogues of known accuracy are a necessity for determining and demonstrating the residual level of systematic errors in any wavemeter.

Reasonably accurate, extensive spectral catalogues of sufficient density to be of general use in the visible have been generated by Fourier-transform spectrometry for the molecules $I_2$ (Gerstenkorn and Luc 1978) and isotopically pure $^{130}Te_2$ (Cariou and Luc 1980), and for atomic thorium (Palmer and Engelman 1983) and uranium (Palmer et al. 1980, 1981). In addition, less extensive but more accurate lists of selected lines of iodine (Hlousek and Fairbank 1983; Bönsch 1985), helium (Hlousek et al. 1983) and the alkalis potassium (Lorenzen et al. 1981; Thompson et al. 1983), rubidium (S. A. Lee et al. 1978; Stoicheff and Weinberger 1979; Sansonetti and Weber 1985) and cesium (Lorenzen and Niemax 1983; O'Sullivan and Stoicheff 1983) have been published. Finally, for ultimate accuracy, there are the reference lines recommended as secondary length standards by the Conférence Général des Poids et Mesures (CGPM) in the new definition of the meter (Metrologia 1984).

Diatomic iodine, which covers the spectral range 500–676 nm, provides the most convenient of the commonly used reference spectra. The iodine spectrum can readily be observed in a cell at room temperature by monitoring either

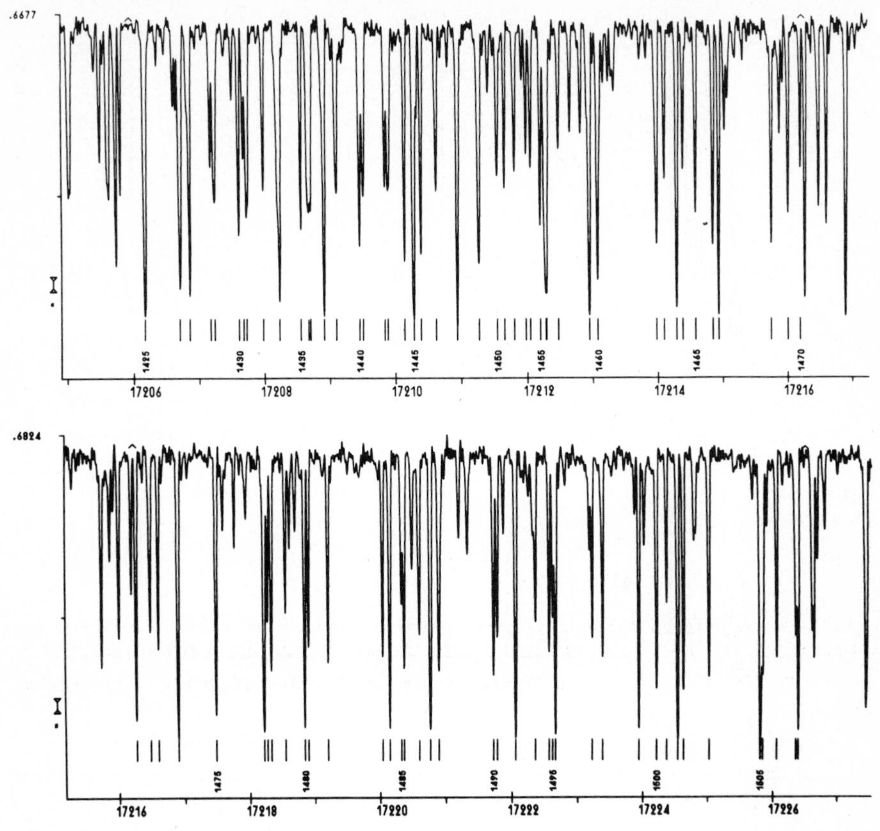


**Fig. 6.1.** A sample spectrum from the iodine atlas (Gerstenkorn and Luc 1978). The atlas catalogues the absorption lines of diatomic iodine from 500 to 676 nm

direct laser absorption or laser-excited fluorescence, and usually there will be at least several absorption lines within a laser scan range of one wavenumber. A sample spectrum from the published iodine atlas (Gerstenkorn and Luc 1978) is shown in Fig. 6.1. The tabulated wavenumbers have been found to have an offset of $0.0056\,cm^{-1}$ (Gerstenkorn and Luc 1979), but after correction are estimated to be accurate to $\pm 0.002\,cm^{-1}$ (about $10^{-7}$ of the optical frequency), although the asymmetry of the absorption lines due to the unresolved underlying hyperfine structure may degrade that accuracy somewhat for laser applications. Iodine cells suitable for providing accurate reference spectra are commercially available[1].

The $^{130}Te_2$ spectrum from 420 to 541 nm is similar in density to, and nicely complements the spectrum of iodine, but tellurium requires operation at elevated temperatures of 500°–600 °C. The accuracy of the tellurium atlas is probably comparable to that of the iodine atlas, but there are no published cor-

[1] Opthos Instruments, Inc., 17805 Caddy Dr., Rockville, MD 20855, USA.

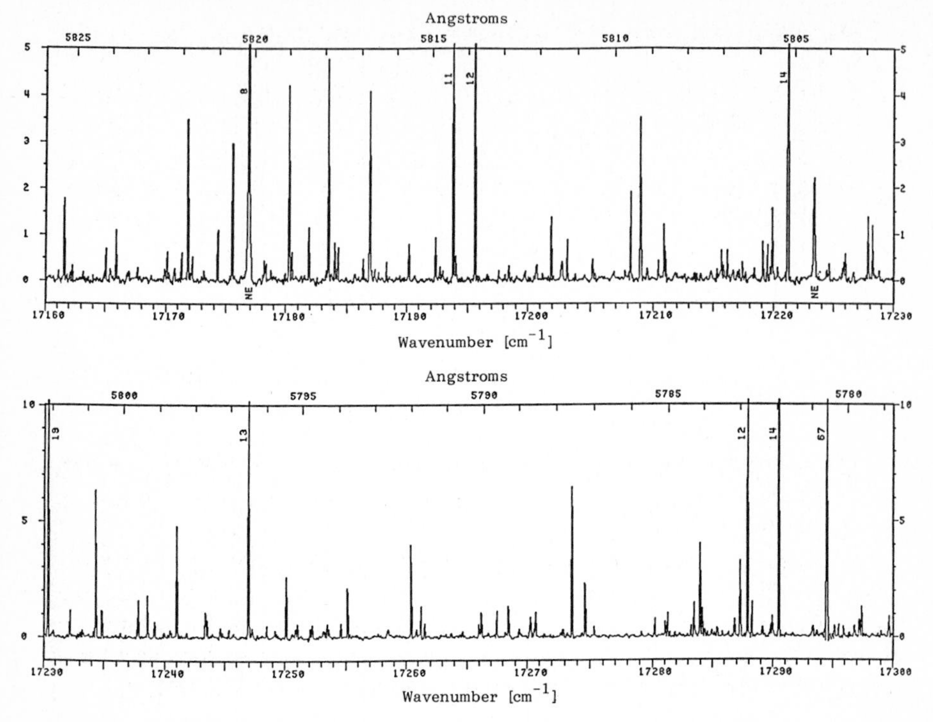


**Fig. 6.2.** A sample spectrum from the uranium atlas (Palmer et al. 1980). The atomic uranium spectrum provides known reference lines from 385 to 909 nm

roborating measurements. Optical cells of isotopically pure $^{130}Te_2$ are also available commercially[1].

The uranium and thorium atomic spectra are considerably less dense than the molecular spectra of iodine and tellurium, which can be a problem for the initial determination of the wavelength with lasers of limited continuous tuning range. On the other hand, the individual lines are generally very symmetric, and the spectra cover much broader regions. The uranium atlas extends from 385 to 909 nm, and the thorium atlas from 278 to 1350 nm. Figure 6.2 shows a sample spectrum from the uranium atlas. Both uranium and thorium sources are available commercially as hollow cathode lamps[2]. Absorption lines in these lamps can be observed conveniently by optogalvanic spectroscopy (R. B. Green et al. 1976; King et al. 1977; King and Schenck 1978). The estimated accuracy for the atlases of uranium and thorium is about $\pm 0.003\ \mathrm{cm}^{-1}$ (depending on line strength), and a number of the lines have been independently measured to substantially higher accuracy (Palmer et al. 1981; Sansonetti and Weber 1984).

[2] Westinghouse Industrial and Government Tube Division, Westinghouse Circle, Horseheads, NY 14845, USA.

## 6.2 Wavelength Measurements Based on Spectrographs

The classical instrument for determining the wavelength of any light source is the spectrograph. Small grating spectrographs are still commonly used for this purpose in laser labs since they have the advantage of broad spectral coverage and are reasonably simple to operate. Moreover, for broad sources they not only give information about the wavelength of the line center but also about the line shape of the source. They are deficient primarily in their rather limited resolution and accuracy. The resolution of a typical quartermeter monochromator is about 0.01 nm (about $10^{-5}$ in the visible), and the accuracy is somewhat worse unless a reference spectrum is included.

A small echelle-based wavemeter with a simple He-Ne reference laser and a photodiode-array readout has been developed (Morris and McIlrath 1979) and is commercially available[3]. This instrument has an estimated accuracy of 0.01 nm, and resolution (using a Fabry-Perot attachment) of 0.001 nm. It does not give the wavelength directly, but requires a separate calculation after each measurement.

A wavemeter based on a self-calibrating grating has been proposed (Hänsch 1977). As illustrated in Fig. 6.3, it incorporates a reference laser and multiple rulings on the grating surface to produce a wavelength scale superimposed on the spectrum being measured. The reference wavelength scale reduces systematic errors, and also simplifies the wavelength determination.

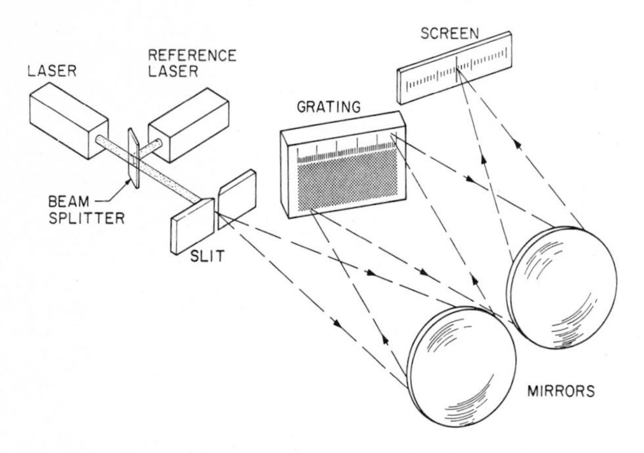


**Fig. 6.3.** Grating spectrograph with self-calibrating grating. Multiple rulings on the grating are illuminated by a reference laser to superimpose a wavelength scale onto any spectrum (Hänsch 1977)

[3] Candela Corporation, 19 Strathmore Rd., Natick, MA 01760, USA.

## 6.3 Wavelength Measurements Based on Interferometry

By far the greatest and most diverse research efforts in wavemeters have been directed at developing instruments employing interferometry. In general, most interferometer-based wavemeters can be conveniently separated into two categories: dynamic and static. Dynamic wavemeters require mechanical motion to change the optical path difference of the interferometer, whereas static wavemeters do not. As a result, dynamic wavemeters are mostly suitable only for cw laser wavelength measurements, whereas static wavemeters may be used with either pulsed or cw lasers. Many of the interferometric wavemeters have been described in a review article by Snyder (1982), and wavemeters suitable for use with pulsed lasers were reviewed by Reiser (1988). An earlier review article by Solomakha and Toropov (1977) described many of the basic problems and approaches.

One of the difficulties encountered by developers of interferometric wavemeters is sensitivity to optical alignment and to wavefront quality. Most wavemeters utilize resolution-enhancing techniques in order to achieve the maximum measurement precision with a limited number of fringes. In effect, these methods measure distances to small fractions of an optical fringe. However, for the measurements to be accurate, the optical alignment and the fringe quality must be very good. In particular this usually implies that both the alignment and the wavefront flatness must be very reproducible under all conditions of use and all wavelengths of the unknown laser. Ishikawa (1986) has described the use of a single-mode polarization-preserving optical fiber to carry both the reference and the unknown laser beam to the entrance port of the wavemeter. This technique ensures good alignment of the two orthogonally polarized laser beams.

An analysis of potential systematic errors in a Michelson interferometer used for wavelength comparison has been given by Monchalin et al. (1981). Dorenwendt and Bönsch (1976) have analyzed the inherent wavefront curvature of a laser beam of finite aperture and Lichten (1986) has analyzed the wavelength-dependent phase shift due to the use of dielectric mirror coatings. Some specific sources of error in dynamic Michelson and static Fabry-Perot wavemeters are also discussed by Castell et al. (1985).

When possible in our discussions of the various types of interferometric wavemeters, we shall report or estimate the demonstrated accuracy and the fractional fringe resolution of each instrument. By demonstrated accuracy, we mean the accuracy in measuring the position of known lines, or of lines later measured by other means. Of course it should be noted that a full test of the true accuracy requires a large number of calibration measurements at wavelengths throughout the relevant spectral region in order to sufficiently sample the effects of arbitrary periodic and dispersion-type systematic errors. Unfortunately, there are only a few lines known with sufficient accuracy to calibrate wavemeters at or below the $10^{-8}$ level (Metrologia 1984; Pollock et al. 1983; Jennings et al. 1983; Bönsch 1983a, b) so most researchers have reported calibration results for only a very few points. The estimated accuracies should therefore be accepted with caution.

We have chosen not to emphasize the frequency or wavelength resolution of the various wavemeters because that depends not only on the fractional fringe resolution but also on the total path change, which is not necessarily a fundamental property of the wavemeter. The fractional fringe resolution, on the other hand, indicates how well the interval between fringes (i.e., the free spectral range) can be subdivided. This value can be converted easily into wavelength resolution by determining the total path difference change of the interferometer, and it simplifies the comparison of wavemeters based on different classes of interferometers.

Many of the interferometric wavemeters incorporate a stabilized reference laser in order to reduce the sensitivity of a measurement to mechanical stability. Accurate measurements at or below the $10^{-9}$ level generally would require an iodine-stabilized He-Ne laser such as has been described by Schweitzer et al. (1973) and Layer (1980). Iodine-stabilized He-Ne lasers are commercially available.[4] For more modest accuracy requirements in the region of $10^{-8}$, simpler stabilized lasers suitable for use as a wavemeter reference have been developed (Balhorn et al. 1972; Bennett et al. 1973; Gordon and Jacobs 1974; Umeda et al. 1980; Baer et al. 1980; Zumberge 1985), and are available from a number of manufacturers.[5]

### 6.3.1 Dynamic Michelson Wavemeters

Basically, a dynamic Michelson wavemeter is a Michelson (or similar) interferometer in which the optical path difference of the two arms can be changed continuously and smoothly. The interferometer is simultaneously illuminated by the laser whose wavelength is to be measured, and by a separate reference laser of known wavelength, as illustrated in Fig. 6.4. The changing path difference causes the fringe intensity for each laser beam at the output port to oscillate at a frequency $f = (1/\lambda) dl/dt$, where $\lambda$ is the laser wavelength in the interferometer medium, and $dl/dt$ is the rate of change of the optical path difference. The ratio of the fringe frequencies of the two lasers is approximately equal to the inverse ratio of their wavelengths. The unknown wavelength can therefore be found by multiplying the known wavelength by the easily measured ratio of the fringe frequencies.

Dynamic Michelson wavemeters are usually not suitable for pulsed lasers, which often have pulsewidths and duty cycles too small to allow the wavelength to be determined from samples fringes. For cw lasers, however, the performance of even very simple Michelson wavemeters is generally quite good. And since dynamic fringe-counting wavemeters are reasonably easy to construct, they are the kind most commonly found in spectroscopy laboratories.

---

[4] Frazier Precision Instrument Co., Inc., 210 Oakmont Ave., Gaithersburg, MD 20877, USA.

[5] Coherent, Inc., 3210 Porter Dr., P.O. Box 10321, Palo Alto, Ca 94304, USA; Laboratory for Science, 2821 9th St., Berkeley, CA 94710, USA; Spectra-Physics Inc., 1250 W. Middlefield Rd., Mountain View, CA 94042, USA; Spindler & Hoyer GmbH & Co., Königsallee 23, Postfach 3353, D-3400 Göttingen, FRG; Newport Corporation, P.O. Box 8020, Fountain Valley, CA 92728.

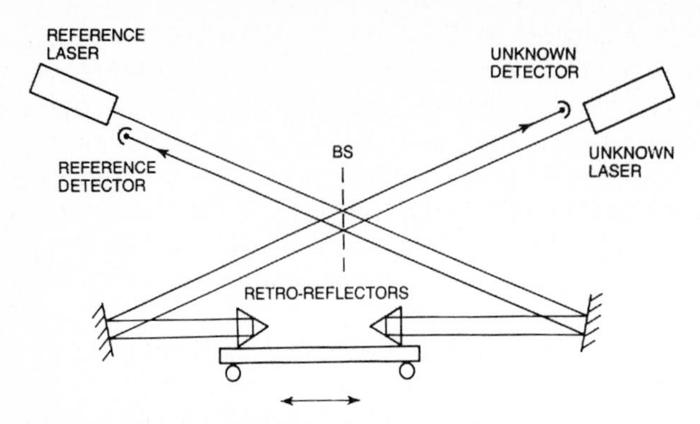


**Fig. 6.4.** Scheme of a dynamic Michelson wavemeter, changing the path length in both arms simultaneously with back-to-back corner cubes on a rolling cart (Snyder 1982)

The earliest dynamic Michelson wavemeters (Bukovskii et al. 1974; F. V. Kowalski et al. 1976, 1977; Hall and Lee 1976; S. A. Lee and Hall 1977; Rowley et al. 1977) generally had total retro-reflector motion of several tens of centimeters.[6] A variety of methods have been used to enhance the resolution over the $\sim 10^{-6}$ available from simple fringe counting.

The wavemeter of Bukovskii et al. (1974) used a single moving corner cube, which gave a total path change near $10^6$ fringes in the visible. The fringes could be subdivided to about $10^{-2}$ by using a zero-phase or fringe coincidence technique in which the fringe counting gate opened and closed only when the fringes from the two lasers were in phase. Limited evidence at that time suggested an accuracy below $10^{-7}$. More-recent results from a similar instrument (Kahane et al. 1983) have demonstrated measurement reproducibility to about $10^{-8}$.

An extension of the fringe coincidence approach is to use the many coincidences that typically occur during a single long scan of the interferometer path length in order to make a large number of separate wavelength measurements with reduced resolution (Ishikawa et al. 1986). At the end of the scan, the multiple measurements are averaged, thus increasing the single-measurement wavelength resolution. This approach offers faster wavelength updates (1 – 3 s) with moderate resolution, and a high resolution average at the end of the scan (50 s). In addition, bad measurements due to fringe dropout (e.g., due to bubbles in a dye laser jet) and other noise can be deleted before averaging. However, since the averaging process improves the resolution only as the square root of the number of measurements, there will be a net loss of resolution compared to a single measurement utilizing the full interferometer path change. This instrument was used to compare iodine-stabilized lasers at 612 nm and 633 nm, and showed a reproducibility of $4 \times 10^{-10}$.

---

[6] A commercial dynamic Michelson wavemeter (Coherent Optics model 404), which was briefly on the market in the early 1970s, had only a few millimeters of mirror motion. However, this instrument only offered resolution of about 0.1 nm ($10^{-5}$).

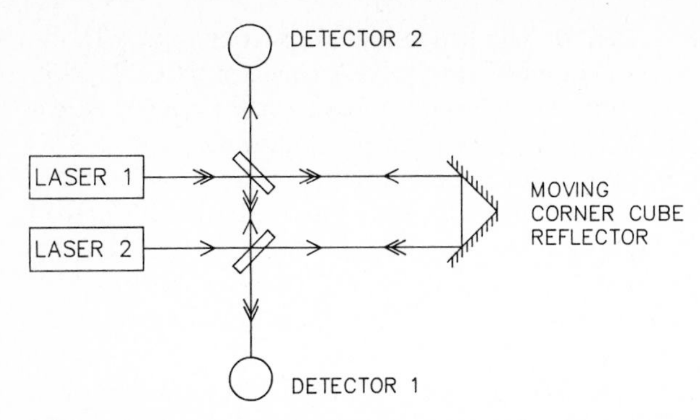


**Fig. 6.5.** A dynamic Michelson wavemeter allowing the same optical path for counter-propagating beams from the reference and unknown lasers (F. V. Kowalski et al. 1976)

Kowalski et al. (1976, 1977) used an optical arrangement similar to that of Bukovskii et al., except that counter-propagating reference and unknown laser beams followed the same optical path within the interferometer (Fig. 6.5). This geometry offers good immunity from various possible perturbations such as mirror vibrations. The moving corner cube traveled on an air track, which provided relatively smooth and virtually frictionless motion. The path difference change available gave about $2.5 \times 10^6$ fringes, and electronic frequency quadrupling was used to subdivide fringes by four. An accuracy of $6 \times 10^{-8}$ based on calibration at a few known points was reported.

Hall and Lee (1976; S. A. Lee and Hall 1977), in the instrument they named the "lambda-meter", further increased the path difference change to about $6.5 \times 10^6$ fringes by using dual back-to-back corner cubes on a rolling cart to change both arms of the interferometer simultaneously, and by increasing the physical distance moved. They also introduced phase-locked frequency multiplication to subdivide the fringes by as much as 100. Their reported accuracy for a number of measurements was $2 \times 10^{-7}$.

Other Michelson wavemeters, also using phase-locked fringe subdivision, have since been developed in other laboratories (Petley and Morris 1978; F. V. Kowalski et al. 1978; Cachenaut et al. 1979; Nagai et al. 1980). Recent results by Braun et al. (1987) have shown good performance, with fringe division by 128 and measurement of a number of lines of the iodine spectrum agreeing with the atlas of Gerstenkorn and Luc (1978) to $5 \times 10^{-8}$. A commercial dynamic Michelson wavemeter using phase-locked frequency multiplication to subdivide fringes by 16 is available[7]. This instrument offers accuracy in the $10^{-6}$ range.

Another technique for subdividing fringes from a dynamic Michelson was reported by Bennett and Gill (1980). In their digital interpolation technique, the fringe fraction is found to about 1% by measuring the time delay between the reference and unknown fringe signal zero crossings at the beginning and

[7] Burleigh Instruments, Inc., Burleigh Park, Fishers, NY 14453, USA.

at the end of a measurement period. The fringe fraction is combined with the integer fringe number, which is counted separately. A single point calibration using the orange He-Ne laser line at 612 nm is the basis for their reported accuracy of a few parts in $10^8$. A similar approach by J. Kowalski et al. (1985) included additional logic to correct for fringe dropout.

A Michelson-type wavemeter in which a rotating fused silica parallelepiped changes the optical path length has been reported by Docchio et al. (1985). This instrument uses fractional fringe counting circuitry (Bennett and Gill 1980) to subdivide the fringes by about 50. The reported accuracy for a number of $Ar^+$ laser lines was a few parts in $10^6$.

A tilt-compensated (Steel 1967) Michelson wavemeter has been described (Xia et al. 1981) that is so insensitive to transverse motion of the corner cube that a model train could be used to provide the motion. Since the optical configuration couples light from the interferometer back into the two illuminating lasers, attenuation or optical isolation may be necessary to avoid destabilizing them. A similar wavemeter reported by Beigang et al. (1982) used phase-locked frequency multiplication to subdivide fringes by 10. Accuracy results have not been reported.

Snyder et al. (1981) reported a method for increasing the subdivision of a fringe by several orders of magnitude using an electronic instrument called a "frequency meter". This instrument (Snyder 1981 b) uses a real-time digital averaging algorithm to increase the fringe resolution by an amount proportional to the square root of the number of fringes in a measurement interval. Fringe subdivision by $10^5$ has been achieved with this instrument, but accuracy tests have not been completed.

Finally, in view of the obvious similarities of a dynamic Michelson wavemeter and a Fourier transform spectrometer, it is perhaps surprising that a Fourier transform wavemeter has only recently been reported (Junttila et al. 1987). This approach offers the advantage of operation with multimode lasers, which are generally a problem for the usual wavemeters, as well as with single-mode lasers. In order to do the Fast Fourier Transform (FFT) on a personal computer, Junttila et al. have chosen to reduce the data set by undersampling the scanning fringes, at the cost of some additional software complexity in order to alleviate the effects of spectral aliasing.

### 6.3.2 Dynamic Fabry-Perot Wavemeters

The principal difference between Michelson and Fabry-Perot interferometers is that the former is a two-beam interferometer whereas the latter is a multibeam interferometer. By exploiting the interference of many waves, which have made different numbers of round trips, a Fabry-Perot interferometer can provide much sharper and narrower interference fringes relative to the fringe spacing.

Two wavemeters based on a scanning Fabry-Perot interferometer have been described. Pole et al. (1980) compare the time interval between fringes from a reference laser with the time interval between fringes from the unknown laser when both illuminate a linearly scanning spherical Fabry-Perot interferometer. The fringe interval could be subdivided by $10^5$, so that a scan over only one or two fringes provided a resolution of $10^{-5}$. This resolution could be further improved by averaging over a number of measurements.

Salimbeni and Pole (1980) also use a scanning Fabry-Perot with a reference laser, but the fringe ratio is determined by an electronic coincidence technique similar to that described by Bukovskii et al. (1974). They report a fringe interval subdivision of 100 and a scan of $\sim 10^5$ fringes. Reported accuracy, based on a single-point calibration, is about $10^{-7}$.

An unusual application of a Fabry-Perot interferometer to the measurement of laser wavelengths was reported by DeVoe and Brewer (1984). This method takes advantage of the extreme sensitivity with which the conversion of a laser frequency modulation into an amplitude modulation can be detected (Bjorklund 1980; Hall et al. 1981). With the help of dual-frequency modulation, a laser frequency can be locked to a Fabry-Perot cavity, while the cavity mode spacing is simultaneously locked to an rf source. In the absence of phase dispersion in the mirror coatings, the optical frequency is then an integer multiple of the rf frequency, which can be measured by conventional counting techniques. This approach may ultimately offer accuracy competitive with direct frequency-measuring techniques, although accuracy tests have not yet been reported.

### 6.3.3 Static Michelson Wavemeters

A static Michelson wavemeter has been developed by Juncar and Pinard (1975, 1982; Jacquinot et al. 1977). This instrument, which is called the "sigma-meter" and illustrated in Fig. 6.6, polarization-encodes the interference phase

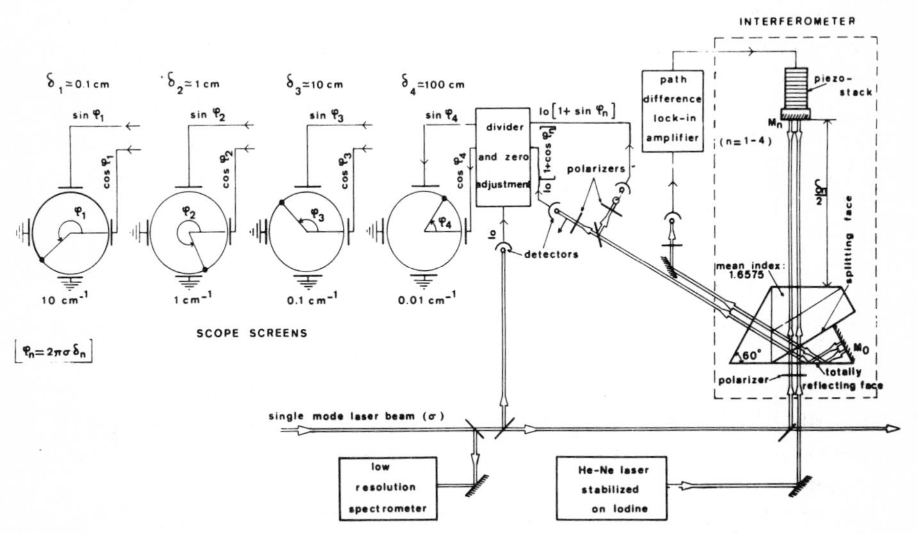


**Fig. 6.6.** Scheme of a fourfold static Michelson interferometer ("sigma-meter"), with polarization encoding (Juncar and Pinard 1975)

of each of four Michelson interferometers of fixed path differences ranging from 0.1 cm to 100 cm. The fringe resolution for each interferometer is about $10^{-2}$ fringes. The length of each of the four interferometers is independently controlled using an iodine-stabilized He-Ne laser to reduce thermal drifts. An accuracy of $10^{-7}$ has been reported for a series of measurements of several known neon lines using a cw dye laser. Because the wavemeter is static, it can also be used with pulsed lasers.

### 6.3.4 Static Fabry-Perot Wavemeters

Byer et al. (1977) reported the development of a static wavelength meter based on multiple Fabry-Perot interferometers with free spectral ranges varying by decades from $10\,\mathrm{cm}^{-1}$ to $0.01\,\mathrm{cm}^{-1}$. The interferometer lengths are periodically calibrated using a reference laser. Pulsed or cw laser fringe patterns from the interferometers are recorded by photodiode arrays or vidicon tubes and analyzed by computer. Fringe interval subdivision by 100 has been achieved, but accuracy results have not been reported. Other static Fabry-Perot wavemeters have also been reported, with similar characteristics (Konishi et al. 1981; Fischer et al. 1981).

A wavemeter based on the Fabry-Perot interferometer is available as an accessory to one commercial cw dye laser.[8] This instrument, called the Autoscan wavemeter, has two static, optically parallel Fabry-Perot interferometers about 2.3 cm long and differing in length by about 5%. During a scan of the laser, the digitally recorded fringes of the Fabry-Perot pair form a "vernier" pattern with an effective free spectral range of 150 GHz. After the laser scan is completed, a measurement of the separation of the transmission peaks of the two interferometers determines the wavelength of each point in the scan with a resolution of about 25 MHz (equivalent to subdividing each

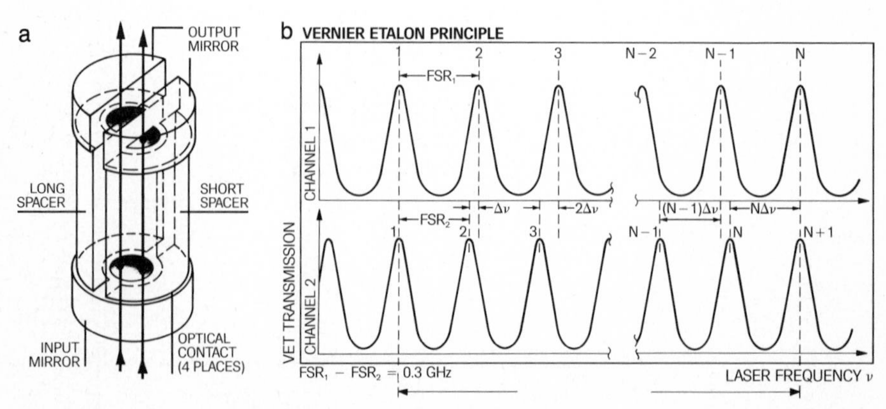


**Fig. 6.7 a, b.** Fixed vernier etalon used in the Autoscan wavemeter, a subsystem of a commercial microcomputer-controlled cw ring dye laser (Courtesy: Coherent, Inc., Palo Alto, CA)

[8] Coherent, Inc., 3210 Porter Dr., P.O.Box 10321, Palo Alto, CA 94304, USA.

Fabry-Perot fringe interval by 250). The wavemeter also uses the optical rotary power of a pair of quartz crystals to measure the approximate wavelength with sufficient accuracy to determine the order numbers of the Fabry-Perot interferometer pair. A schematic of this wavemeter is shown in Fig. 6.7. Reported accuracy is a few times $10^{-7}$.

One unusual feature of the Autoscan wavemeter is that the digital processing algorithm used requires a laser scan of about 15 GHz in order to sufficiently sample the fringes of the two Fabry-Perot interferometers. This is in contrast to the other interferometric wavemeters discussed here, which can measure the wavelength of a fixed-wavelength source. These other wavemeters either sample a spatially dispersed fringe pattern, or else change the optical path difference and sequentially sample the time-dependent fringes.

### 6.3.5 Fizeau Wavemeters

A static wavemeter based on the Fizeau interferometer has been developed by Snyder (1977, 1979a, b, 1981a) for use with either pulsed or cw lasers. The Fizeau wavemeter (Fig. 6.8) typically consists of a pair of uncoated optical flats separated by $\sim 0.1$ cm, and with a wedge of about three arc minutes ($\sim 9 \times 10^{-4}$ rad) between the surfaces. Collimated laser light reflected from the wedge produces a pattern of straight, parallel fringes, which is recorded by a linear photodiode array for processing by a microcomputer. The processing algorithm (Snyder 1980) is a three-step bootstrap that first uses the measured fringe period to find the fractional order number and estimate the wavelength, then uses the estimated wavelength to find the integer order number, and finally uses the order number together with the measured fringe phase to calculate the exact wavelength. Fringe recording and processing time is below 100 ms per measurement. Measurement accuracy requires initial calibration and depends on the stability of the interferometer, since a reference laser is not used.

The fringe resolution reported by Snyder for the Fizeau wavemeter is below $10^{-3}$ for typical cw lasers, and slightly above $10^{-3}$ for most pulsed lasers. The

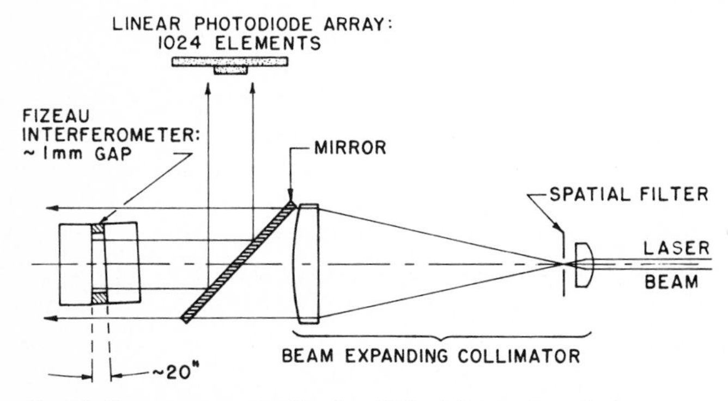


**Fig. 6.8.** Fizeau wavemeter (Snyder 1977). A linear photodiode array monitors the fringe pattern obtained in reflection from an uncoated wedge interferometer

interferometer order number is typically about 3000 at 633 nm. Accuracy tests on nine lines of an argon ion laser (Morris et al. 1984) and with a cw dye laser referenced to some two dozen iodine lines indicate an uncertainty near $3 \times 10^{-7}$. Other Fizeau wavemeters using a similar design as reported by Miller (1982) and Hackel et al. (1986) demonstrated similar performance. A commercial version of the Fizeau wavemeter[9] offers $\sim 10^{-6}$ accuracy.

Lawrence Livermore National Laboratory is currently using three Fizeau wavemeters to monitor and control the wavelengths of several pulsed dye lasers used in the laser isotope separation program (Hackel 1989). In their approach, each laser is coupled into a single-mode optical fiber, which acts as a transport system as well as a spatial filter. The desired laser light is input to the Fizeau wavemeter by means of a computer controlled fiber-optic switch, which can connect up to 18 separate fibers to the single input fiber of the wavemeter. They calibrate their wavemeters over the spectral range 560–750 nm using the iodine atlas (Gerstenkorn and Luc 1978, 1979) and report an accuracy of $\sim 1.5 \times 10^{-7}$ and a reproducibility slightly below $10^{-7}$.

A recent version of the Fizeau wavemeter using a solid wedge has been developed by Reiser and Lopert (1988). This instrument is substantially less complex than the original Fizeau wavemeters because it has no vacuum requirements, but is somewhat less accurate because of the dispersion of the solid wedge. They calibrated their wavemeter against a dynamic Michelson wavemeter, and demonstrated on accuracy of $\sim 2 \times 10^{-6}$.

Reiser et al. (1988) have recently demonstrated the ability to measure laser linewidths by analyzing the contrast of the Fizeau fringes. They calculate that this approach works for linewidths ranging approximately from the free spectral range (FSR) of the Fizeau interferometer down to 0.5% of the FSR. For the typical 1 mm spacing of the majority of these instruments, the linewidth measuring range extends from 900 MHz to over 100 GHz.

One problem of the early versions of the Fizeau wavemeter was the error in the fringe pattern introduced by the combination of wavefront curvature and shear of the light incident on the detector array. Snyder (1981a; Morris et al. 1984) solved this by choosing an optical geometry in which the detector array was located at a plane of "zero shear". In this plane, there is no shear of the two beams reflected from the wedge faces, and the fringe pattern is to first order independent of wavefront curvature. The small amount of dispersion introduced by the zero shear geometry could be corrected either with software or by introduction of a compensating plate, as is done in the commercial version[9]. Gardner (1983, 1985) developed a more compact geometry that reduced the wavefront shear by placing the photodiode array in close proximity to the wedge. This approach gave slightly degraded accuracy near $10^{-6}$. Gray et al. (1986) developed an instrument of comparable accuracy that used a different geometry. In this approach the orientation of the detector array about the direction of the incident illumination compensates for the curvature-in-

[9] Lasertechnics, Inc., 5500 Wilshire Ave. NE, Albuquerque, NM 87113, USA.

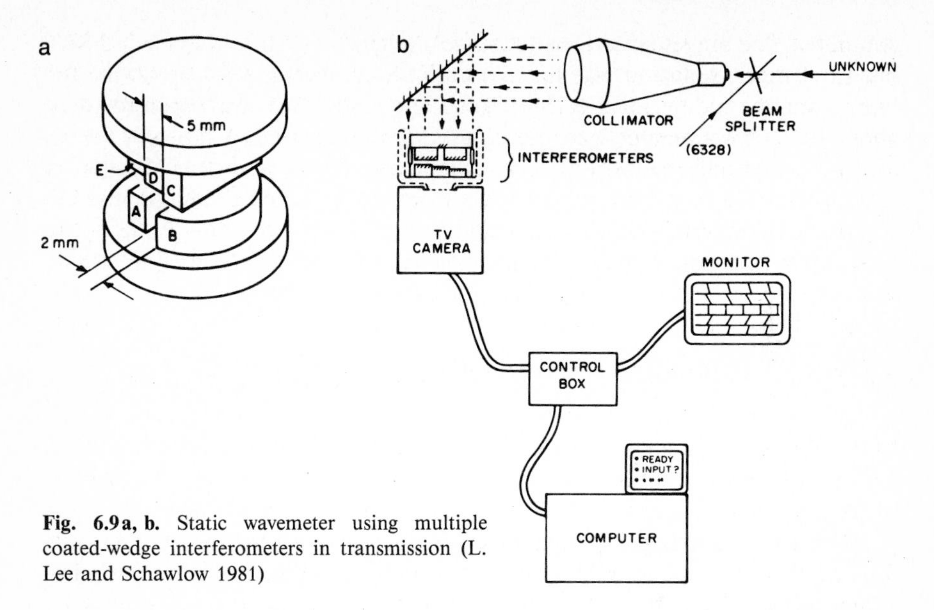


**Fig. 6.9 a, b.** Static wavemeter using multiple coated-wedge interferometers in transmission (L. Lee and Schawlow 1981)

duced fringe shifts. However, as shown by Gardner (1986), the geometry of Gray et al. requires extraordinary stability of the angle of incidence of the light on the interferometer.

A more serious problem in using the Fizeau wavemeter with typical pulsed lasers has been the relatively high error rate of the bootstrap processing algorithm in the presence of excessive spatial noise or transverse mode structure in the laser beam (Morris et al. 1984). In this mode of failure, the algorithm estimate gives the wrong integer order number, and therefore calculates a wavelength in error by the $\sim 5\ \mathrm{cm}^{-1}$ free spectral range of the Fizeau interferometer. Lasers which generate relatively "clean" output beams (i.e., $TEM_{00}$) do not appear to cause this problem.

A different type of Fizeau wavemeter using coated wedge surfaces to give multiple-beam fringes was described by L. Lee and Schawlow (1981). As shown in Fig. 6.9, their wavemeter contains a compact assembly of five Fizeau interferometers with spacings ranging from 0.0005 cm to 1.3 cm, and a common wedge angle of 20 arcsec. The interferometers are coated with aluminum, giving 80% reflectivity. All five interferometers are read out simultaneously in transmission by a single vidicon tube, and the pattern is analyzed by a microcomputer. The wavemeter is calibrated before use with a He-Ne reference laser. The reported fringe subdivision by 80 gives an instrumental resolution near $3 \times 10^{-7}$ for the longest interferometer. Reported accuracy for a few calibration points is $10^{-6}$.

Other wavemeters using coated Fizeau interferometers have been reported by C. Cahen et al. (1981) and Volkov et al. (1982). In the instrument of Volkov et al., four coated Fizeau interferometers are simultaneously read by photodiode arrays, and the recorded fringe patterns are analyzed by a microcomputer. The longest interferometer has a length of 4 cm (0.125 $cm^{-1}$ FSR), but the fringe resolution was not reported. Accuracy for tests on about 100 lines from the iodine atlas (Gerstenkorn and Luc 1978) was reported to be about $10^{-7}$. A wavemeter incorporating both an uncoated (two-beam) Fizeau and a coated (multi-beam) Fizeau interferometer was described by Cotnoir et al. (1985). In this instrument the fringes from the uncoated Fizeau are used to determine the wavelength, and the fringe pattern from the coated Fizeau gives directly the laser line shape in the manner described by Westling et al. (1984).

## 6.4 Direct Frequency Measurements

Generally the most accurate determinations of laser wavelengths have been achieved by frequency measurements, which in effect are direct comparisons of infrared or optical frequencies with microwave frequencies (Evenson et al. 1983). Unlike wavelength comparisons, frequency comparisons are relatively insensitive to diffraction effects, alignment errors, and imperfection in optical components.

Elaborate frequency synthesis chains can be designed to compare selected laser frequencies directly with the primary microwave cesium frequency standard, which is four orders of magnitude more reproducible than the former standard of length based on $^{86}$Kr. The success of such techniques led to the redefinition of the meter as "the length of the path traveled by light in vacuum during a time interval of 1/299 792 458 of a second" (Metrologia 1984). In their extension of direct frequency measurements for the first time into the visible spectral region, Pollock et al. (1983) and Jennings et al. (1983) have determined the frequencies of two molecular iodine transitions, with a total fractional uncertainty of a few parts in $10^{10}$. Following the scheme shown in Fig. 6.10, Pollock et al. employed a W-Ni point-contact metal-insulator-metal (MIM) diode to observe an rf heterodyne signal between the sum of three harmonics of one $CO_2$ laser and two harmonics of another $CO_2$ laser, with a 2.3 μm (130 THz) color center laser acting as a transfer oscillator. Nonlinear frequency doubling crystals were then used to compare the color center laser frequency with that of a 1.15 μm He-Ne laser, and finally the doubled 1.15 μm light was compared with the frequency of a dye laser locked to the iodine transition at 576 nm. Jennings et al. (1983) used a similar approach except that a He-Ne plasma was used instead of a nonlinear crystal to synthesize light at 633 nm (Klement'ev et al. 1976; Chebotayev et al. 1976) for comparison with a He-Ne laser locked to a convenient iodine transition. While such complex frequency chains are hardly suitable for routine use in the laboratory, they promise to establish a number of well-known reference line at visible frequencies, which can then serve as secondary standards for optical frequency measurements.

An interesting alternative to a conventional optical frequency synthesizer described above is the recently proposed (McIntyre and Hänsch 1989) novel optical frequency divider and synthesizer chain illustrated in Fig. 6.11. Each

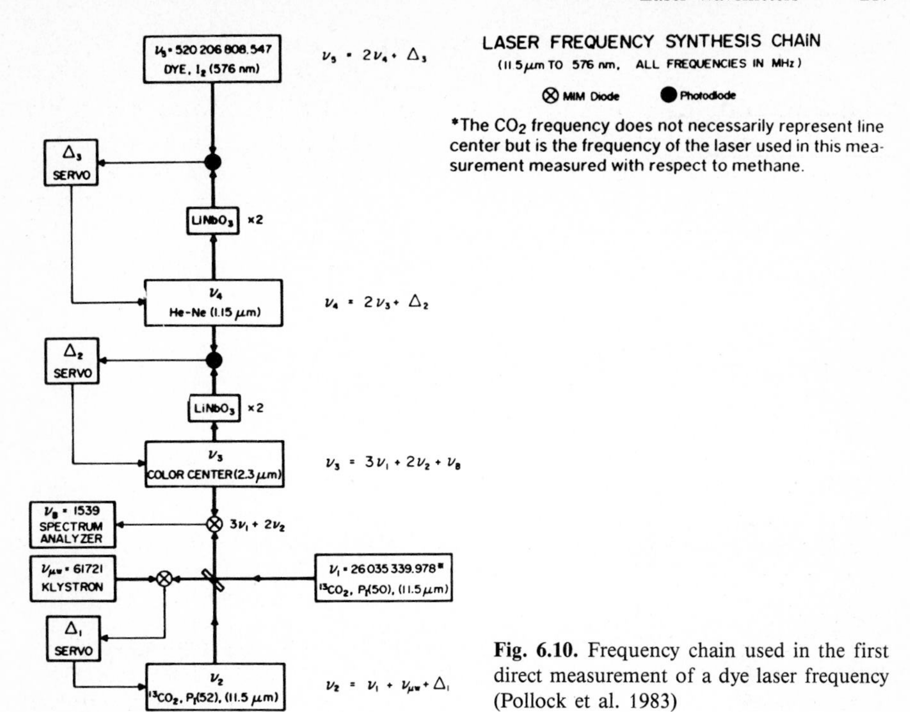


**Fig. 6.10.** Frequency chain used in the first direct measurement of a dye laser frequency (Pollock et al. 1983)

stage (Fig. 6.11 a) can be compared to a differential gear box. It receives two input frequencies $f_1$ and $f_2$, and it contains a tunable laser whose frequency $f_3$ is held at the midpoint of the two input frequencies by phase-locking the sum frequency $f_1 + f_2$ to the second harmonic of $f_3$. Several such stages can be connected in cascade (Fig. 6.11 b), successively halving the difference between two optical frequencies until the beat signal becomes accessible to a microwave counter. If the first stage uses a stable laser of frequency $f_0$ and its second harmonic as the two input frequencies, the difference frequency after $n$ stages will be $f_0/2^n$. A connection between 500 nm and the 9 GHz cesium frequency standard would require 16 stages. Although the scheme may appear complex, many of the stages can use identical components since the laser frequencies rapidly converge towards a limiting value which can be chosen either for convenience or in order to synthesize a desired optical frequency. The use of diode lasers would allow a compact system.

The task of measuring the frequency of a laser is of course much simplified if a reference laser is available that oscillates at a nearby known frequency. For regions within a few gigahertz of the reference laser, the rf difference frequency between the known and unknown lasers can be measured directly by observing a heterodyne beat signal with a fast photodetector, followed by a conventional frequency counter. Frequency-offset locking (Barger and Hall 1969) can be used to control and scan the frequency of a tunable laser with extreme precision.

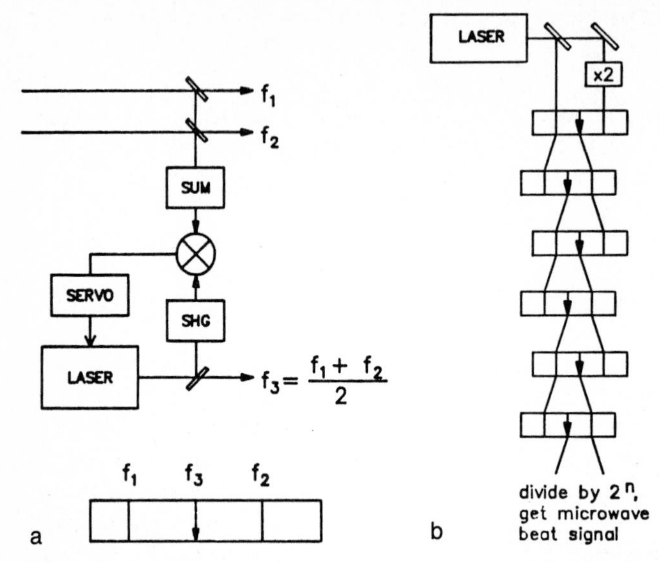


**Fig. 6.11 a, b.** Scheme for an optical frequency divider and synthesizer (McIntyre and Hänsch 1989). **(a)** Typical divide-by-two stage contains a laser whose frequency $f_3$ is locked halfway between two input frequencies $f_1$ and $f_2$. **(b)** Divide-by-$2^n$ chain to get from laser frequency to a microwave frequency

Optical difference frequency measurements have been extended to several terahertz by superposing two visible laser beams and microwave or far-infrared radiation on a MIM point-contact diode (Daniel et al. 1981; Drullinger 1983). For frequency mixing over somewhat smaller ranges, GaAs Schottky point diodes seem to offer superior sensitivity and stability, and they have been used for precise frequency-controlled operation of one single-mode ring dye laser with respect to another at a frequency difference of 243 GHz (Bergquist and Daniel 1984).

Very recently a novel type of electro-optic light modulator has generated optical sidebands 72 GHz from the carrier frequency of a 633 nm He-Ne laser (Kallenbach et al. 1989). This modulation frequency is several times higher than previously demonstrated with cw electro-optic modulators. An electro-optic crystal is placed as a resonant etalon in an open microwave Fabry-Perot resonator, and the laser beam follows a zigzag path inside the crystal due to total internal reflection so that phase matching is achieved and transit time limitations are overcome. Such a modulator can operate at any microwave frequency from a few tens of gigahertz up, limited only by the availability of millimeter-wave sources. It can facilitate the measurement of large optical difference-frequencies by shifting the beat frequency so that slower, more sensitive photodetectors can be used.

It has also been proposed that optical difference-frequencies of similar magnitude could be measured without an extremely fast photodetector by taking advantage of the far reaching comb of exactly equidistant modes of an actively mode-locked (Eckstein et al. 1978) or frequency-modulated (Hänsch and Wong 1980) dye laser.

Other new schemes will certainly continue to emerge, such as the proposed synchrotron-based optical frequency divider (Wineland 1979). The inherent accuracy of frequency metrology suggests that such methods may eventually dominate the interferometric wavemeters as rf and laser technology continues to develop.

## 6.5 Conclusion

We have attempted to outline the rapidly growing field of laser wavemeters. For completeness we have included a number of topics related to wavemeters and wavelength metrology, so that the laser scientist who needs to satisfy a wavelength measurement requirement is able to get some idea of what methods are already available. We hope that this review will solve problems for most researchers, and encourage others to study and resolve some of the many remaining problems.

*Acknowledgements.* One of us (TWH) acknowledges support by the National Science Foundation under Grant No. NSF PHY 83-08721. An earlier version of this paper was completed while one of us (JJS) was at the National Bureau of Standards and the other (TWH) was at Stanford University.

# References


Abakumov, G.A., A.P. Simonov, V.V. Fadeev, L.A. Kharitonov, R.V. Khokhlov: JETP Lett. **9**, 9 (1969a)

Abakumov, G.A., A.P. Simonov, V.V. Fadeev, M.A. Kasymdganov, L.A. Kharitonov, R.V. Khokhlov: Opto-Electron. **1**, 205 (1969b)

Ackerman, J.R., B.E. Kohler, D. Huppert, P.M. Rentzepis: J. Chem. Phys. **77**, 3967 (1982)

Alekseev, V.A., I.V. Antonov, V.E. Korobov, S.A. Mikhnov, V.S. Prokudin, B.V. Skortsov: Sov. J. Quantum Electron. **1**, 643 (1972)

Andreoni, P., P. Benetti, C.A. Sacchi: Appl. Phys. **7**, 61 (1975)

Anikiev, Yu.G., R.G. Vdovchenko, G.S.N. Telepin: Sov. J. Quantum Electron. **5**, 1 (1976)

Anliker, P., H.R. Lüthi, W. Seelig, J. Steinger, H.P. Weber, S. Leutwyler, E. Schumacher, L. Wöste: IEEE J. QE-**13**, 547 (1977)

Antonov, V.S., K.L. Hohla: Appl. Phys. B **32**, 9 (1983)

Aristov, A.V., V.A. Kuzin: Opt. Spectrosc. **30**, 77 (1971)

Aristov, A.V., E.N. Viktorova, D.A. Kozlovskii, V.A. Kuzin: Opt. Spectrosc. **28**, 293 (1970)

Armstrong, J.A.: Appl. Phys. Lett. **10**, 16 (1967)

Arthurs, E.G., D.J. Bradley, B. Liddy, F. O'Neill, A.G. Roddie, W. Sibbett, W.E. Sleat: Proc. Int. Conf. on High Speed Photography (Nice 1972a)

Arthurs, E.G., D.J. Bradley, A.G. Roddie: Appl. Phys. Lett. **20**, 125 (1972b)

Arvan, Kh.L., N.E. Zaitseva: Opt. Spectrosc. **10**, 137 (1961)

Aussenegg, F., A. Leitner: Opt. Commun. **32**, 121 (1980)

Aussenegg, F., J. Schubert: Phys. Lett. **30A**, 488 (1969)

Baer, T., F.V. Kowalski, J.L. Hall: Appl. Opt. **19**, 3173 (1980)

Balhorn, R., H. Kunzman, F. Lebowsky: Appl. Opt. **11**, 742 (1972)

Baltakov, F.N., B.A. Barikhin: Sov. J. Quantum Electron. **5**, 456 (1976)

Baltakov, F.N., B.A. Barikhin, V.G. Kornilov, S.A. Mikhnov, A.N. Rubinov, L.V. Sukhanov: Sov. Phys. – Tech. Phys. **17**, 1161 (1973)

Baltakov, F.N., B.A. Barikhin, L.V. Sukhanov: Sov. J. Quantum Electron. **4**, 537 (1974)

Baranova, E.G.: Opt. Spectrosc. **18**, 230 (1965)

Barger, R.L., J.L. Hall: Phys. Rev. Lett. **22**, 4 (1969)

Bass, M., T.F. Deutsch: Appl. Phys. Lett. **11**, 89 (1967)

Bass, M., J.I. Steinfeld: IEEE J. QE-**4**, 53 (1968)

Bass, M., T.F. Deutsch, M.J. Weber: Appl. Phys. Lett. **13**, 120 (1968)

Bastow, S.J., M.H. Dunn: Opt. Commun. **35**, 259 (1980)

Bates, B., D.J. Bradley, T. Kohno, H.W. Yates: Rev. Sci. Instrum. **34**, 476 (1968)

Becker, R.S.: *Theory and Interpretation of Fluorescence and Phosphorescence* (Wiley-Interscience, New York 1969)

Beigang, R., K. Lücke, A. Timmermann, P.J. West, D. Frölich: Opt. Commun. **42**, 19 (1982)

Bennett, S.J., P. Gill: J. Phys. E **13**, 174 (1980)

Bennett, S.J., R.E. Ward, D.C. Wilson: Appl. Opt. **12**, 1406 (1973)

Berg, E., L. Klynning, H. Martin: Opt. Commun. **17**, 320 (1976)

Bergman, A., R. David, J. Jortner: Opt. Commun. **4**, 431 (1972)

Bergquist, J.C., H. Daniel: Opt. Commun. **48**, 327 (1984)

Bethune, D.S.: Appl. Opt. **20**, 1897 (1981)

Birks, J.B.: *Photophysics of Aromatic Molecules* (Wiley-Interscience, London 1970)

Birks, J.B.: Chem. Phys. Lett. **17**, 370 (1972)
Bjorkholm, J.E., C.V. Shank: IEEE J. QE-**8**, 833 (1972a)
Bjorkholm, J.E., C.V. Shank: Appl. Phys. Lett. **20**, 306 (1972b)
Bjorkholm, J.E., F.C. Damen, J. Shah: Opt. Commun. **4**, 283 (1971)
Bjorklund, G.C.: Opt. Lett. **5**, 154 (1980)
Bloom, A.L.: J. Opt. Soc. Am. **64**, 447 (1974)
Bloomfield, L.A., H. Gerhardt, T.W. Hänsch, S.C. Rand: Opt. Commun. **42**, 247 (1982)
Boiteux, M., O. De Witte: Appl. Opt. **9**, 514 (1970)
Bonneau, R., J. Faure, J. Joussot-Dubien: Chem. Phys. Lett. **2**, 65 (1968)
Bönsch, G.: Appl. Opt. **22**, 3414 (1983a)
Bönsch, G.: Metrologia **19**, 93 (1983b)
Bönsch, G.: IEEE Trans. IM-**34**, 248 (1985)
Bor, Z.: Opt. Commun. **29**, 103 (1979)
Bor, Z.: IEEE J. QE-**16**, 517 (1980)
Bor, Z., B. Rácz: Appl. Opt. **24**, 1910 (1985)
Bor, Z., S. Szatmári, Alexander Müller: Appl. Phys. B**32**, 101 (1983)
Borisevich, N.A., I.I. Kalosa, V.A. Talkacev: Zh. Prikl. Spektrosk. **19**, 1108 (1973)
Born, M., E. Wolf: *Principles of Optics*, 4th ed. (Pergamon, New York 1970)
Bowman, M.R., A.J. Gibson, M.C. Sandford: Joint Conference on Lasers and Optoelectronics (Southampton 1969a)
Bowman, M.R., A.J. Gibson, M.C. Sandford: Nature **221**, 456 (1969b)
Boyd, G.D., J.P. Gordon: Bell System Tech. J. **40**, 489 (1961)
Brackmann, U.: *Lambdachrome Laser Dyes* (Lambda Physik, Göttingen 1986)
Bradley, D.J.: Phys. Bull. **21**, 116 (1970)
Bradley, D.J.: Opto-Electronics **6**, 25 (1974)
Bradley, D.J., A.J.F. Durrant: Phys. Lett. **27**A, 73 (1968)
Bradley, D.J., F. O'Neill: J. Opto-Electron. **1**, 69 (1969)
Bradley, D.J., G.M. Gale, M. Moore, P.D. Smith: Phys. Lett. **26**A, 378 (1968a)
Bradley, D.J., A.J.F. Durrant, G.M. Gale, M. Moore, P.D. Smith: IEEE J. QE-**4**, 707 (1968b)
Bradley, D.J., A.J.F. Durrant, F. O'Neill, B. Sutherland: Phys. Lett. **30**A, 535 (1969)
Bradley, D.J., G.M. Gale, P.D. Smith: J. Phys. B**3**, 11 (1970a)
Bradley, D.J., G.M. Gale, P.D. Smith: Nature **225**, 719 (1970b)
Braun, M., J. Maier, H. Liening: J. Phys. E**20**, 1247 (1987)
Braverman, L.W.: Appl. Phys. Lett. **27**, 602 (1975)
Brock, E.G., P. Czavinsky, E. Hormats, H.C. Nedderman, D. Stripe, F. Unterleitner: J. Chem. Phys. **35**, 759 (1961)
Broida, H.P., S.C. Haydon: Appl. Phys. Lett. **16**, 142 (1970)
Broude, V.L., V.S. Mashkevich, A.F. Prikhot'ko, N.F. Prokopyuk, M.S. Soskin: Sov. Phys. – Solid State **4**, 2182 (1963)
Brown, R.T.: IEEE J. QE-**11**, 800 (1975)
Bücher, H., H. Kuhn: Chem. Phys. Lett. **6**, 183 (1970)
Bücher, H., W. Chow: Appl. Phys. **13**, 267 (1977)
Bukovskii, B.L., L.G. Vasil'eva, L.A. Sakaeva, Y.F. Tomashevskii, A.K. Toropov, Y.A. Federov: Instrum. Exp. Tech. (Engl. Transl.) **3**, 175 (1974)
Burlamacchi, P., R. Pratesi: Appl. Phys. Lett. **22**, 7 (1973a)
Burlamacchi, P., R. Pratesi: Appl. Phys. Lett. **22**, 334 (1973b)
Burlamacchi, P., R. Pratesi: Appl. Phys. Lett. **23**, 475 (1973c)
Burlamacchi, P., R. Pratesi: Appl. Phys. Lett. **28**, 124 (1976)
Burlamacchi, P., R. Salimbeni: Opt. Commun. **17**, 6 (1976)
Burlamacchi, P., R. Pratesi, R. Salimbeni: Opt. Commun. **11**, 109 (1974)
Burlamacchi, P., R. Pratesi, R. Salimbeni: Appl. Opt. **14**, 1311 (1975a)
Burlamacchi, P., R. Pratesi, L. Ronchi: Appl. Opt. **14**, 79 (1975b)
Burlamacchi, P., R. Pratesi, U. Vanni: Appl. Opt. **15**, 2684 (1976)
Buettner, A.V., B.B. Snavely, O.G. Peterson: *Molecular Luminescence*, ed. by E.C. Lim (Benjamin, New York 1969) p. 403

Byer, R.L., J. Paul, M.D. Duncan: "A Wavelength Meter", in *Laser Spectroscopy III*, ed. by J.L. Hall, J.L. Carlsten, Springer Ser. Opt. Sci., Vol. 7 (Springer, Berlin, Heidelberg 1977) p. 414

Cachenaut, J., C. Man, P. Cerez, F. Stoeckel, A. Jourdan, F. Hartman: Rev. Phys. Appl. **14**, 685 (1979)
Cahen, C., J.P. Jegou, J. Pelon, P. Gildwarg, J. Porteneuve: Rev. Phys. Appl. **16**, 353 (1981)
Calkins, J., E. Colley, J. Hazle: Opt. Commun. **42**, 275 (1982)
Cariou, J., P. Luc: *Atlas du Spectre d'Absorption de le Molécule de Tellure, Partie 2: 18500–21200 cm$^{-1}$; Partie 5: 21100–23800 cm$^{-1}$, Temperature: 510 °C.* Technical Report, Laboratoire Aime-Cotton, Orsay, France (1980)
Cassard, P., P.B. Corkum, A.J. Alcock: Appl. Phys. **25**, 17 (1981)
Castell, R., W. Demtröder, A. Fischer, R. Kullmer, H. Weickenmeier, K. Wickert: Appl. Phys. B**38**, 1 (1985)
Chan, C.K., S.O. Sari: Appl. Phys. Lett. **25**, 403 (1974)
Chandross, E.A., R.E. Visco: J. Am. Chem. Soc. **86**, 5350 (1964)
Chang, M.S., P. Burlamacchi, C. Hu, J.R. Whinnery: Appl. Phys. Lett. **20**, 313 (1972)
Chebotayev, V.P., V.M. Klementyev, Y.A. Matyugin: Appl. Phys. **11**, 163 (1976)
Chiu, P.H., S. Hsu, S.J.C. Box, H.-S. Kwok: IEEE J. QE-**20**, 652 (1984)
Chou, P., T.J. Aartsma: J. Phys. Chem. **90**, 721 (1986)
Cirkel, H.J., L. Ringwelski, F.P. Schäfer: Z. Phys. Chem. NF **81**, 158 (1972)
Claesson, S., L. Lindquist: Arkiv Kemi **12**, 1 (1958)
Clobes, A.R., M.J. Brienza: Appl. Phys. Lett. **21**, 265 (1972)
Colour Index: Third Edition Vol. 5 (The Society of Dyers and Colourists, Bradford, England 1971)
Cotnoir, L.J., T.D. Wilkerson, M. Dombrowski, R.H. Kagann, C.L. Korb, G.K. Schwemmer, H. Walden: In Technical Digest, Topical Meeting on Tunable Solid-State Lasers, FC5-1 (1985)
Couillaud, B., T.W. Hänsch: Opt. Commun. **35**, 441 (1980)
Couillaud, B., L.A. Bloomfield, J.E. Lawler, A. Siegel, T.W. Hänsch: Opt. Commun. **35**, 359 (1980)
Cox, A.J., G.W. Scott: Appl. Opt. **18**, 532 (1979)
Cox, A.J., C.D. Merritt, G.W. Scott: Appl. Phys. Lett. **40**, 664 (1982)
Crozet, P., Y. Meyer: C.R. Acad. Sci. **271**, 718 (1970)
Crozet, P., B.S. Kirkiacharian, C. Soula, Y.H. Meyer: J. Chim. Phys. **68**, 1388 (1971)

Dal Pozzo, P., R. Polloni, O. Svelto: Appl. Phys. **6**, 381 (1975)
Daniel, H., M. Steiner, H. Walther: Appl. Phys. B **26**, 19 (1981)
Danielmeyer, H.G., W.G. Nilsen: Appl. Phys. Lett. **16**, 124 (1970)
Decker, C.D., T.S. Fahlen, J. Falk: J. Appl. Phys. **46**, 2308 (1975)
DeMaria, A.J., D.A. Stetser, H. Heyman: Appl. Phys. Lett. **8**, 22 (1966)
Demas, J.N., G.A. Crosby: J. Phys. Chem. **75**, 991 (1971)
Derkacheva, L.D., A.I. Sokolovskaya: Opt. Spectrosc. **25**, 244 (1968)
Derkacheva, L.D., A.I. Krymova, A.F. Vompe, I.I. Levkoev: Opt. Spectrosc. **25**, 404 (1968a)
Derkacheva, L.D., A.I. Krymova, V.I. Malyshev, A.S. Markin: JETP Lett. **7**, 362 (1968b)
Deryugin, L.N., I.V. Cheremiskin, T.K. Chekhlova: Sov. J. Quantum Electron. **5**, 439 (1976a)
Deryugin, L.N., O.I. Ovcharenko, V.E. Sotin, T.K. Chekhlova: Sov. J. Quantum Electron. **5**, 1129 (1976b)
Deutsch, T.F., M. Bass: IEEE J. QE-**5**, 260 (1969)
DeVoe, R.G., R.G. Brewer: Phys. Rev A**30**, 2827 (1984)
Dienes, A., C.V. Shank, A.M. Trozzolo: Appl. Phys. Lett. **17**, 189 (1970)
Dienes, A., E.P. Ippen, C.V. Shank: Appl. Phys. Lett. **19**, 258 (1971a)
Dienes, A., E.P. Ippen, C.V. Shank: unpublished (1971b)
Dienes, A., E.P. Ippen, C.V. Shank: IEEE J. QE-**8**, 388 (1972)
Dinev, S.G., I.G. Koprinkov, K.V. Stamenov, K.A. Stankov, C. Radzewicz: Opt. Commun. **32**, 313 (1980)
Docchio, F., F.P. Schäfer, J. Jethwa, J. Jasny: J. Phys. E**18**, 845 (1985)
Dorenwendt, K., G. Bönsch: Metrologia **12**, 57 (1976)
Dorsinville, R.: Opt. Commun. **26**, 419 (1978)
Dorsinville, R., M.M. Denariez-Roberge: Opt. Commun. **24**, 31 (1978)

Drake, J.M., R.I. Morse: Opt. Commun. **12**, 132 (1974)
Drexhage, K.H.: unpublished results (1971)
Drexhage, K.H.: Design of Laser Dyes, 7th Intl. Quantum Electronics Conference (Montreal 1972a)
Drexhage, K.H.: unpublished results (1972b)
Drexhage, K.H.: Laser Focus **9** (3), 35 (1973a)
Drexhage, K.H.: private communication (1973b)
Drexhage, K.H., G.A. Reynolds: unpublished results (1972)
Drexhage, K.H., G.A. Reynolds: IEEE J. QE-**10**, 695 (1974)
Drexhage, K.H., G.H. Hawks, G.A. Reynolds: unpublished results (1972)
Drullinger, R.E., K.M. Evenson, D.A. Jennings, J.C. Bergquist, L. Burkins, H. Daniel: Appl. Phys. Lett. **42**, 137 (1983)
Duarte, F.J., J.A. Piper: Opt. Commun. **35**, 100 (1980)
Duarte, F.J., J.A. Piper: Opt. Commun. **43**, 303 (1982)

Eckstein, J.N., A.I. Ferguson, T.W. Hänsch: Phys. Rev. Lett. **40**, 847 (1978)
Einstein, A.: Ann. Physik **33**, 1275 (1910)
Eranian, A., P. Dezauzier, O. de Witte: Opt. Commun. **7**, 150 (1973)
Erickson, L.E., A. Szabo: Appl. Phys. Lett. **18**, 433 (1971)
Ernsting, N.P., M. Asimov, F.P. Schäfer: Chem. Phys. Lett. **91**, 231 (1982)
Erskine, D.J., A.J. Taylor, C.L. Tang: J. Chem. Phys. **80**, 5338 (1984)
Evans, J.W.: J. Opt. Soc. Am. **39**, 229 (1949)
Evenson, K.M., D.A. Jennings, F.R. Petersen, J.S. Wells, R.E. Drullinger: "Optical Frequency Synthesis Spectroscopy," in *Laser-Cooled and Trapped Atoms,* ed. by W.D. Phillips, Special Publication 653 (1983) p. 27
Evtuhov, V., A.E. Siegman: Appl. Opt. **4**, 142 (1965)
Exciton Laser Dye Catalog (Exciton Inc., Dayton, OH 1989)

Fan, B., T.K. Gustafson: Appl. Phys. Lett. **28**, 202 (1976)
Farmer, G.I., B.G. Huth, L.M. Taylor, M.R. Kagan: Appl. Phys. Lett. **12**, 136 (1968)
Farmer, G.I., B.G. Huth, L.M. Taylor, M.R. Kagan: Appl. Opt. **8**, 363 (1969)
Ferguson, J., A.W.H. Mau: Chem. Phys. Lett. **14**, 245 (1972a)
Ferguson, J., A.W.H. Mau: Chem. Phys. Lett **17**, 543 (1972b)
Ferrar, C.M.: IEEE J. QE-**5**, 621 (1969a)
Ferrar, C.M.: IEEE J. QE-**5**, 550 (1969b)
Ferrar, C.M.: Appl. Phys. Lett. **20**, 419 (1972)
Ferrar, C.M.: Appl. Phys. Lett. **23**, 548 (1973)
Ferrar, C.M.: Appl. Opt. **13**, 1992 (1974)
Fischer, A., R. Kullmer, W. Demtröder: Opt. Commun. **39**, 277 (1981)
Flach, R., I.S. Shahin, W.M. Yen: Appl. Opt. **13**, 2095 (1974)
Flamant, P., Y.H. Meyer: Appl. Phys. Lett. **19**, 491 (1971)
Fork, R.L., Z. Kaplan: Appl. Phys. Lett. **20**, 472 (1972)
Fork, R.L., B.I. Greene, C.V. Shank: Appl. Phys. Lett. **38**, 671 (1981)
Fork, R.L., C.V. Shank, R.T. Yen: Appl. Phys. Lett. **41**, 223 (1982)
Fork, R.L., C.H. Brito Cruz, P.C. Becker, C.V. Shank: Opt. Lett. **12**, 483 (1987)
Förster, Th.: *Fluoreszenz organischer Verbindungen* (Vandenhoeck and Ruprecht, Göttingen 1951)
Förster, Th.: Discuss. Faraday Soc. **27**, 7 (1959)
Förster, Th.: Angew. Chemie **81**, 364; internat. ed. **8**, 333 (1969)
Förster, Th., E. König: Z. Elektrochem. **61**, 344 (1957)
Förster, Th., B. Selinger: Z. Naturforsch. **19a**, 38 (1964)
Försterling, H.D., H. Kuhn: *Physikalische Chemie in Experimenten.* Ein Praktikum (Verlag Chemie, Weinheim/Bergstr. 1971) p. 373
Försterling, H.D., W. Huber, H. Kuhn, H.H. Martin, A. Schweig, F.F. Seelig, W. Stratmann: In *Optische Anregung organischer Syteme*, ed. by W. Foerst (Verlag Chemie, Weinheim 1966) p. 55

Fowler, M.C., W.H. Glenn: Appl. Opt. **15**, 2624 (1976)
Friesem, A.A., U. Ganiel, G. Neumann: Appl. Phys. Lett. **23**, 249 (1973a)
Friesem, A.A., U. Ganiel, G. Neumann, D. Peri: Opt. Commun. **9**, 149 (1973b)
Frigo, N.I., H. Mahr, T. Daly: IEEE J. QE-**13**, 101 (1977)
Frölich, D., L. Stein, H.W. Schröder, H. Welling: Appl. Phys. **11**, 97 (1976)
Furumoto, H.W., H.L. Ceccon: Appl. Opt. **8**, 1613 (1969a)
Furumoto, H.W., H.L. Ceccon: J. Appl. Phys. **40**, 4204 (1969b)
Furumoto, H.W., H.L. Ceccon: IEEE J. QE-**6**, 262 (1970)

Gacoin, P., P. Flamant: Opt. Commun. **5**, 351 (1972)
Gardner, J.L.: Opt. Lett. **8**, 91 (1983)
Gardner, J.L.: Appl. Opt. **24**, 3570 (1985)
Gardner, J.L.: Appl. Opt. **25**, 3799 (1986)
Gerhardt, H., A. Timmermann: Opt. Commun. **21**, 343 (1977)
Gerstenkorn, S., P. Luc: *Atlas du Spectre d'Absorption de la Molécule d'Iode 14800–20000 cm$^{-1}$*, Technical Report (CNRS, Paris 1978)
Gerstenkorn, S., P. Luc: Rev. Phys. Appl. **14**, 791 (1979)
Gibbs, W.E.K., H.A. Kellog: IEEE J. QE-**4**, 293 (1968)
Giordmaine, J.A., P.M. Rentzepis, S.L. Shapiro, K.W. Wecht: Appl. Phys. Lett. **11**, 216 (1967)
Glenn, W.H., M.J. Brienza, A.J. DeMaria: Appl. Phys. Lett. **12**, 54 (1968)
Gordon, S.K., S.F. Jacobs: Appl. Opt. **13**, 231 (1974)
Gray, D.F., K.A. Smith, F.B. Dunning: Appl. Opt. **25**, 1339 (1986)
Green, J.M., J.P. Hohimer, F.K. Tittel: Opt. Commun. **7**, 349 (1973a)
Green, J.M., J.P. Hohimer, F.K. Tittel: Opt. Commun. **9**, 407 (1973b)
Green, R.B., R.A. Keller, G.G. Luther, P.K. Schenck, J.C. Travis: Appl. Phys. Lett. **29**, 727 (1976)
Gregg, D.W., S.J. Thomas: IEEE J. QE-**5**, 302 (1969)
Gregg, D.W., M.R. Querry, J.B. Marling, S.J. Thomas, C.V. Dobler, N.J. Davies, J.F. Belew: IEEE J. QE-**6**, 270 (1970)
Gronau, B., E. Lippert, W. Rapp: Ber. Bunsenges. Phys. Chem. **76**, 432 (1972)
Grove, R.E., F.Y. Wu, L.A. Hackel, D.G. Youmans, S. Ezekiel: Appl. Phys. Lett. **23**, 442 (1973)
Gutfeld, R.J. von, B. Welber, E.E. Tynan: IEEE J. QE-**6**, 532 (1970)

Hackel, R.P.: private communication (1989)
Hackel, R.P., M. Feldman, J. Baker, R.D. Paris, J.M. Tampico, T.J. Kauppila: "Pulsed, Multiple Dye Laser Oscillator System with Accurate Wavelength Control," in Technical Digest, Conference on Lasers and Electro-Optics (1986) p. 266
Hall, J.L., S.A. Lee: Appl. Phys. Lett. **29**, 367 (1976)
Hall, J.L., L. Hollberg, T. Baer, H.G. Robinson: Appl. Phys. Lett. **39**, 680 (1981)
Hambenne, J.B., M. Sargent III: IEEE J. QE-**11**, 90 (1975)
Hammond, P.R., R.S. Hughes: Nature Phys. Sci. **231**, 59 (1971)
Hanna, D.C., P.A. Karkkainen, R. Wyatt: Opt. Quantum Electron. **7**, 115 (1975)
Hänsch, T.W.: Appl. Opt. **11**, 895 (1972)
Hänsch, T.W.: "A Self-calibrating Grating", in *Laser Spectroscopy III*, ed. by J.L. Hall, J.L. Carlsten, Springer Ser. Opt. Sci., Vol. 7 (Springer, Berlin, Heidelberg 1977) p. 423
Hänsch, T.W., N.C. Wong: Metrologia **16**, 39 (1980)
Hänsch, T.W., M. Pernier, A.L. Schawlow: IEEE J. QE-**7**, 45 (1971a)
Hänsch, T.W., F. Varsanyi, A.L. Schawlow: Appl. Phys. Lett. **18**, 108 (1971b)
Hänsch, T.W., I.S. Shahin, A.L. Schawlow: Phys. Rev. Lett. **27**, 707 (1971c)
Hänsch, T.W., A.L. Schawlow, P. Toschek: IEEE J. QE-**9**, 553 (1973)
Hartig, W., H. Walther: Appl. Phys. **1**, 171 (1973)
Hausser, K.W., R. Kuhn, E. Kuhn: Z. Physik. Chemie B **29**, 417 (1935)
Held, M.: Paper presented at the 8th Conference on High Speed Photography (Stockholm 1968)
Heller, C.A., J.L. Jernigan: Appl. Opt. **16**, 61 (1977)
Heller, E.J., R.C. Brown: J. Chem. Phys. **29**, 3336 (1983)
Hercher, M., H.A. Pike: Opt. Commun. **3**, 65 (1971a)
Hercher, M., H.A. Pike: Opt. Commun. **3**, 346 (1971b)

Hercher, M., H.A. Pike: IEEE J. QE-**7**, 473 (1971c)
Hercher, M., B.B. Snavely: Conference on Coherence and Quantum Optics (Rochester 1972)
Hercules, D.M.: Science **145**, 808 (1964)
Hill, K.O., A. Watanabe: Opt. Commun. **5**, 389 (1972)
Hirth, A., J. Faure, R. Schoenenberger: C.R. Acad. Sci. **274B**, 747 (1972)
Hirth, A., K. Vollrath, J. Faure, D. Lougnot: Opt. Commun. **7**, 339 (1973a)
Hirth, A., K. Vollrath, J.-P. Fouassier: Opt. Commun. **9**, 139 (1973b)
Hirth, A., K. Vollrath, J.Y. Allain: Opt. Commun. **20**, 347 (1977)
Hlousek, L., W.M. Fairbank Jr.: Opt. Lett. **8**, 322 (1983)
Hlousek, L., S.A. Lee, W.M. Fairbank Jr.: Phys. Rev. Lett. **50**, 328 (1983)
Hoffnagle, J., L.P. Roesch, N. Schlumpf, A. Weis: Opt. Commun. **42**, 267 (1982)
Holtom, G., O. Teschke: IEEE J. QE-**10**, 577 (1974)
Horobin, R.W., L.B. Murgatroyd: Stain Technol. **44**, 297 (1969)
Hsu, S.C., H.S. Kwok: Appl. Opt. **25**, 470 (1986)
Hüffer, W., R. Schieder, H. Telle, R. Raue, W. Brinkwerth: Opt. Commun. **28**, 353 (1979)
Hüffer, W., R. Schieder, H. Telle, R. Raue, W. Brinkwerth: Opt. Commun. **33**, 85 (1980)
Hutcheson, L.D., R.S. Hughes: IEEE J. QE-**10**, 462 (1974a)
Hutcheson, L.D., R.S. Hughes: Appl. Opt. **13**, 1395 (1974b)
Huth, B.G.: Appl. Phys. Lett. **16**, 185 (1970)
Huth, B.G., G.I. Farmer: IEEE J. QE-**4**, 427 (1968)
Huth, B.G., G.I. Farmer, M.R. Kagan: J. Appl. Phys. **40**, 5145 (1969)

Inomata, H., A.I. Carswell: Opt. Commun. **22**, 278 (1977)
Ippen, E.P., C.V. Shank: Appl. Phys. Lett. **21**, 301 (1972a)
Ippen, E.P., C.V. Shank: unpublished (1972b)
Ippen, E.P., C.V. Shank: Appl. Phys. Lett. **27**, 488 (1975)
Ippen, E.P., C.V. Shank, A. Dienes: IEEE J. QE-**7**, 178 (1971)
Ippen, E.P., C.V. Shank, A. Dienes: Appl. Phys. Lett. **21**, 348 (1972)
Ishida, Y., N. Iwasaki, K. Asaumi, T. Yajima, Y. Maruyama: Appl. Phys. B**38**, 159 (1985)
Ishikawa, J.: Appl. Opt. **25**, 3013 (1986)
Ishikawa, J., N. Ito, K. Tanaka: Appl. Opt. **25**, 639 (1986)
Ishizaka, S., S. Kotani: Chem. Phys. Lett. **117**, 251 (1985)

Jacquinot, P., P. Juncar, J. Pinard: "Motionless Michelson for High Precision Laser Frequency Measurements: The Sigmameter", in *Laser Spectroscopy III*, ed. by J.L. Hall, J.L. Carlsten, Springer Ser. Opt. Sci., Vol. 7 (Springer, Berlin, Heidelberg 1977) p. 417
Jain, R.K., J.P. Heritage: Appl. Phys. Lett. **32**, 41 (1978)
Jakobi, H., H. Kuhn: Z. Elektrochem. Ber. Bunsenges. Phys. Chem. **66**, 46 (1962)
Jarrett, S.M., J.F. Young: Opt. Lett. **4**, 176 (1979)
Jennings, D.A., C.R. Pollock, F.R. Petersen, R.E. Drullinger, K.M. Evenson, J.S. Wells: Opt. Lett. **8**, 136 (1983)
Jethwa, J., F.P. Schäfer: Appl. Phys. **4**, 299 (1974)
Jethwa, J., F.P. Schäfer, J. Jasny: IEEE J. QE-**14**, 119 (1978)
Jethwa, J., S.St. Anufrik, F. Docchio: Appl. Opt. **21**, 2778 (1982)
Johnson, A.M., W.M. Simpson: J. Opt. Soc. Am. B**2**, 619 (1985)
Johnston, T.F., Jr.: In *Encyclopedia of Physical Science and Technology*, Vol. 14 (Academic, New York 1987)
Johnston, T.F., Jr., W. Proffitt: IEEE J. QE-**16**, 483 (1980)
Johnston, T.F., Jr., R.H. Brady, W. Proffitt: Appl. Opt. **21**, 2307 (1982)
Juncar, P., J. Pinard: Opt. Commun. **14**, 438 (1975)
Juncar, P., J. Pinard: Rev. Sci. Instrum. **53**, 939 (1982)
Junttila, M., B. Ståhlberg, E. Kyrö, T. Veijola, J. Kauppinen: Rev. Sci. Instrum. **58**, 1180 (1987)

Kagan, M.R., G.I. Farmer, B.G. Huth: Laser Focus **4** (9), 26 (1968)
Kahane, A., M.S. O'Sullivan, N.M. Sanford, B.P. Stoicheff: Rev. Sci. Instrum. **54**, 1138 (1983)

Kallenbach, R., B. Scheumann, C. Zimmermann, D. Meschede, T.W. Hänsch: to be published in Appl. Phys. Lett.
Kaminow, I.P., H.P. Weber, E.A. Chandross: Appl. Phys. Lett. **18**, 497 (1971)
Karl, N.: Phys. Status Solidi (a) **13**, 651 (1972)
Kato, D., T. Sato: Opt. Commun, **5**, 134 (1972)
Keller, R.A.: IEEE J. QE-**6**, 411 (1970)
Kellogg, R.E.: J. Luminesc. **1**, **2**, 435 (1970)
Keszthelyi, C.P.: Appl. Opt. **14**, 1710 (1975)
King, D.S., P.K. Schenck: Laser Focus **14**(3), 50 (1978)
King, D.S., P.K. Schenck, K.C. Smyth, J.C. Travis: Appl. Opt. **16**, 2617 (1977)
Kittrell, C., R.A. Bernheim: Opt. Commun. **19**, 5 (1976)
Klauder, J.R., M.A. Duguay, J.A. Giordmaine, S.L. Shapiro: Appl. Phys. Lett. **13**, 174 (1968)
Klauminzer, G.K.: US Patent no. **4**,127,828 (1978)
Klement'ev, V.M., Y.A. Matyugin, V.P. Chebotaev: JETP Lett. **24**, 5 (1976)
Knibbe, H., D. Rehm, A. Weller: Ber. Bunsenges. Phys. Chem. **72**, 257 (1968)
Kogelnik, H.: Bell Syst. Tech. J. **44**, 455 (1965)
Kogelnik, H., T. Li: Appl. Opt. **5**, 1550 (1966)
Kogelnik, H., C.V. Shank: Appl. Phys. Lett. **18**, 152 (1971)
Kogelnik, H., C.V. Shank, T.P. Sosnowski, A. Dienes: Appl. Phys. Lett. **16**, 499 (1970)
Kogelnik, H., E.P. Ippen, A. Dienes, C.V. Shank: IEEE J. QE-**8**, 373 (1972)
Kohlmannsperger, J.: Z. Naturforsch. **24a**, 1547 (1969)
Kohn, R.L., C.V. Shank, E.P. Ippen, A. Dienes: Opt. Commun. **3**, 177 (1971)
Kolbin, I.I., O.I. Ovcharenko, V.E. Sotin, I.V. Cheremiskin: Sov. J. Quantum Electron. **5**, 860 (1976)
Kommandeur, J.: Recueil Rev. **102**, 421 (1983)
Kong, J.J., S.S. Lee: IEEE J. QE-**17**, 439 (1981)
Konishi, N., T. Suzuki, Y. Taira, H. Kato, T. Kasuya: Appl. Phys. **25**, 311 (1981)
Kopainsky, B., W. Kaiser, F.P. Schäfer: Chem. Phys. Lett. **56**, 458 (1978)
Kopainsky, K.: Appl. Phys. **8**, 229 (1975)
Kotani, H., O. Nakagawa, M. Kawabe, K. Masuda: Jpn. J. Appl. Phys. **15**, 1581 (1976)
Kotzubanov, V.D., Yu. V. Naboikin, L.A. Ogurtsova, A.P. Podgornyi, F.S. Pokrovskaya: Opt. Spectrosc. **25**, 406 (1968a)
Kotzubanov, V.D., L.Ya. Malkes, Yu.V. Naboikin, L.A. Ogurtsova, A.P. Podgornyi, F.S. Pokrovskaya: Bull. Acad. Sci. USSR, Phys. Ser. **32**, 1357 (1968b)
Kowalski, F.V., R.T. Hawkins, A.L. Schawlow: J. Opt. Soc. Am. **66**, 965 (1976)
Kowalski, F.V., W. Demtröder, A.L. Schawlow: "Digital Wavemeter for cw Lasers", in *Laser Spectroscopy III*, ed. by J.L. Hall, J.L. Carlsten, Springer Ser. Opt. Sci., Vol. 7 (Springer, Berlin, Heidelberg 1977) p. 412
Kowalski, F.V., R.E. Teets, W. Demtröder, A.L. Schawlow: J. Opt. Soc. Am. **68**, 1611 (1978)
Kowalski, J., R. Neuman, S. Noehte, R. Schwarzwald, H. Suhr, G. zu Putlitz: Opt. Commun. **53**, 141 (1985)
Kozlov, N.T., Yu.S. Protasov: Sov. Phys. – Dokl. **20**, 500 (1975/6)
Kuhl, J., H. Telle, R. Schieder, U. Brinkmann: Opt. Commun. **24**, 251 (1978)
Kuhn, H.: Chimia **9**, 237 (1955)
Kuhn, H.: *Progress in the Chemistry of Organic Natural Products*, ed. by D.L. Zechmeister, Vol. 16 (Springer, Wien 1959) p. 17
Kuizenga, D.J.: Appl. Phys. Lett. **19**, 260 (1971)

Labhart, H.: Helv. Chim. Acta **47**, 2279 (1964)
Lamm, M.E., D.M. Neville Jr.: J. Phys. Chem. **69**, 3872 (1965)
Larsson, R., B. Nordén: Acta Chem. Scand. **24**, 2583 (1970)
Laubereau, A., A. Seilmeier, W. Kaiser: Chem. Phys. Lett. **36**, 232 (1975)
Layer, H.P.: IEEE Trans. IM-**29**, 358 (1980)
Leduc, M., C. Weisbuch: Opt. Commun. **26**, 78 (1978)
Lee, L., A.L. Schawlow: Opt. Lett. **6**, 610 (1981)

Lee, S.A., J.L. Hall: "A Traveling Michelson Interferometer with Phase-Locked Fringe Interpolation", in *Laser Spectroscopy III*, ed. by J.L. Hall, J.L. Carlsten, Springer Ser. Opt. Sci., Vol. 7 (Springer, Berlin, Heidelberg 1977) p. 421
Lee, S.A., J. Helmcke, J.L. Hall, B.P. Stoicheff: Opt. Lett. **3**, 141 (1978)
Leeb, W.R.: Appl. Phys. **6**, 267 (1975)
Lempicki, A., H. Samelson: Lasers, ed. by A.K. Levine, Vol. 1 (Marcel Dekker, New York 1966) p. 181
Leonhardt, H., A. Weller: *Luminescence of Organic and Inorganic Materials*, ed. by H.P. Kallmann, G.M. Spruch (Wiley, New York 1962) p. 74
Letouzey, J.P., S.O. Sari: Appl. Phys. Lett. **23**, 311 (1973)
Leutwyler, S., E. Schumacher, L. Wöste: Opt. Commun. **19**, 197 (1976)
Levenson, M.D., G.L. Eesley: IEEE J. QE-**12**, 259 (1976)
Levshin, L.V., V.K. Gorshkov: Opt. Spectrosc. **10**, 401 (1961)
Lichten, W.: J. Opt. Soc. Am. A**3**, 909 (1986)
Lidholt, L.R., W.W. Wladimiroff: Opto-Electron. **2**, 21 (1970)
Lin, C.: J. Appl. Phys. **46**, 4076 (1975a)
Lin, C.: IEEE J. QE-**10**, 602 (1975b)
Lin, C.: IEEE J. QE-**11**, 61 (1975c)
Lin, C., A. Dienes: Opt. Commun. **9**, 21 (1973a)
Lin, C., A. Dienes: J. Appl. Phys. **44**, 5050 (1973b)
Liphardt, Bo., Be. Liphardt, W. Lüttke: Opt. Commun. **38**, 207 (1981)
Lippert, E.: Z. Elektrochem. **61**, 962 (1957)
Littman, M.G.: Opt. Lett. **3**, 138 (1978)
Littman, M.G., H.J. Metcalf: Appl. Opt. **17**, 2224 (1978)
Lorenzen, C., K. Niemax: Z. Phys. A**311**, 249 (1983)
Lorenzen, C., K. Niemax, L.R. Pendrill: Opt. Commun. **39**, 370 (1981)
Lotem, H., R.T. Lynch Jr.: Appl. Phys. Lett. **27**, 344 (1975)
Lyot, B.: C.R. Acad. Sci. **197**, 1593 (1933)

Mack, M.E.: Appl. Phys. Lett. **15**, 166 (1969)
Mack, M.E., O.B. Northam, L.G. Crawford: J. Opt. Soc. Am. **66**, 1108 (1976)
Maeda, M.: *Laser Dyes* (Academic, New York 1984)
Maeda, M., Y. Miyazoe: Jpn. J. Appl. Phys. **11**, 692 (1972)
Maeda, M., O. Uchino, E. Doi, K. Watanabe, Y. Miyazoe: IEEE J. QE-**13**, 65 (1977)
Magde, D., M.W. Windsor: Chem. Phys. Lett. **24**, 144 (1974a)
Magde, D., M.W. Windsor: Chem. Phys. Lett. **27**, 31 (1974b)
Magyar, G., H.J. Schneider-Muntau: Appl. Phys. Lett. **20**, 406 (1972)
Maier, J.P., A. Seilmeier, A. Laubereau, W. Kaiser: Chem. Phys. Lett. **46**, 527 (1977)
Malley, M.M., G. Mourou: Opt. Commun. **10**, 323 (1974)
Marason, E.G.: Opt. Commun. **37**, 57 (1981)
Marling, J.B., D.W. Gregg, S.J. Thomas: IEEE J. QE-**6**, 570 (1970a)
Marling, J.B., D.W. Gregg, L. Wood: Appl. Phys. Lett. **17**, 527 (1970b)
Marling, J.B., L.L. Wood, D.W. Gregg: IEEE J. QE-**7**, 498 (1971)
Marling, J.B., J.G. Hawley, E.M. Liston, W.B. Grant: Appl. Opt. **13**, 2317 (1974)
Marowsky, G.: IEEE J. QE-**9**, 245 (1973a)
Marowsky, G.: Rev. Sci. Instrum. **44**, 890 (1973b)
Marowsky, G.: Appl. Phys. **2**, 213 (1973c)
Marowsky, G., K. Kaufmann: IEEE J. QE-**12**, 207 (1976)
Marowsky, G., L. Ringwelski, F.P. Schäfer: Z. Naturforsch. **27**a, 711 (1972)
Marowsky, G., F.K. Tittel, F.P. Schäfer: Opt. Commun. **13**, 100 (1975)
Marth, K.: Diplomarbeit, Universität Marburg (1967)
Matsuzawa, H., S. Suganomata, H. Inaba: Jpn. J. Appl. Phys. **15**, 1155 (1976)
McFarland, B.B.: Appl. Phys. Lett. **10**, 208 (1967)
McIntyre, D.H., T.W. Hänsch: to be published (1989)
Measures, R.M.: Appl. Opt. **13**, 1121 (1974)
Measures, R.M.: Appl. Opt. **14**, 909 (1975)

Melhuish, W. H.: J. Opt. Soc. Am. **52**, 1256 (1962)
Metrologia **19**, 163 (1984); "Documents concerning the new definition of the metre"
Migus, A., C. V. Shank, E. P. Ippen, R. L. Fork: IEEE J. QE-**18**, 101 (1982)
Miller, C. K.: Wavelength Meter for Pulsed Laser Applications. Technical Report SAND81-0310, Sandia National Laboratories, Albuquerque, NM (1982)
Miyazoe, Y., M. Maeda: Appl. Phys. Lett. **12**, 206 (1968)
Miyazoe, Y., M. Maeda: Opto-Electron. **2**, 227 (1970)
Monchalin, J., M. J. Kelly, J. E. Thomas, N. A. Kurnit, A. Szöke, F. Zernike, P. H. Lee, A. Javan: Appl. Opt. **20**, 736 (1981)
Morantz, D. J.: Proc. of the Symposium on Optical Masers (Polytechnic Press, Brooklyn 1963) p. 491
Morantz, D. J., B. G. White, A. J. C. Wright: Phys. Rev. Lett. **8**, 23 (1962)
Morey, W. W., W. H. Glenn: IEEE J. QE-**12**, 311 (1976)
Moriarty, A., W. Heaps, D. D. Davis: Opt. Commun. **16**, 324 (1976)
Morris, M. B., T. J. McIlrath: Appl. Opt. **18**, 4145 (1979)
Morris, M. B., T. J. McIlrath, J. J. Snyder: Appl. Opt. **23**, 3862 (1984)
Morrow, T., M. Quinn: Private communication (1973)
Mourou, G.: IEEE J. QE-**11**, 1 (1975)
Mourou, G., M. M. Malley: Opt. Commun. **11**, 282 (1974)
Mourou, G., T. Sizer: Opt. Commun. **41**, 47 (1982)
Müller, Alexander: Z. Naturforsch. **23a**, 946 (1968)
Müller, A., E. Pflüger: Chem. Phys. Lett. **2**, 155 (1968)
Müller, A., U. Sommer: Ber. Bunsenges. Phys. Chem. **73**, 819 (1969)
Myer, J. A., I. Itzkan, E. Kierstead: Nature **225**, 544 (1970)
Myers, S. A.: Opt. Commun. **4**, 187 (1971)

Naboikin, Yu. V., L. A. Ogurtsova, A. P. Podgornyi, F. S. Pokrovskaya, V. I. Grigoryeva, B. M. Krasovitskii, L. M. Kutsyna, V. G. Tishchenko: Opt. Spectrosc. **28**, 528 (1970)
Nagai, K., K. Kawaguchi, C. Yamada, K. Hayakawa, Y. Takagi, E. Hirota: J. Mol. Spectrosc. **84**, 197 (1980)
Nair, L. G.: Appl. Phys. **20**, 97 (1979)
Nair, L. G., K. Dasgupta: IEEE J. QE-**16**, 111 (1980)
Nakashima, M., J. A. Sousa, R. C. Clapp: Nature Phys. Sci. **235**, 16 (1972)
Nakato, Y., N. Yamamoto, H. Tsubomura: Chem. Phys. Lett. **2**, 57 (1968)
Nakatsuka, H., D. Grischkowsky: Opt. Lett. **6**, 13 (1981)
Neporent, B. S., V. B. Shilov: Opt. Spectrosc. **30**, 576 (1971)
New, G. H. C., D. H. Rea: J. Appl. Phys. **47**, 3107 (1976)
Nitzan, A., J. Jortner: Theor. Chim. Acta **29**, 97 (1973a)
Nitzan, A., J. Jortner: Theor. Chim. Acta **30**, 217 (1973b)
Nouchi, G.: J. Chim. Phys. **66**, 548 (1969)
Novak, J. R., M. W. Windsor: J. Chem. Phys. **47**, 3075 (1967)
Novak, J. R., M. W. Windsor: Proc. R. Soc. London A **308**, 95 (1968)

O'Bryan, C. L. III, M. Sargent III: Phys. Rev. A**8**, 3071 (1973)
Okada, M., S. Shimizu, S. Ieiri: Appl. Opt. **14**, 917 (1975)
Okada, M., K. Takizawa, S. Ieiri: Appl. Opt. **15**, 472 (1976)
O'Neill, F.: Opt. Commun. **6**, 360 (1972)
Opower, H., W. Kaiser: Phys. Lett. **21**, 638 (1966)
Ornstein, M. H., V. E. Derr: Appl. Opt. **13**, 2100 (1974)
Osborne, A. D., G. Porter: Proc. R. Soc. London A **284**, 9 (1965)
O'Sullivan, M. S., B. P. Stoicheff: Can. J. Phys. **61**, 940 (1983)

Palmer, B. A., R. Engelman Jr.: Atlas of the Thorium Spectrum, Technical Report LA-9615, Los Alamos National Laboratory (1983)
Palmer, B. A., R. A. Keller, R. Engelman Jr.: An Atlas of Uranium Emission Intensities in a Hollow Cathode Discharge, Technical Report LA-8251-MS, Los Alamos National Laboratory (1980)

Palmer, B.A., R.A. Keller, F.V. Kowalski, J.L. Hall: J. Opt. Soc. Am. **71**, 948 (1981)
Pappalardo, R., H. Samelson, A. Lempicki: Appl. Phys. Lett. **16**, 267 (1970a)
Pappalardo, R., H. Samelson, A. Lempicki: IEEE J. QE-**6**, 716 (1970b)
Parker, C.A.: *Photoluminescence of Solutions* (Elsevier, Amsterdam 1968)
Parker, C.A., C.G. Hatchard: Trans. Faraday Soc. **57**, 1894 (1961)
Pease, A.A., W.M. Pearson: Appl. Opt. **16**, 57 (1977)
Penzkofer, A., W. Falkenstein: Chem. Phys. Lett. **44**, 547 (1976)
Penzkofer, A., W. Falkenstein, W. Kaiser: Chem. Phys. Lett. **44**, 82 (1976)
Periasamy, N., Z. Bor: Opt. Commun. **39**, 298 (1981)
Personov, R.I., V.V. Solochmov: Opt. Spectrosc. **23**, 317 (1967)
Peterson, N.C., M.J. Kurylo, W. Braun, A.M. Bass, R.A. Keller: J. Opt. Soc. Am. **61**, 746 (1971)
Peterson, O.G., B.B. Snavely: Appl. Phys. Lett. **12**, 238 (1968)
Peterson, O.G., S.A. Tuccio, B.B. Snavely: Appl. Phys. Lett. **17**, 245 (1970)
Peterson, O.G., J.P. Webb, W.C. McColgin, J.H. Eberly: J. Appl. Phys. **42**, 1917 (1971)
Petley, B.W., K. Morris: Opt. Quantum Electron. **10**, 277 (1978)
Pike, C.T.: Opt. Commun. **10**, 14 (1974)
Pike, H.A.: Ph. D. thesis, University of Rochester, available from University Microfilms, Ann Arbor, Mich., USA (1971)
Pilloff, H.S.: Appl. Phys. Lett **21**, 339 (1972)
Pole, R.V., A.J. Spiekerman, T.W. Hänsch: IBM J. Res. Dev. **24**, 85 (1980)
Polland, H.J., T. Elsaesser, A. Seilmeier, W. Kaiser: Appl. Phys. B**32**, 53 (1983)
Pollock, C.R., D.A. Jennings, F.R. Petersen, J.S. Wells, R.E. Drullinger, E.C. Beaty, K.M. Evenson: Opt. Lett. **8**, 133 (1983)
Porter, G., M.W. Windsor: Proc. R. Soc. London A **245**, 238 (1958)
Pringsheim, P.: *Fluorescence and Phosphorescence* (Interscience, New York 1949)
Prior, Y.: Rev. Sci. Instrum. **50**, 259 (1979)

Rabinowitch, E., L.F. Epstein: J. Am. Chem. Soc. **63**, 69 (1941)
Rácz, B., Z. Bor, S. Szatmári, G. Szabó: Opt. Commun. **36**, 399 (1981)
Ramette, R.W., E.B. Sandell: J. Am. Chem. Soc. **78**, 4872 (1956)
Rautian, S.G., I.I. Sobel'man: Opt. Spectrosc. **10**, 65 (1961)
Rebane, K.K., V.V. Khizhnyakov: Opt. Spectrosc. **14**, 262 (1963)
Reiser, C.: "Modern Pulsed Wavemeters," in *Pulsed Single-Frequency Lasers: Technology and Applications*, Proc. SPIE **912**, 214 (1988)
Reiser, C., R.B. Lopert: Appl. Opt. **27**, 3656 (1988)
Reiser, C., P. Esherick, R.B. Lopert: Opt. Lett. **13**, 981 (1988)
Rentzepis, P.M., V.E. Bondybey: J. Chem. Phys. **80**, 4727 (1984)
Rentzepis, P.M., C.J. Mitschele, A.C. Saxman: Appl. Phys. Lett. **17**, 122 (1970)
Reynolds, G.A., K.H. Drexhage: unpublished results (1972)
Ricard, D., J. Ducuing: IEEE J. QE-**10**, 745 (1974)
Ricard, D., J. Ducuing: J. Chem. Phys. **62**, 3616 (1975)
Ricard, D., W.H. Lowdermilk, J. Ducuing: Chem. Phys. Lett. **16**, 617 (1972)
Ringwelski, L., F.P. Schäfer: unpublished results, (1970)
Robinson, G.W., R.P. Frosch: J. Chem. Phys. **38**, 1187 (1963)
Roess, D.: J. Appl. Phys. **37**, 2004 (1966)
Rohatgi, K.K., G.S. Singhal: J. Phys. Chem. **70**, 1695 (1966)
Rohatgi, K.K., A.K. Mukhopadhyay: Photochem. Photobiol. **14**, 551 (1971)
Romanek, K.M., O. Hildebrand, E. Göbel: Opt. Commun. **21**, 16 (1977)
Rosker, M.J., F.W. Wise, C.L. Tang: Phys. Rev. Lett. **57**, 321 (1986)
Rowley, W.R.C., K.C. Shotton, P.T. Woods: "A Simple Moving-Carriage Interferometer for 1 in $10^7$ Wavelength Intercomparison, and a Servocontrolled Fabry-Perot System for 3 in $10^{11}$ Accuracy," in *Laser Spectroscopy III*, ed. by J.L. Hall, J.L. Carlsten, Springer Ser. Opt. Sci., Vol. 7 (Springer, Berlin, Heidelberg 1977) p. 425
Rubinov, A.N., V.A. Mostovnikov: J. Appl. Spectrosc. **7**, 223 (1967)
Rubinov, A.N., V.A. Mostovnikov: Bull. Acad. Sci. USSR, Phys. Ser. **32**, 1348 (1968)
Runge, P.K.: Opt. Commun. **4**, 195 (1971)

Runge, P.K.: Opt. Commun. **5**, 311 (1972)
Runge, P.K., R. Rosenberg: IEEE J. QE-**8**, 910 (1972)

Saikan, S.: Appl. Phys. **17**, 41 (1978)
Salimbeni, R., R.V. Pole: Opt. Lett. **5**, 39 (1980)
Salomon, C., D. Hils, J.L. Hall: J. Opt. Soc. Am. B**5**, 1576 (1988)
Sansonetti, C.J., K. Weber: J. Opt. Soc. Am. B**1**, 361 (1984)
Sansonetti, C.J., K. Weber: J. Opt. Soc. Am. B**2**, 1385 (1985)
Sargent, M. III: Appl. Phys. **9**, 127 (1976)
Schäfer, F.P.: Invited Paper at Int. Quantum Electron. Conf., Miami, Fla. (1968)
Schäfer, F.P.: Conference on Nonlinear Optics, Belfast (1969)
Schäfer, F.P.: Angew. Chem. **82**, 25; Int. Ed. Engl. **9**, 9 (1970)
Schäfer, F.P.: Laser Chem. **3**, 265 (1983)
Schäfer, F.P.: Appl. Phys. B**39**, 1 (1986)
Schäfer, F.P., H. Müller: Opt. Commun., **2**, 407 (1971)
Schäfer, F.P., L. Ringwelski: Z. Naturforsch. **28a**, 792 (1973)
Schäfer, F.P., W. Schmidt, J. Volze: Appl. Phys. Lett. **9**, 306 (1966)
Schäfer, F.P., W. Schmidt, K. Marth: Phys. Lett. **24A**, 280 (1967)
Schäfer, F.P., W. Schmidt, J. Volze, K. Marth: Ber. Bunsenges. Phys. Chem. **72**, 328 (1968)
Schäfer, F.P., Z. Bor, W. Lüttke, B. Lipphardt: Chem. Phys. Lett. **56**, 455 (1978)
Schäfer, F.P., F.-G. Zhang, J. Jethwa: Appl. Phys. B**28**, 37 (1982)
Schäfer, F.P., Lee Wenchong, S. Szatmári: Appl. Phys. B**32**, 123 (1983)
Schappert, G.T., K.W. Billman, D.C. Burnham: Appl. Phys. Lett. **13**, 124 (1968)
Schawlow, A.L., C.H. Townes: Phys. Rev. **112**, 1940 (1958)
Schearer, L.D.: IEEE J. QE-**11**, 935 (1975)
Scheibe, G.: Z. Elektrochem. **47**, 73 (1941)
Schinke, D.P., R.G. Smith, E.G. Spencer, M.F. Galvin: Appl. Phys. Lett. **21**, 494 (1972)
Schmidt, A.J.: Opt. Commun. **14**, 287 (1975a)
Schmidt, A.J.: Opt. Commun. **14**, 294 (1975b)
Schmidt, W.: Laser **2**, 47 (1970)
Schmidt, W., F.P. Schäfer: Z. Naturforsch. **22a**, 1563 (1967)
Schmidt, W., F.P. Schäfer: Phys. Lett. **26A**, 558 (1968)
Schmidt, W., N. Wittekindt: Appl. Phys. Lett. **20**, 71 (1972)
Schröder, H.W., H. Welling, B. Wellegehausen: Appl. Phys. **1**, 347 (1973)
Schröder, H.W., L. Stein, D. Frölich, B. Fugger, H. Welling: Appl. Phys. **14**, 377 (1977)
Schweitzer, W.G., E.G. Kessler, R.D. Deslattes, H.P. Layer, J.R. Whetstone: Appl. Opt. **12**, 2927 (1973)
Scott, G.W., S.G.-Z. Shen, A.J. Cox: Rev. Sci. Instrum. **55**, 358 (1984)
Seilmeier, A., P.O.J. Scherer, W. Kaiser: Chem. Phys. Lett. **105**, 140 (1984)
Selwyn, J.E., J.I. Steinfeld: J. Phys. Chem. **76**, 762 (1972)
Sevchenko, A.N., L.G. Pikulik, L.F. Gladchenko, A.D. Das'ko: J. Appl. Spectrosc. **8**, 556 (1968a)
Sevchenko, A.N., A.A. Kovalev, V.A. Pilipovich, Yu. V. Razvin: Sov. Phys. – Dokl. **13**, 226 (1968b)
Shank, C.V.: Rev. Mod. Phys. **41**, 649 (1975)
Shank, C.V., E.P. Ippen: Appl. Phys. Lett. **24**, 373 (1974)
Shank, C.V., A. Dienes, A.M. Trozzolo, J.A. Myer: Appl. Phys. Lett. **16**, 405 (1970a)
Shank, C.V., A. Dienes, W.T. Silfvast: Appl. Phys. Lett. **17**, 307 (1970b)
Shank, C.V., J.E. Bjorkholm, H. Kogelnik: Appl. Phys. Lett. **18**, 395 (1971)
Shank, C.V., E.P. Ippen, O. Teschke: Chem. Phys. Lett. **45**, 291 (1977)
Shilov, V.B., B.S. Neporent, G.V. Lukomskii, A.G. Spiro, G.N. Antonevich: Sov. J. Quantum Electron. **5**, 1024 (1976)
Shingh, S.: *Stimulated Raman Scattering,* CRC Handbook of Lasers, ed. by R.J. Pressley (CRC, Cleveland 1971)
Shoshan, I., N.N. Danon, U.P. Oppenheim: J. Appl. Phys. **48**, 4495 (1977)
Shpolski, E.V.: Usp. Fiz. Nauk **77**, 321 and **80**, 255 (1962)
Siegman, A.E.: Opt. Commun. **5**, 200 (1972)
Siegman, A.E., D.W. Phillion, D.J. Kuizenga: Appl. Phys. Lett. **21**, 345 (1972)

Silfvast, W.T., O.R. Wood II: Appl. Phys. Lett. **26**, 447 (1975a)
Silfvast, W.T., O.R. Wood II: Digest of Technical Papers 1D, IEEE/OSA Conference on Laser Engineering and Applications (1975b)
Simon, P., J. Klebniczki, G. Szabó: Opt. Commun. **56**, 359 (1986)
Smith, W.V., P.P. Sorokin: *The Laser* (McGraw-Hill, New York 1966) p. 74
Smolskaya, T.I., A.N. Rubinov: Opt. Spectrosc. **31**, 235 (1971)
Snavely, B.B.: Proc. IEEE **57**, 1374 (1969)
Snavely, B.B., O.G. Peterson: IEEE J. QE-**4**, 540 (1968)
Snavely, B.B., F.P. Schäfer: Phys. Lett. **28A**, 728 (1969)
Snavely, B.B., O.G. Peterson, R.F. Reithel: Appl. Phys. Lett. **11**, 275 (1967)
Snyder, J.J.: "Fizeau Wavelength Meter", in *Laser Spectroscopy III*, ed. by J.L. Hall, J.L. Carlsten, Springer Ser. Opt. Sci., Vol. 7 (Springer, Berlin, Heidelberg 1977) p. 419
Snyder, J.J.: US Patent no. 4,173,442 (1979a)
Snyder, J.J.: Sov. J. Quantum Electron. **9**, 959 (1979b)
Snyder, J.J.: Appl. Opt. **19**, 1223 (1980)
Snyder, J.J.: "Fizeau Wavemeter", in *Los Alamos Conference on Optics 81*, ed. by D.L. Liebenberg, Proc. SPIE **288**, 258 (1981a)
Snyder, J.J.: "An Ultra-high Resolution Frequency Meter," in Proc. of the 35th Annual Symposium on Frequency Control (1981b) p. 464
Snyder, J.J.: Laser Focus **18**(5), 55 (1982)
Snyder, J.J., T. Baer, L. Hollberg, J.L. Hall: IEEE J. QE-**17**, 176 (1981)
Soep, B.: Opt. Commun. **1**, 433 (1970)
Soep, B., A. Kellmann, M. Martin, L. Lindqvist: Chem. Phys. Lett. **13**, 241 (1972)
Soffer, B.H., J.W. Linn: J. Appl. Phys. **39**, 5859 (1968)
Soffer, B.H., B.B. McFarland: Appl. Phys. Lett. **10**, 266 (1967)
Solomakha, D.A., A.K. Toropov: Sov. J. Quantum Electron. **7**(8), 929 (1977)
Sorokin, P.P.: Sci. Am. **220**(2), 30 (1969)
Sorokin, P.P., J.R. Lankard: IBM J. Res. Dev. **10**, 162 (1966)
Sorokin, P.P., J.R. Lankard: IBM J. Res. Dev. **11**, 148 (1967)
Sorokin, P.P., W.H. Culver, E.C. Hammond, J.R. Lankard: IBM J. Res. Dev. **10**, 401 (1966a)
Sorokin, P.P., W.H. Culver, E.C. Hammond, J.R. Lankard: IBM J. Res. Dev. **10**, 428 (1966b)
Sorokin, P.P., J.R. Lankard, E.C. Hammond, V.L. Moruzzi: IBM J. Res. Dev. **11**, 130 (1967)
Sorokin, P.P., J.R. Lankard, V.L. Moruzzi, E.C. Hammond: J. Chem. Phys. **48**, 4726 (1968)
Sorokin, P.P., J.R. Lankard, V.L. Moruzzi: Appl. Phys. Lett. **15**, 179 (1969)
Spaeth, M.L., D.P. Bortfeld: Appl. Phys. Lett. **9**, 179 (1966)
Speiser, S., N. Shakkour: Appl. Phys. B**38**, 3112 (1985)
Srinivasan, R.: IEEE J. QE-**5**, 552 (1969)
Steel, W.H. *Interferometry* (Cambridge University Press, Cambridge 1967)
Steer, R.P., V. Ramamurthy: Acc. Chem. Res. **21**, 380 (1988)
Stepanov, B.I., A.N. Rubinov: Sov. Phys. – Usp. **11**, 304 (1968)
Stepanov, B.I., A.N. Rubinov, V.A. Mostovnikov: J. Appl. Spectrosc. **7**, 116 (1967a)
Stepanov, B.I., A.N. Rubinov, V.A. Mostovnikov: JETP Lett. **5**, 117 (1967b)
Stephens, R.L., W.F. Hug: Laser Focus **8**(6), 38 (1972)
Steyer, B., F.P. Schäfer: Opt. Commun. **10**, 219 (1974a)
Steyer, B., F.P. Schäfer: IEEE J. QE-**10**, 736 (1974b)
Stix, M.S., E.P. Ippen: IEEE J. QE-**19**, 520 (1983)
Stockman, D.L.: Proc. of the ONR Conf. on Organic Lasers, Doc. no. AD 447468 (Defence Documentation Center for Scientific and Technical Information, Cameron Station, Alexandria, Va., USA 1964)
Stockman, D.L., W.R. Mallory, F.K. Tittel: Proc. IEEE **52**, 318 (1964)
Stoicheff, B.P., E. Weinberger: Can. J. Phys. **57**, 2143 (1979)
Stoilov, Yu. Yu.: Appl. Phys. B**33**, 63 (1984)
Streifer, W., P. Saltz: IEEE J. QE-**9**, 563 (1973)
Strickler, S.J., R.A. Berg: J. Chem. Phys. **37**, 814 (1962)
Strizhnev, V.S.: Sov. J. Quantum Electron. **5**, 119 (1976)
Strome, F.C. Jr., S.A. Tuccio: Opt. Commun. **4**, 58 (1971)

Strome, F.C. Jr., J.P. Webb: Appl. Opt. **10**, 1348 (1971)
Sutton, D.G., G.A. Capelle: Appl. Phys. Lett. **29**, 563 (1976)
Szabó, G., Z. Bor, Alexander Müller: Appl. Phys. B**31**, 53 (1983)
Szabó, G., B. Rácz, Z. Bor, B. Nikolaus, Alexander Müller: In Ultrafast Phenomena IV, ed. by D.H. Auston, K.B. Eisenthal, Springer Ser. Chem. Phys., Vol. 38 (Springer, Berlin, Heidelberg 1984) p. 60
Szatmári, S., F.P. Schäfer: Opt. Commun. **49**, 281 (1984)
Szatmári, S., F.P. Schäfer: Appl. Phys. B**46**, 305 (1988)

Tang, C.L., H. Statz, G. deMars: J. Appl. Phys. **34**, 2289 (1963)
Taylor, A.J., D.J. Erskine, C.L. Tang: Chem. Phys. Lett. **103**, 430 (1984)
Telle, J.M., C.L. Tang: Appl. Phys. Lett **24**, 85 (1974a)
Telle, J.M., C.L. Tang: Opt. Commun. **11**, 251 (1974b)
Telle, J.M., C.L. Tang: Appl. Phys. Lett. **26**, 572 (1975)
Terenin, A.N., V.L. Ermolaev: Usp. Fiz. Nauk **58**, 37 (1956)
Teschke, O., J.R. Whinnery, A. Dienes: IEEE J. QE-**12**, 513 (1976)
Thiel, E., C. Zander, K.H. Drexhage: Opt. Commun. **62**, 171 (1987)
Thompson, D.C., M.S. O'Sullivan, B.P. Stoicheff, G. Xu: Can. J. Phys. **61**, 949 (1983)
Tuccio, S.A., F.C. Strome Jr.: Appl. Opt. **11**, 64 (1972)
Tuccio, S.A., K.H. Drexhage, G.A. Reynolds: Opt. Commun. **7**, 248 (1973)
Turek, C.A., J.T. Yardley: IEEE J. QE-**7**, 102 (1971)
Turner, J.J., E.I. Moses, C.L. Tang: Appl. Phys. Lett. **27**, 441 (1975)
Turro, N.J.: *Molecular Photochemistry* (Benjamin, New York 1965)

Umeda, N., M. Tsukiji, H. Takasaki: Appl. Opt. **19**, 442 (1980)

Valdmanis, J.A., R.L. Fork, J.P. Gordon: Opt. Lett. **10**, 131 (1985)
Varga, P., P.G. Kryukov, V.F. Kuprishov, Yu. V. Senatskii: JETP Lett. **8**, 307 (1968)
Viktorova, E.N., I.A. Gofman: Russ. J. Phys. Chem. **39**, 1416 (1965)
Volkov, S.Y., V.I. Pelipenko, V.V. Smirnov: Sov. J. Quantum Electron. **12**, 380 (1982)
Volze, J.: Dissertation, Marburg (1969)
Vrehen, Q.H.F.: Opt. Commun. **3**, 144 (1971)
Vrehen, Q.H.F., A.J. Breimer: Opt. Commun. **4**, 416 (1972)

Wallenstein, R., T.W. Hänsch: Opt. Commun. **14**, 353 (1975)
Walther, H., J.L. Hall: Appl. Phys. Lett. **17**, 239 (1970)
Wang, G.: Opt. Commun. **10**, 149 (1974)
Wang, G., J.P. Webb: IEEE J. QE-**10**, 722 (1974)
Watanabe, A., H. Saito, Y. Ishida, M. Nakamoto, T. Yajima: Opt. Commun. **71**, 301 (1989)
Webb, J.P., W.C. McColgin, O.G. Peterson, D.L. Stockman, J.H. Eberly: J. Chem. Phys. **53**, 4227 (1970)
Weber, H.P.: Phys. Lett. **27A**, 321 (1968)
Weber, H.P., R. Ulrich: Appl. Phys. Lett. **19**, 38 (1971)
Weber, M.J., M. Bass: IEEE J. QE-**5**, 175 (1969)
Wehry, E.L.: "Structural and Environmental Factors in Fluorescence" in *Fluorescence – Theory, Instrumentation, and Practice,* ed. by G.G. Guilbault (Dekker, New York 1967)
Wellegehausen, B., H. Welling, R. Beigang: Appl. Phys. **3**, 387 (1974)
Wellegehausen, B., L. Laepple, H. Welling: Appl. Phys. **6**, 335 (1975)
Weller, A.: Z. Phys. Chemie NF **18**, 163 (1958)
Westling, L.A., M.G. Raymer, J.J. Snyder: J. Opt. Soc. Am. B**1**, 150 (1984)
Weysenfeld, C.H.: Appl. Opt. **13**, 2816 (1974)
Wieder, I.: Appl. Phys. Lett. **21**, 318 (1972)
Wineland, D.J.: J. Appl. Phys. **50**, 2528 (1979)
Winter, E., G. Veith, A.J. Schmidt: Opt. Commun. **25**, 87 (1978)
Wu, C.-Y., J.R. Lombardi: Opt. Commun. **7**, 233 (1973)
Wyatt, R.: Opt. Commun. **26**, 429 (1978)
Wyatt, R., E. Marinero: Appl. Phys. **25**, 297 (1981)

Xia, H., S.V. Benson, T.W. Hänsch: Laser Focus **17**(3), 54 (1981)

Yamada, K., K. Miyazaki, T. Hasama, T. Sato: Appl. Opt. **25**, 634 (1986)
Yamagishi, A., H. Inaba: Opt. Commun. **16**, 223 (1976)
Yamaguchi, G., F. Endo, S. Murakawa, S. Okamura, C. Yamanaka: Jpn. J. Appl. Phys. **7**, 179 (1968)
Yao, J.Q.: Appl. Phys. Lett. **41**, 136 (1982)
Yarborough, J.M.: Appl. Phys. Lett. **24**, 629 (1974)
Yarborough, J.M., J. Hobart: Post-deadline paper at CLEA Meeting (1973)
Yariv, A.: *Quantum Electronics* (Wiley, New York 1967) Chap. 15 (1967)
Yguerabide, J.: Rev. Sci. Instrum. **39**, 1048 (1968)

Zanker, V., E. Miethke: Z. Phys. Chem. NF **12**, 13 (1957a)
Zanker, V., E. Miethke: Z. Naturforsch. **12a**, 385 (1957b)
Zeidler, G.: J. Appl. Phys. **42**, 884 (1971)
Zhang, F.G., F.P. Schäfer: Appl. Phys. B**26**, 211 (1981)
Zumberge, M.A.: Appl. Opt. **24**, 1902 (1985)

# Subject Index